'HE STANDARD MODEL OF
E PHYSICS

d Edition

; is the mathematical theory that describes
teractions between leptons and quarks, the

graduate textbook provides a concise but
Model. It has been updated to account for
eractions, and the observations on matter–
ar that neutrinos are not mass-less, and this
e phenomena and the theory that describes
; in the theory of strong interactions and of
c clearly develops the theoretical concepts
ractions of leptons and quarks to the strong

e introduction to the Standard Model for
ch chapter ends with problems, and hints to
d of the book. The mathematical treatments
nd more sophisticated mathematical ideas
s. This title, first published in 2007, has
)lication on Cambridge Core.

:ENWOOD are theoreticians working in the
iversity of Bristol. They have published two
niversity Press, *Electricity and Magnetism*
Physics, now in its second edition (2001).

AN INTRODUCTION TO THE STANDARD MODEL OF PARTICLE PHYSICS

Second Edition

W. N. COTTINGHAM and D. A. GREENWOOD
University of Bristol, UK

CAMBRIDGE
UNIVERSITY PRESS

CAMBRIDGE
UNIVERSITY PRESS

Shaftesbury Road, Cambridge CB2 8EA, United Kingdom

One Liberty Plaza, 20th Floor, New York, NY 10006, USA

477 Williamstown Road, Port Melbourne, VIC 3207, Australia

314–321, 3rd Floor, Plot 3, Splendor Forum, Jasola District Centre, New Delhi – 110025, India

103 Penang Road, #05-06/07, Visioncrest Commercial, Singapore 238467

Cambridge University Press is part of Cambridge University Press & Assessment, a department of the University of Cambridge.

We share the University's mission to contribute to society through the pursuit of education, learning and research at the highest international levels of excellence.

www.cambridge.org
Information on this title: www.cambridge.org/9781009401722

DOI: 10.1017/9781009401685

First published 2007
5th printing 2013
Reissued as OA 2023

A catalogue record for this publication is available from the British Library.

ISBN 978-1-009-40172-2 Hardback
ISBN 978-1-009-40170-8 Paperback

Contents

Preface to the second edition

In the eight years since the first edition, the Standard Model has not been seriously discredited as a description of particle physics in the energy region (<2 TeV) so far explored. The principal discovery in particle physics since the first edition is that neutrinos carry mass. In this new edition we have added chapters that extend the formalism of the Standard Model to include neutrino fields with mass, and we consider also the possibility that neutrinos are Majorana particles rather than Dirac particles.

The Large Hadron Collider (LHC) is now under construction at CERN. It is expected that, at the energies that will become available for experiments at the LHC ($\sim$20 TeV), the physics of the Higgs field will be elucidated, and we shall begin to see 'physics beyond the Standard Model'. Data from the 'B factories' will continue to accumulate and give greater understanding of CP violation. We are confident that interest in the Standard Model will be maintained for some time into the future.

Cambridge University Press have again been most helpful. We thank Miss V. K. Johnson for secretarial assistance. We are grateful to Professor Dr J. G. Körner for his corrections to the first edition, and to Professor C. Davies for her helpful correspondence.

Preface to the second edition

Preface to the first edition

The 'Standard Model' of particle physics is the result of an immense experimental and inspired theoretical effort, spanning more than fifty years. This book is intended as a concise but accessible introduction to the elegant theoretical edifice of the Standard Model. With the planned construction of the Large Hadron Collider at CERN now agreed, the Standard Model will continue to be a vital and active subject.

The beauty and basic simplicity of the theory can be appreciated at a certain 'classical' level, treating the boson fields as true classical fields and the fermion fields as completely anticommuting. To make contact with experiment the theory must be quantised. Many of the calculations of the consequences of the theory are made in quantum perturbation theory. Those we present are for the most part to the lowest order of perturbation theory only, and do not have to be renormalised. Our account of renormalisation in Chapter 8 is descriptive, as is also our final Chapter 19 on the anomalies that are generated upon quantisation.

A full appreciation of the success and significance of the Standard Model requires an intimate knowledge of particle physics that goes far beyond what is usually taught in undergraduate courses, and cannot be conveyed in a short introduction. However, we attempt to give an overview of the intellectual achievement represented by the Model, and something of the excitement of its successes. In Chapter 1 we give a brief résumé of the physics of particles as it is qualitatively understood today. Later chapters developing the theory are interspersed with chapters on the experimental data. The amount of supporting data is immense and so we attempt to focus only on the most salient experimental results. Unless otherwise referenced, experimental values quoted are those recommended by the Particle Data Group (1996).

The mathematical background assumed is that usually acquired during an undergraduate physics course. In particular, a facility with the manipulations of matrix algebra is very necessary; Appendix A provides an *aide-mémoire*. Principles of symmetry play an important rôle in the construction of the model, and Appendix B is a self-contained account of the group theoretic ideas we use in describing these symmetries. The mathematics we require is not technically difficult, but the reader must accept a gradually more abstract formulation of physical theory than that presented at undergraduate level. Detailed derivations that would impair the flow of the text are often set as problems (and outline solutions to these are provided).

The book is based on lectures given to beginning graduate students at the University of Bristol, and is intended for use at this level and, perhaps, in part at least, at senior undergraduate level. It is not intended only for the dedicated particle physicist: we hope it may be read by physicists working in other fields who are interested in the present understanding of the ultimate constituents of matter.

We should like to thank the anonymous referees of Cambridge University Press for their useful comments on our proposals. The Department of Physics at Bristol has been generous in its encouragement of our work. Many colleagues, at Bristol and elsewhere, have contributed to our understanding of the subject. We are grateful to Mrs Victoria Parry for her careful and accurate work on the typescript, without which this book would never have appeared.

Notation

Position vectors in three-dimensional space are denoted by $\mathbf{r} = (x, y, z)$, or $\mathbf{x} = (x^1, x^2, x^3)$ where $x^1 = x$, $x^2 = y$, $x^3 = z$.

A general vector $\mathbf{a}$ has components (a^1, a^2, a^3), and $\hat{\mathbf{a}}$ denotes a unit vector in the direction of $\mathbf{a}$.

Volume elements in three-dimensional space are denoted by $\mathrm{d}^3\mathbf{x} = \mathrm{d}x\mathrm{d}y\mathrm{d}z = \mathrm{d}x^1\mathrm{d}x^2\mathrm{d}x^3$.

The coordinates of an event in four-dimensional time and space are denoted by $x = (x^0, x^1, x^2, x^3) = (x^0, \mathbf{x})$ where $x^0 = ct$.

Volume elements in four-dimensional time and space are denoted by $\mathrm{d}^4x = \mathrm{d}x^0\mathrm{d}x^1\mathrm{d}x^2\mathrm{d}x^3 = c\,\mathrm{d}t\,\mathrm{d}^3\mathbf{x}$.

Greek indices μ, ν, λ, ρ take on the values 0, 1, 2, 3.

Latin indices *i, j, k, l* take on the space values 1, 2, 3.

Pauli matrices

We denote by σ^μ the set $(\sigma^0, \sigma^1, \sigma^2, \sigma^3)$ and by $\tilde{\sigma}^\mu$ the set $(\sigma^0, -\sigma^1, -\sigma^2, -\sigma^3)$, where

$$\sigma^0 = \mathbf{I} = \begin{pmatrix} 1 & 0 \\ 0 & 1 \end{pmatrix}, \quad \sigma^1 = \begin{pmatrix} 0 & 1 \\ 1 & 0 \end{pmatrix}, \quad \sigma^2 = \begin{pmatrix} 0 & -\mathrm{i} \\ \mathrm{i} & 0 \end{pmatrix}, \quad \sigma^3 = \begin{pmatrix} 1 & 0 \\ 0 & -1 \end{pmatrix},$$

$$(\sigma^1)^2 = (\sigma^2)^2 = (\sigma^3)^2 = \mathbf{I}; \quad \sigma^1\sigma^2 = \mathrm{i}\sigma^3 = -\sigma^2\sigma^1, \text{ etc.}$$

Chiral representation for γ-matrices

$$\gamma^0 = \begin{pmatrix} \mathbf{0} & \mathbf{I} \\ \mathbf{I} & \mathbf{0} \end{pmatrix}, \quad \gamma^{\mathrm{i}} = \begin{pmatrix} \mathbf{0} & \sigma^{\mathrm{i}} \\ -\sigma^{\mathrm{i}} & \mathbf{0} \end{pmatrix},$$

$$\gamma^5 = \mathrm{i}\gamma^0\gamma^1\gamma^2\gamma^3 = \begin{pmatrix} -\mathbf{I} & \mathbf{0} \\ \mathbf{0} & \mathbf{I} \end{pmatrix}.$$

Quantisation ($\hbar = c = 1$)

$$(E, \mathbf{p}) \to (\mathrm{i}\partial/\partial t, -\mathrm{i}\boldsymbol{\nabla}), \text{ or } p^\mu \to \mathrm{i}\partial^\mu.$$

For a particle carrying charge q in an external electromagnetic field,

$$(E, \mathbf{p}) \to (E - q\phi, \mathbf{p} - q\mathbf{A}), \text{ or } p^\mu \to p^\mu - qA^\mu,$$
$$\mathrm{i}\partial^\mu \to (\mathrm{i}\partial^\mu - qA^\mu) = \mathrm{i}(\partial^\mu + \mathrm{i}qA^\mu).$$

Field definitions

$$Z_\mu = W_\mu{}^3 \cos\theta_\mathrm{w} - B_\mu \sin\theta_\mathrm{w},$$
$$A_\mu = \mathrm{W}_\mu{}^3 \sin\theta_\mathrm{w} + B_\mu \cos\theta_\mathrm{w},$$

where $\sin^2\theta_\mathrm{w} = 0.2315(4)$

$$g_2 \sin\theta_\mathrm{w} = g_1 \cos\theta_\mathrm{w} = e, \quad G_\mathrm{F} = g_2{}^2/(4\sqrt{2}M_\mathrm{w}{}^2).$$

Glossary of symbols

$\mathbf{A}$	electromagnetic vector potential Section 4.3
A^μ	electromagnetic four-vector potential
$A^{\mu\nu}$	field strength tensor Section 11.3
A_FB	forward–backward asymmetry Section 15.2
a	wave amplitude Section 3.5
$a, a^\dagger$	boson annihilation, creation operator
$\mathbf{B}$	magnetic field
B^μ	gauge field Section 11.1
$B^{\mu\nu}$	field strength tensor Section 11.2
$b, b^\dagger$	fermion annihilation, creation operator
$\mathbf{D}$	isospin doublet Section 16.6
$d, d^\dagger$	antifermion annihilation, creation operator
d_k	($k = 1,2,3$) down-type quark field
$\mathbf{E}$	electric field
E	energy
$e, e_\mathrm{L}, e_\mathrm{R}$	electron Dirac, two-component left-handed, right-handed field
$F^{\mu\nu}$	electromagnetic field strength tensor Section 4.1
f	radiative corrections factor Sections 15.1, 17.4
f_{abc}	structure constants of *SU(3)* Section B.7
G^μ	gluon matrix gauge field
$\mathrm{G}^{\mu\nu}$	gluon field strength tensor
G_F	Fermi constant Section 9.4

$g^{\mu\nu}$	metric tensor
g	strong coupling constant Section 16.1
g_1, g_2	electroweak coupling constants
H	Hamiltonian Section 3.1
$h(x)$	Higgs field
$\mathcal{H}$	Hamiltonian density Section 3.3
$\mathbf{I}$	isospin operator Sections 1.5, 16.6
$\mathbf{J}$	electric current density Section 4.1
$\mathbf{J}$	total angular momentum operator
J	Jarlskog constant Section 14.3
J^{μ}	lepton number current Section 12.4
$\mathbf{j}$	probability current Section 7.1
j^{μ}	lepton current Section 12.2
K	string tension Section 17.1
$\mathbf{k}$	wave vector
$\mathbf{L}$	lepton doublet Section 12.1
L	Lagrangian Section 3.1
$\mathcal{L}$	Lagrangian density Section 3.3
l^3	normalisation volume Section 3.5
$\mathbf{M}$	left-handed spinor transformation matrix Section B.6
M	proton mass Section D.1
m	mass
$\mathbf{N}$	right-handed spinor transformation matrix Section B.6
N	number operator Section C.1
$\hat{\mathrm{O}}$	quantum operator
$\mathbf{P}$	total field momentum
$\mathbf{p}$	momentum
Q^2	$= -q_{\mu}q^{\mu}$
$\mathbf{q}$	quark colour triplet
q^{μ}	energy–momentum transfer
$\mathbf{R}$	rotation matrix Section B.2
$\mathbf{S}$	spin operator
S	action Section 3.1
s	square of centre of mass energy
T^{μ}_{ν}	energy–momentum tensor Section 3.6
$\mathbf{U}$	unitary matrix
u_k	$(k = 1, 2, 3)$ up-type quark field
$u_{\mathrm{L}}, u_{\mathrm{R}}$	two-component left-handed, right-handed spinors Section 6.1
u_{+}, u_{-}	Dirac spinors Section 6.3
$\mathbf{V}$	Kobayashi–Maskawa matrix Section 14.2

V	normalisation volume
$\mathbf{v}$	velocity
v	$= \lvert\mathbf{v}\rvert$
$v_{\mathrm{L}}, v_{\mathrm{R}}$	two-component left-handed, right-handed spinors
v_+, v_-	Dirac spinors Section 6.4
$\mathbf{W}^{\mu}$	matrix of vector gauge field Section 11.1
$\mathrm{W}^{\mu\nu}$	field strength tensor Section 11.2
$W^1_{\mu}, W^2_{\mu}, W^+_{\mu}, W^-_{\mu}$	fields of W boson
Z_{μ}	field of Z boson
$\alpha(Q^2)$	effective fine structure constant Section 16.3
$\alpha_s(Q^2)$	effective strong coupling constant Section 16.3
α_{latt}	lattice coupling constant Section 17.1
α^i	Dirac matrix Section 5.1
β	Dirac matrix Section 5.1
β	$= v/c$
Γ	width of excited state, decay rate
γ^{μ}	Dirac matrix Section 5.5
γ	$= (1 - \beta^2)^{-1/2}$
δ	Kobayashi–Maskawa phase Section 14.3
$\boldsymbol{\varepsilon}$	polarisation unit vector Section 4.7
ε	helicity index
θ	boost parameter: $\tanh\theta = \beta$, $\cosh\theta = \gamma$ Section 2.1, phase angle, scattering angle, scalar potential Section 4.3, gauge parameter field Section 10.2
θ_{W}	Weinberg angle
Λ^{-1}	confinement length Section 16.3
Λ_{latt}	lattice parameter Section 17.1
λ_a	matrices associated with $SU(3)$ Section B.7
$\mu, \mu_{\mathrm{L}}, \mu_{\mathrm{R}}$	muon Dirac, two-component left-handed, right-handed field
$\nu_{e\mathrm{L}}, \nu_{\mu\mathrm{L}}, \nu_{\tau\mathrm{L}}$	electron neutrino, muon neutrino, tau neutrino field
Π	momentum density Section 3.3
ρ	electric charge density
$\rho(E)$	density of final states at energy E
Σ	spin operator acting on Dirac field Section 6.2
τ	mean life
$\tau, \tau_{\mathrm{L}}, \tau_{\mathrm{R}}$	tau Dirac, two-component left-handed, right-handed field
Φ	complex scalar field Section 3.7

ϕ	real scalar field Section 2.3, scalar potential Section 4.1, gauge parameter field Section 10.2
ϕ_0	vacuum expectation value of the Higgs field
χ	gauge parameter field Section 4.3, scalar field Section 10.3
ψ	four-component Dirac field
ψ_L, ψ_R	two-component left-handed, right-handed spinor field
$\bar{\psi}$	$\psi^\dagger\gamma^o$ Section 5.5
ω	frequency

1

The particle physicist's view of Nature

1.1 Introduction

It is more than a century since the discovery by J. J. Thomson of the electron. The electron is still thought to be a structureless point particle, and one of the elementary particles of Nature. Other particles that were subsequently discovered and at first thought to be elementary, like the proton and the neutron, have since been found to have a complex structure.

What then are the ultimate constituents of matter? How are they categorised? How do they interact with each other? What, indeed, should we ask of a mathematical theory of elementary particles? Since the discovery of the electron, and more particularly in the last sixty years, there has been an immense amount of experimental and theoretical effort to determine answers to these questions. The present Standard Model of particle physics stems from that effort.

The Standard Model asserts that the material in the Universe is made up of elementary fermions interacting through fields, of which they are the sources. The particles associated with the interaction fields are bosons.

Four types of interaction field, set out in Table 1.1., have been distinguished in Nature. On the scales of particle physics, gravitational forces are insignificant. The Standard Model excludes from consideration the gravitational field. The quanta of the electromagnetic interaction field between electrically charged fermions are the massless photons. The quanta of the weak interaction fields between fermions are the charged W^+ and W^- bosons and the neutral Z boson, discovered at CERN in 1983. Since these carry mass, the weak interaction is short ranged: by the uncertainty principle, a particle of mass M can exist as part of an intermediate state for a time $\hbar/Mc^2$, and in this time the particle can travel a distance no greater than $\hbar c/Mc$. Since $M_W \approx 80\,\text{GeV}/c^2$ and $M_Z \approx 90\,\text{GeV}/c^2$, the weak interaction has a range $\approx 10^{-3}$ fm.

Table 1.1. *Types of interaction field*

Interaction field	Boson	Spin
Gravitational field	'Gravitons' postulated	2
Weak field	W^+, W^-, Z particles	1
Electromagnetic field	Photons	1
Strong field	'Gluons' postulated	1

The quanta of the strong interaction field, the gluons, have zero mass and, like photons, might be expected to have infinite range. However, unlike the electromagnetic field, the gluon fields are *confining*, a property we shall be discussing at length in the later chapters of this book.

The elementary fermions of the Standard Model are of two types: *leptons* and *quarks*. All have spin $\frac{1}{2}$, in units of $\hbar$, and in isolation would be described by the Dirac equation, which we discuss in Chapters 5, 6 and 7. Leptons interact only through the electromagnetic interaction (if they are charged) and the weak interaction. Quarks interact through the electromagnetic and weak interactions and also through the strong interaction.

1.2 The construction of the Standard Model

Any theory of elementary particles must be consistent with special relativity. The combination of quantum mechanics, electromagnetism and special relativity led Dirac to the equation now universally known as the *Dirac equation* and, on quantising the fields, to quantum field theory. Quantum field theory had as its first triumph quantum electrodynamics, QED for short, which describes the interaction of the electron with the electromagnetic field. The success of a post-1945 generation of physicists, Feynman, Schwinger, Tomonaga, Dyson and others, in handling the infinities that arise in the theory led to a spectacular agreement between QED and experiment, which we describe in Chapter 8.

The Standard Model, like the QED it contains, is a theory of interacting fields. Our emphasis will be on the beauty and simplicity of the theory, and this can be understood at a certain 'classical' level, treating the boson fields as true classical fields, and the fermion fields as completely anticommuting. To make a judgement of the success of the model in describing the data, it is necessary to quantise the fields, but to keep this book concise and accessible, results beyond the lowest orders of perturbation theory will only be quoted.

The construction of the Standard Model has been guided by principles of symmetry. The mathematics of symmetry is provided by group theory; groups of

Table 1.2. *Leptons*

	Mass (MeV/c^2)	Mean life (s)	Electric charge
Electron e^-	0.5110	∞	$-e$
Electron neutrino ν_e	$< 3 \times 10^{-6}$		0
Muon μ^-	105.658	2.197×10^{-6}	$-e$
Muon neutrino ν_μ			0
Tau τ^-	1777	$(291.0 \pm 1.5) \times 10^{-15}$	$-e$
Tau neutrino ν_τ			0

For neutrino masses see Chapter 20.

particular significance in the formulation of the Model are described in Appendix B. The connection between symmetries and physics is deep. *Noether's theorem* states, essentially, that for every continuous symmetry of Nature there is a corresponding conservation law. For example, it follows from the presumed homogeneity of space and time that the Lagrangian of a closed system is invariant under uniform translations of the system in space and in time. Such transformations are therefore symmetry operations on the system. It may be shown that they lead, respectively, to the laws of conservation of momentum and conservation of energy. Symmetries, and symmetry breaking, will play a large part in this book.

In the following sections of this chapter, we remind the reader of some of the salient discoveries of particle physics that the Standard Model must incorporate. In Chapter 2 we begin on the mathematical formalism we shall need in the construction of the Standard Model.

1.3 Leptons

The known leptons are listed in Table 1.2.. The Dirac equation for a charged massive fermion predicts, correctly, the existence of an *antiparticle* of the same mass and spin, but opposite charge, and opposite magnetic moment relative to the direction of the spin. The Dirac equation for a neutrino ν allows the existence of an antineutrino $\bar{\nu}$.

Of the charged leptons, only the electron e^- carrying charge $-e$ and its antiparticle e^+, are stable. The muon μ^- and tau τ^- and their antiparticles, the μ^+ and τ^+, differ from the electron and positron only in their masses and their finite lifetimes. They appear to be elementary particles. The experimental situation regarding small neutrino masses has not yet been clarified. There is good experimental evidence that the e, μ and τ have different neutrinos ν_e, ν_μ and ν_τ associated with them.

It is believed to be true of all interactions that they preserve electric charge. It seems that in its interactions a lepton can change only to another of the same type,

Table 1.3. *Properties of quarks*

Quark	Electric charge (e)	Mass ($\times c^{-2}$)
Up u	2/3	1.5 to 4 MeV
Down d	−1/3	4 to 8 MeV
Charmed c	2/3	1.15 to 1.35 GeV
Strange s	−1/3	80 to 130 MeV
Top t	2/3	169 to 174 GeV
Bottom b	−1/3	4.1 to 4.4 GeV

and a lepton and an antilepton of the same type can only be created or destroyed together. These laws are exemplified in the decay

$$\mu^- \to \nu_\mu + e^- + \bar{\nu}_e.$$

Apart from neutrino oscillations (see Chapters 19–21). This *conservation of lepton number*, antileptons being counted negatively, which holds for each separate type of lepton, along with the conservation of electric charge, will be apparent in the Standard Model.

1.4 Quarks and systems of quarks

The known quarks are listed in Table 1.3.. In the Standard Model, quarks, like leptons, are spin $\frac{1}{2}$ Dirac fermions, but the electric charges they carry are 2e/3, −e/3. Quarks carry quark number, antiquarks being counted negatively. The net quark number of an isolated system has never been observed to change. However, the number of different types or *flavours* of quark are not separately conserved: changes are possible through the weak interaction.

A difficulty with the experimental investigation of quarks is that an isolated quark has never been observed. Quarks are always confined in compound systems that extend over distances of about 1 fm. The most elementary quark systems are *baryons* which have net quark number three, and *mesons* which have net quark number zero. In particular, the proton and neutron are baryons. Mesons are essentially a quark and an antiquark, bound transiently by the strong interaction field. The term *hadron* is used generically for a quark system.

The proton basically contains two *up* quarks and one *down* quark (uud), and the neutron two down quarks and one up (udd). The proton is the only stable baryon. The neutron is a little more massive than the proton, by about 1.3 MeV/c^2, and in free space it decays to a proton through the weak interaction: $n \to p + e^- + \bar{\nu}_e$, with a mean life of about 15 minutes.

All mesons are unstable. The lightest mesons are the π-mesons or 'pions'. The electrically charged π^+ and π^- are made up of $(\mathrm{u\bar{d}})$ and $(\mathrm{\bar{u}d})$ pairs, respectively, and the neutral π^0 is either $\mathrm{u\bar{u}}$ or $\mathrm{d\bar{d}}$, with equal probabilities; it is a coherent superposition $(\mathrm{u\bar{u}} - \mathrm{d\bar{d}})/\sqrt{2}$ of the two states. The π^+ and π^- have a mass of $139.57\,\mathrm{MeV}/c^2$ and the π^0 is a little lighter, $134.98\,\mathrm{MeV}/c^2$. The next lightest meson is the η ($\approx 547\,\mathrm{MeV}/c^2$), which is the combination $(\mathrm{u\bar{u}} + \mathrm{d\bar{d}})/\sqrt{2}$ of quark–antiquark pairs orthogonal to the π^0, with some $\mathrm{s\bar{s}}$ component.

1.5 Spectroscopy of systems of light quarks

As will be discussed in Chapter 16, the masses of the u and d quarks are quite small, of the order of a few MeV/c^2, closer to the electron mass than to a meson or baryon mass. A u or d quark confined within a distance $\approx$ 1 fm has, by the uncertainty principle, a momentum $p \approx \hbar/(1\mathrm{fm}) \approx 200\,\mathrm{MeV}/c$, and hence its energy is $E \approx pc \approx 200\,\mathrm{MeV}$, almost independent of the quark mass. All quarks have the same strong interactions. As a consequence, the physics of light quark systems is almost independent of the quark masses. There is an approximate $SU(2)$ isospin symmetry (Section 16.6), which is evident in the Standard Model.

The symmetry is not exact because of the different quark masses and different quark charges. The symmetry breaking due to quark mass differences prevails over the electromagnetic. In all cases where two particles differ only in that a d quark is substituted for a u quark, the particle with the d quark is more massive. For example, the neutron is more massive than the proton, even though the mass, $\sim 2\,\mathrm{MeV}/c^2$, associated with the electrical energy of the charged proton is far greater than that associated with the (overall neutral) charge distribution of the neutron. We conclude that the d quark is heavier than the u quark.

The evidence for the existence of quarks came first from nucleon spectroscopy. The proton and neutron have many excited states that appear as resonances in photon–nucleon scattering and in pion–nucleon scattering (Fig. 1.1). Hadron states containing light quarks can be classified using the concept of isospin. The u and d quarks are regarded as a doublet of states $|\mathrm{u}\rangle$ and $|\mathrm{d}\rangle$, with $I = 1/2$ and $I_3 = +1/2$, $-1/2$, respectively. The total isospin of a baryon made up of three u or d quarks is then $I = 3/2$ or $I = 1/2$. The isospin 3/2 states make up multiplets of four states almost degenerate in energy but having charges $2e$(uuu), e(uud), 0(udd), $-e$(ddd). The $I = 1/2$ states make up doublets, like the proton and neutron, having charges e(uud) and 0(udd). The electric charge assignments of the quarks were made to comprehend this baryon charge structure.

Energy level diagrams of the $I = 3/2$ and $I = 1/2$ states up to excitation energies of 1 GeV are shown in Fig. 1.2. The energy differences between states in a multiplet are only of the order of 1 MeV and cannot be shown on the scale of the figure. The

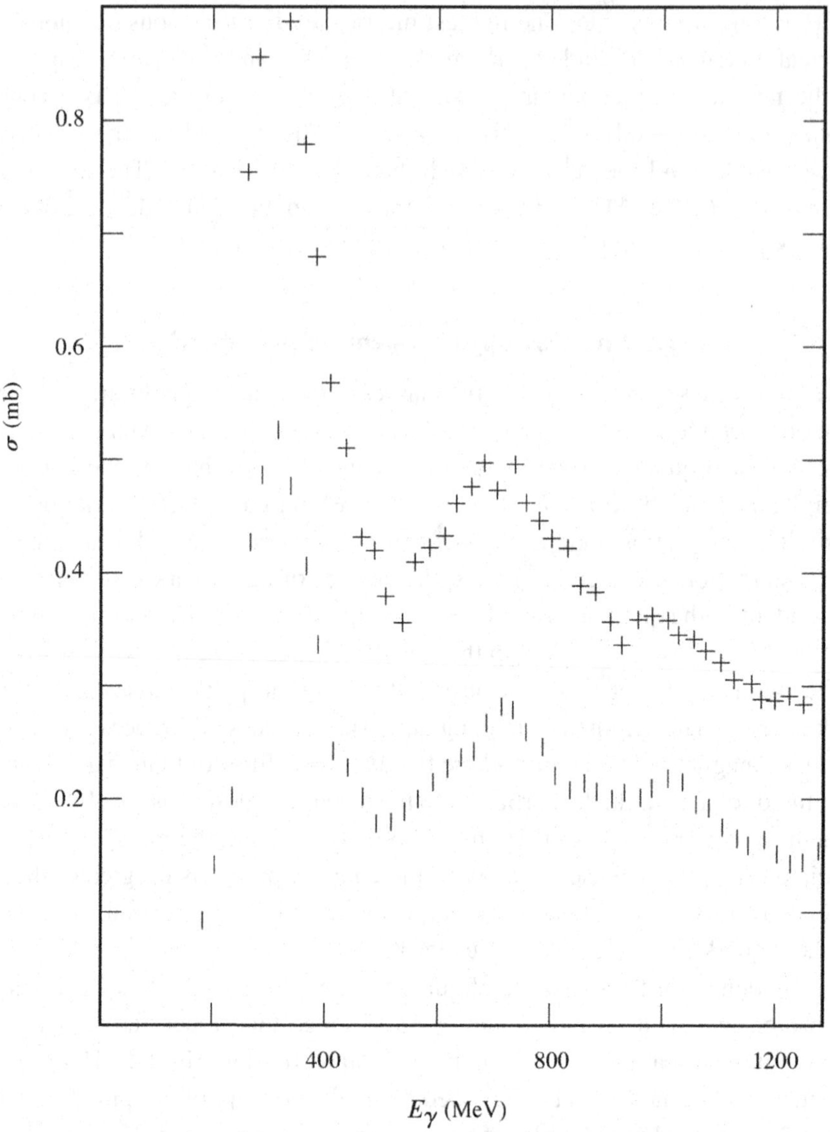

Figure 1.1 The photon cross-section for hadron production by photons on protons (dashes) and deuterons (crosses). The difference between these cross-sections is approximately the cross-section for hadron production by photons on neutrons. (After Armstrong *et al.* (1972).)

widths Γ of the excited states are however quite large, of the order of 100 MeV, corresponding to mean lives $\tau = \hbar/\Gamma \sim 10^{-23}$s. The excited states are all energetic enough to decay through the strong interaction, as for example $\Delta^{++} \to \mathrm{p} + \pi^+$ (Fig. 1.3).

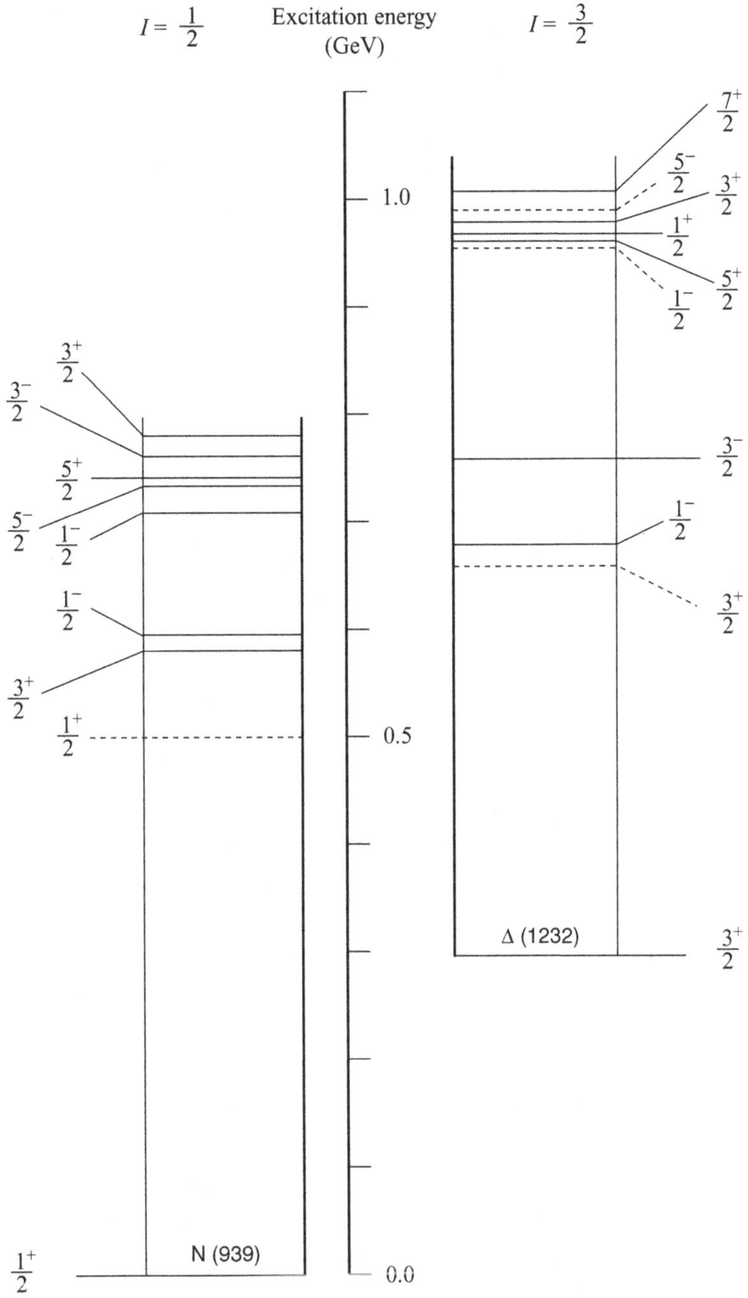

Figure 1.2 An energy-level diagram for the nucleon and its excited states. The levels fall into two classes: isotopic doublets ($I = 1/2$) and isotopic quartets ($I = 3/2$). The states are labelled by their total angular momenta and parities J^P. The nucleon doublet N(939) is the ground state of the system, the $\Delta(1232)$ is the lowest lying quartet. Within the quark model (see text) these two states are the lowest that can be formed with no quark orbital angular momentum ($L = 0$). The other states designated by unbroken lines have clear interpretations: they are all the next most simple states with $L = 1$ (negative parity) and $L = 2$ (positive parity). The broken lines show states that have no clear interpretation within the simple three-quark model. They are perhaps associated with excited states of the gluon fields.

Table 1.4. *Isospin quantum numbers of light quarks*

Quark	Isospin I	I_3
u	1/2	1/2
$\bar{u}$	1/2	−1/2
d	1/2	−1/2
$\bar{d}$	1/2	1/2
s	0	0
$\bar{s}$	0	0

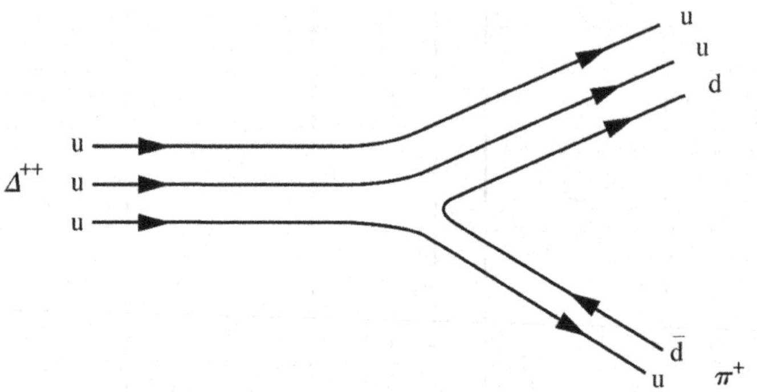

Figure 1.3 A quark model diagram of the decay $\Delta^{++} \to p + \pi^{+}$. The gluon field is not represented in this diagram, but it would be responsible for holding the quark systems together and for the creation of the $d\bar{d}$ pair.

The rich spectrum of the baryon states can largely be described and understood on the basis of a simple 'shell' model of three confined quarks. The lowest states have orbital angular momentum $L = 0$ and positive parity. The states in the next group have $L = 1$ and negative parity, and so on. However, the model has the curious feature that, to fit the data, the states are completely symmetric in the interchange of any two quarks. For example, the Δ^{++}(uuu), which belongs to the lowest $I = 3/2$ multiplet, has $J^{P} = 3/2^{+}$. If $L = 0$ the three quark spins must be aligned $\uparrow\uparrow\uparrow$ in a symmetric state to give $J = 3/2$, and the lowest energy spatial state must be totally symmetric. Symmetry under interchange is not allowed for an assembly of identical fermions! However, there is no doubt that the model demands symmetry, and with symmetry it works very well. The resolution of this problem will be left to later in this chapter. There are only a few states (broken lines in Fig. 1.2) that cannot be understood within the simple shell model.

Mesons made up of light u and d quarks and their antiquarks also have a rich spectrum of states that can be classified by their isospin. Antiquarks have an I_3 of opposite sign to that of their corresponding quark (Table 1.4.). By the rules for the addition of isospin, quark–antiquark pairs have $I = 0$ or $I = 1$. The $I = 0$ states

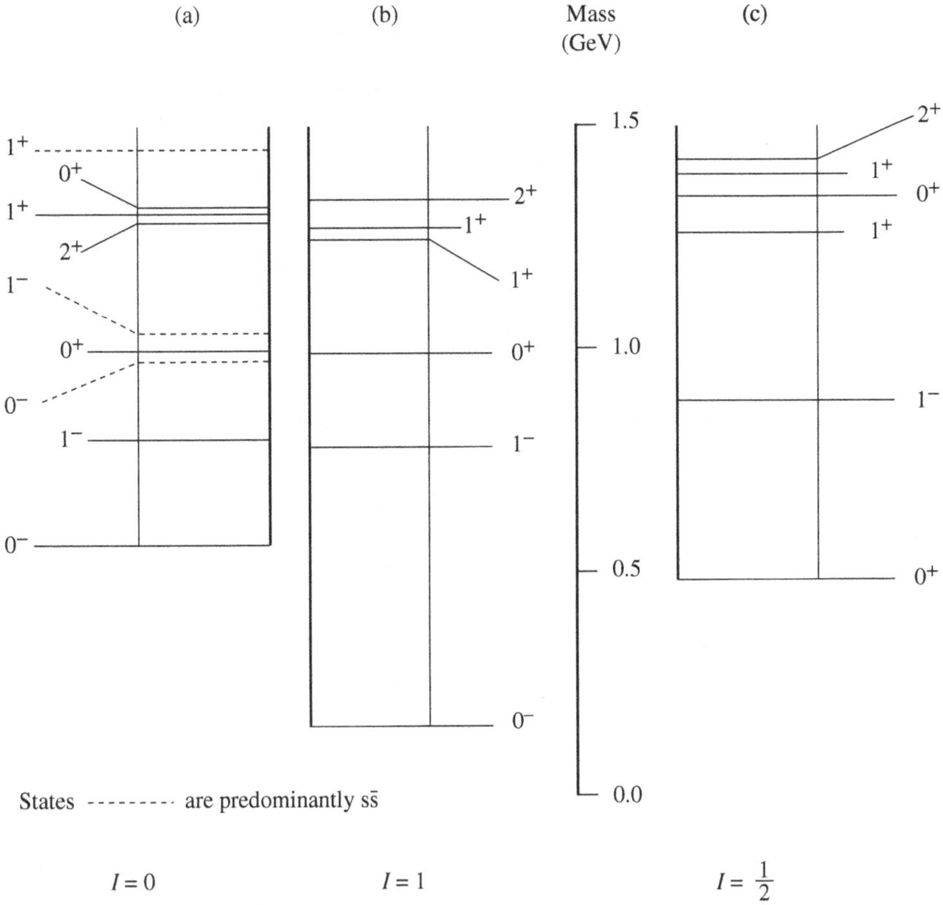

Figure 1.4 States of the quark–antiquark system $u\bar{u}$, $u\bar{d}$, $d\bar{u}$, $d\bar{d}$ form isotopic triplets $(I = 1) : u\bar{d}, (u\bar{u} - d\bar{d})/\sqrt{2}, d\bar{u}$; and also isotopic singlets $(I = 0) : (u\bar{u} + d\bar{d})/\sqrt{2}$. Figure 1.4(a) is an energy-level diagram of the lowest energy isosinglets, including states --- which are interpreted as $s\bar{s}$ states. Figure 1.4(b) is an energy-level diagram of the lowest energy isotriplets. Figure 1.4(c) is an energy-level diagram of the lowest energy K mesons. The K mesons are quark–antiquark systems $u\bar{s}$ and $d\bar{s}$; they are isotopic doublets, as are their antiparticle states $s\bar{u}$ and $s\bar{d}$. Their higher energies relative to the states in Fig. 1.4(b) are largely due to the higher mass of the s over the u and d quarks. The large relative displacement of the 0^+ state is a feature with, as yet, no clear interpretation.

are singlets with charge 0, like the η (Fig. 1.4(a)). The $I = 1$ states make up triplets carrying charge $+e$, 0, $-e$, which are almost degenerate in energy, like the triplet π^+, π^0, π^-.

The spectrum of $I = 1$ states with energies up to 1.5 GeV is shown in Fig. 1.4(b). As in the baryon case the splitting between states in the same isotopic multiplet is only a few MeV; the widths of the excited states are like the widths of the excited baryon states, of the order of 100 MeV. In the lowest multiplet (the pions), the quark–antiquark pair is in an $L = 0$ state with spins coupled to zero. Hence $J^{\mathrm{P}} = 0^-$, since a fermion and antifermion have opposite relative parity (Section 6.4). In the first excited state the spins are coupled to 1 and $J^{\mathrm{P}} = 1^-$. These are the ρ mesons. With $L = 1$ and spins coupled to $S = 1$ one can construct states $2^+, 1^+, 0^+$, and with $L = 1$ and spins coupled to $S = 0$ a state 1^+. All these states can be identified in Fig. 1.4(b).

1.6 More quarks

'Strange' mesons and baryons were discovered in the late 1940s, soon after the discovery of the pions. It is apparent that as well as the u and d quarks there exists a so-called *strange* quark s, and strange particles contain one or more s quarks. An s quark can replace a u or d quark in any baryon or meson to make the strange baryons and strange mesons. The electric charges show that the s quark, like the d, has charge $-e/3$, and the spectra can be understood if the s is assigned isospin $I = 0$.

The lowest mass strange mesons are the $I = 1/2$ doublet, $\mathrm{K}^-(\mathrm{s\bar{u}}$, mass 494 MeV) and $\bar{\mathrm{K}}^{\mathrm{o}}(\mathrm{s\bar{d}}$, mass 498 MeV). Their antiparticles make up another doublet, the $\mathrm{K}^+(\mathrm{u\bar{s}})$ and $\mathrm{K}^{\mathrm{o}}(\mathrm{d\bar{s}})$.

The effect of quark replacement on the meson spectrum is illustrated in Fig. 1.4. Each level in the spectrum of Fig. 1.4(b) has a member $(\mathrm{d\bar{u}})$ with charge $-e$. Figure 1.4(c) shows the spectrum of strange $(\mathrm{s\bar{u}})$ mesons. There is a correspondence in angular momentum and parity between states in the two spectra. The energy differences are a consequence of the s quark having a much larger mass, of the order of 200 MeV.

The excess of mass of the s quark over the u and d quarks makes the s quark in any strange particle unstable to decay by the weak interaction.

Besides the u, d and s quarks there are considerably heavier quarks: the *charmed* quark c (mass $\approx 1.3\ \mathrm{GeV}/c^2$, charge $2e/3$), the *bottom* quark b (mass $\approx 4.3\ \mathrm{GeV}/c^2$, charge $-e/3$), and the *top* quark t (mass $\approx 180\ \mathrm{GeV}/c^2$, charge $2e/3$). The quark masses are most remarkable, being even more disparate than the lepton masses. The experimental investigation of the elusive top quark is still in its infancy, but it seems that three quarks of any of the six known flavours can be bound to form a system of states of a baryon (or three antiquarks to form antibaryon states), and any quark–antiquark pair can bind into mesonic states.

The c and b quarks were discovered in $\mathrm{e^+e^-}$ colliding beam machines. Very prominent narrow resonances were observed in the $\mathrm{e^+e^-}$ annihilation cross-sections. Their widths, of less than 15 MeV, distinguished the meson states responsible from those made up of u, d or s quarks. There are two groups of resonant states.

The group at around 3 GeV centre of mass energy are known as J/ψ resonances, and are interpreted as *charmonium* $c\bar{c}$ states. Another group, around 10 GeV, the Υ (upsilon) resonances, are interpreted as *bottomonium* $b\bar{b}$ states. The current state of knowledge of the $c\bar{c}$ and $b\bar{b}$ energy levels is displayed in Fig. 1.5. We shall discuss these systems in Chapter 17.

The existence of the top quark was established in 1995 at Fermilab, in $\bar{p}p$ collisions.

1.7 Quark colour

Much informative quark physics has been revealed in experiments with e^+e^- colliding beams. We mention here experiments in the range between centre of mass energies 10 GeV and the threshold energy, around 90 GeV, at which the Z boson can be produced.

The e^+e^- annihilation cross-section $\sigma(e^+e^- \to \mu^+\mu^-)$ is comparatively easy to measure, and is easy to calculate in the Weinberg–Salam electroweak theory, which we shall introduce in Chapter 12. At centre of mass energies much below 90 GeV the cross-section is dominated by the electromagnetic process represented by the Feynman diagram of Fig. 1.6. The muon pair are produced 'back-to-back' in the centre of mass system, which for most e^+e^- colliders is the laboratory system. To leading order in the fine-structure constant $\alpha = e^2/(4\pi\varepsilon_0\hbar c)$, the differential cross-section for producing muons moving at an angle θ with respect to unpolarised incident beams is

$$\frac{d\sigma}{d\theta} = \frac{\pi\alpha^2}{2s}(1+\cos^2\theta)\sin\theta \tag{1.1}$$

where s is the square of the centre of mass energy (see Okun, 1982, p. 205). In the derivation of (1.1) the lepton masses are neglected. Integrating with respect to θ, the total cross-section is

$$\sigma = \frac{4\pi\alpha^2}{3s}. \tag{1.2}$$

The quantity $R(E)$ shown in Fig. 1.7 is the ratio

$$R = \frac{\sigma(e^+e^- \to \text{strongly interacting particles})}{\sigma(e^+e^- \to \mu^+\mu^-)}. \tag{1.3}$$

At the lower energies many hadronic states are revealed as resonances, but R seems to become approximately constant, $R \approx 4$, at energies above 10 GeV up to about 40 GeV.

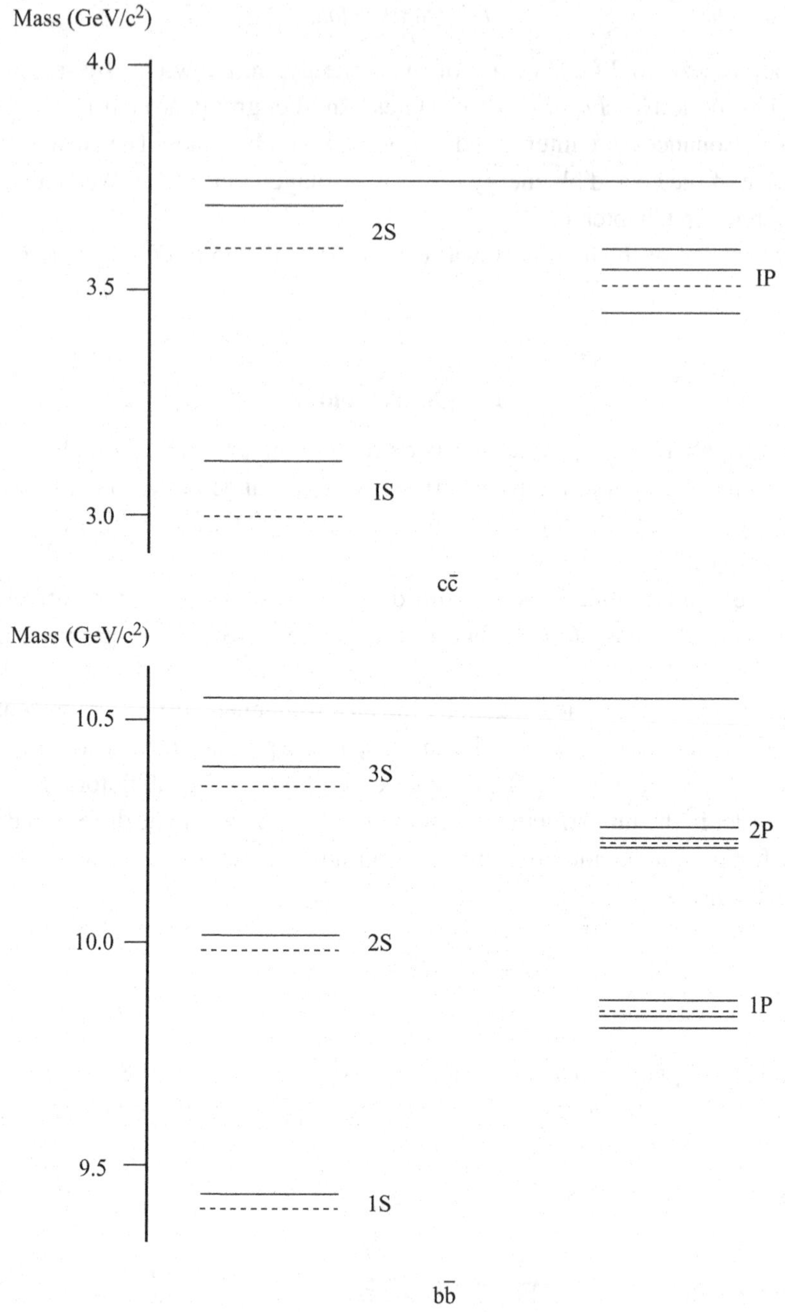

Figure 1.5 Energy-level diagrams for charmonium cc̄ and bottomonium bb̄ states, below the threshold at which they can decay through the strong interaction to meson pairs (for example cc̄ → cū + uc̄). States labelled 1S, 2S, 3S have orbital angular momentum $L = 0$ and the 1P, 2P states have $L = 1$. The intrinsic quark spins can couple to $S = 0$ to give states with total angular momentum $J = L$. These states are denoted by -----; experimentally they are difficult to detect. The intrinsic quark spins can also couple to give $S = 1$. States with $S = 1$ are denoted by —. Spin–orbit coupling splits the P states with $S = 1$ to give rise to states with $J^P = 0^+, 1^+, 2^+$. This spin–orbit splitting is apparent in the figure. All the $S = 1$ states shown have been measured.

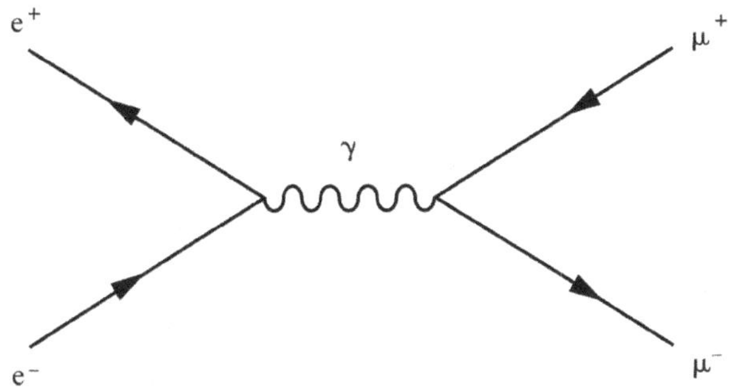

Figure 1.6 The lowest order Feynman diagram (Chapter 8) for electromagnetic $\mu^+ \mu^-$ pair production in e^+e^- collisions.

As fundamental particles, quarks have the same electrodynamics as muons, apart from the magnitude of their electric charge. The Feynman diagrams that dominate the numerator of R in this range 10 GeV to 40 GeV are shown in Fig. 1.8. (The top quark has a mass $\sim$ 174 GeV/c^2 and will not contribute.) For each quark process the formula (1.2) holds, except that e is replaced by the quark's electric charge at the quark vertex, which suggests

$$R = \left(\frac{2}{3}\right)^2 + \left(\frac{1}{3}\right)^2 + \left(\frac{2}{3}\right)^2 + \left(\frac{1}{3}\right)^2 + \left(\frac{1}{3}\right)^2 = \frac{11}{9}. \tag{1.4}$$

This value is too low, by a factor of about 3.

In the Standard Model, the discrepancy is resolved by introducing the idea of quark *colour*. A quark not only has a flavour index, u, d, s, c, b, t, but also, for each flavour, a colour index. There are postulated to be three basic states of colour, say red, green and blue (r, g, b). With three quark colour states to each flavour, we have to multiply the R of (1.4) by 3, to obtain

$$R = \frac{11}{3}, \tag{1.5}$$

which is in excellent agreement with the data of Fig. 1.7.

This invention of colour not only solves the problem of R but, most significantly, solves the problem of the symmetry of the baryon states. We have seen (Section 1.5) that in the absence of any new quantum number baryon states are completely symmetric in the interchange of two quarks. However, if these state functions are multiplied by an antisymmetric colour state function, the overall state becomes antisymmetric, and the Pauli principle is preserved.

Strong support for the mechanism of quark production represented by the Feynman diagrams of Fig. 1.8 is given by other features in the data from $e^+ e^-$ colliders. An $e^+ e^-$ annihilation at high energies produces many hadrons.

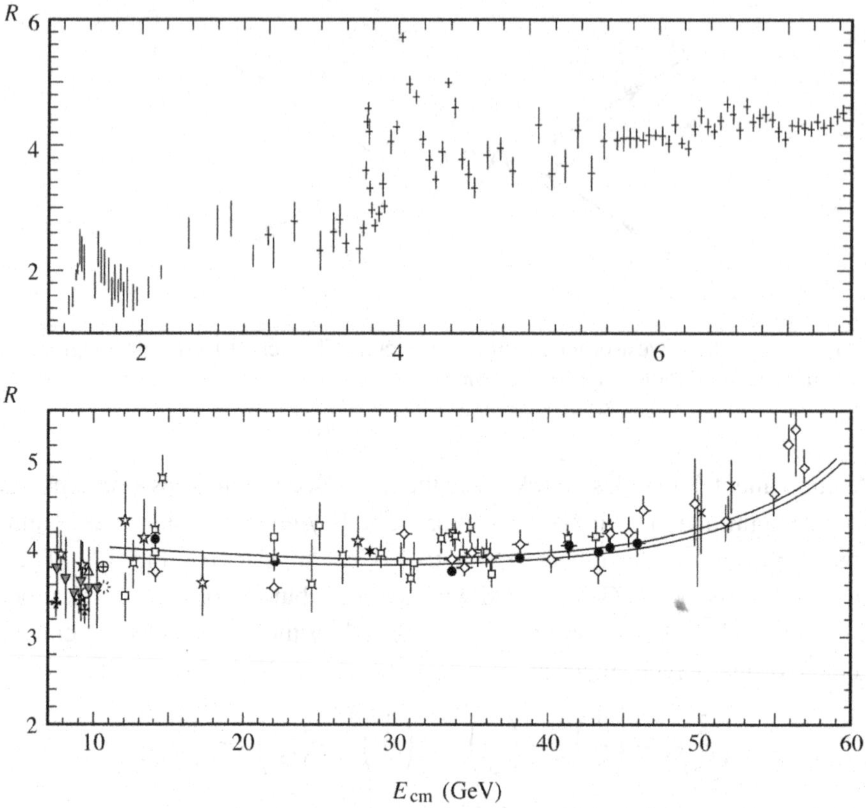

Figure 1.7 Measurements of $R(E)$ from the resonance region 1 GeV < E < 11 GeV into the region 11 GeV < E < 60 GeV, which contains no prominent resonances and no quark–antiquark production threshold. For E > 11 GeV two curves are shown of calculations that take account of quark colour and include electroweak corrections and strong interaction (QCD) effects. (Adapted from Particle Data Group (1996).)

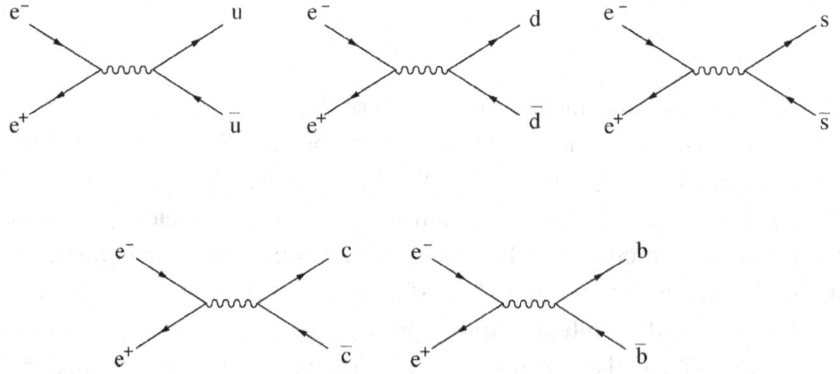

Figure 1.8 The lowest order Feynman diagrams for quark–antiquark pair production in $e^+ e^-$ collisions at energies below the Z threshold.

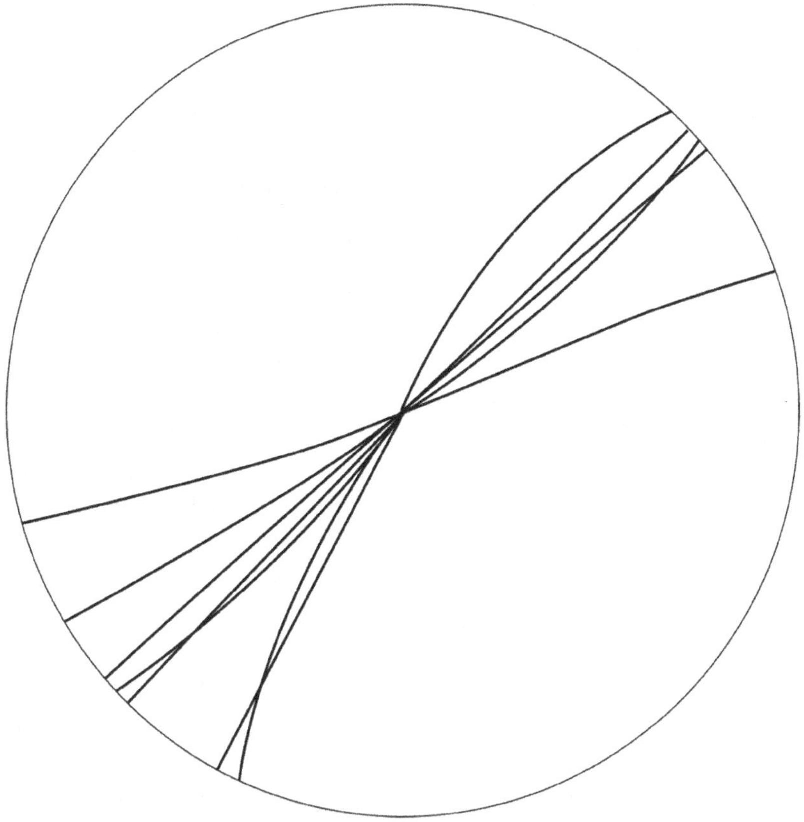

Figure 1.9 An example of an $e^+ e^-$ annihilation event that results in two jets of hadrons. The figure shows the projection of the charged particle tracks onto a plane perpendicular to the axis of the $e^+ e^-$ beams. This figure was taken from an event in the TASSO detector at PETRA DESY.

These are mostly correlated into two back-to-back *jets*. An example is shown in Fig. 1.9. (The charged particle tracks are curved because of the presence of an external magnetic field: the curvature is related to the particle's momentum.) The direction of a jet may be defined as the direction at the point of production of the total momentum of all the hadrons associated with it. The momenta of two back-to-back jets are equal and opposite. The jet directions may be presumed to be the directions of the initial quark–antiquark pair. This interpretation is corroborated by an examination of the angular distribution of the jet directions of two-jet events from many annihilations, with respect to the $e^+ e^-$ beams. The angular distribution is the same as that for muons (equation (1.1)) after allowance has been made for the Z contribution, which becomes significant as the energy for Z production is approached.

The hadron jets result from the original quark and antiquark combining with quark–antiquark pairs generated from the vacuum. The precise details of the processes involved are not yet fully understood.

1.8 Electron scattering from nucleons

There is a clear advantage in using electrons to probe the proton and neutron, since electrons interact with quarks primarily through electromagnetic forces that are well understood: the weak interaction is negligible in the scattering process, except at very high energy and large scattering angle, and the strong interaction is not directly involved.

In the 1950s, experiments at Stanford on nucleon targets at rest in the laboratory revealed the electric charge distribution in the proton and (using scattering data from deuterium targets) the neutron. These early experiments were performed at electron energies $\leq 500\,\text{MeV}$ (Hofstadter *et al.*, 1958). Scattering at higher energies has thrown more light on the behaviour of quarks in the proton. At these energies inelastic electron scattering, which involves meson production, becomes the dominant mode.

At the electron–proton collider HERA at Hamburg, a beam of 30 GeV electrons met a beam of 820 GeV protons head on. Many features of the ensuing electron–proton collisions are well described by the *parton model*, which was introduced by Feynman in 1969. In the parton model each proton in the beam is regarded as a system of sub-particles called *partons.* These are quarks, antiquarks and gluons. Quarks and antiquarks are the particles that carry electric charge. The basic idea of the parton model is that at high energy–momentum transfer Q^2, an electron scatters from an effectively free quark or antiquark and the scattering process is completed before the recoiling quark or antiquark has time to interact with its environment of quarks, antiquarks and gluons. Thus in the calculation of the inclusive cross-section the final hadronic states do not appear.

In the model, at large Q^2 both the electron and the struck quark are deflected through large angles. Figure 1.10 shows an example of an event from the ZEUS detector at HERA. The transverse momentum of the scattered electron is balanced by a jet of hadrons, which can be associated with the recoiling quark. Another jet, the 'proton remnant' jet is confined to small angles with respect to the proton beam. Events like these give further strong support to the parton model.

The success of the parton model in interpreting the data gives added support to the concept of quarks. The parton model is not strictly part of our main theme but, in view of its interest and importance in particle physics, a simple account of the model and its relation to experiment is given in Appendix D.

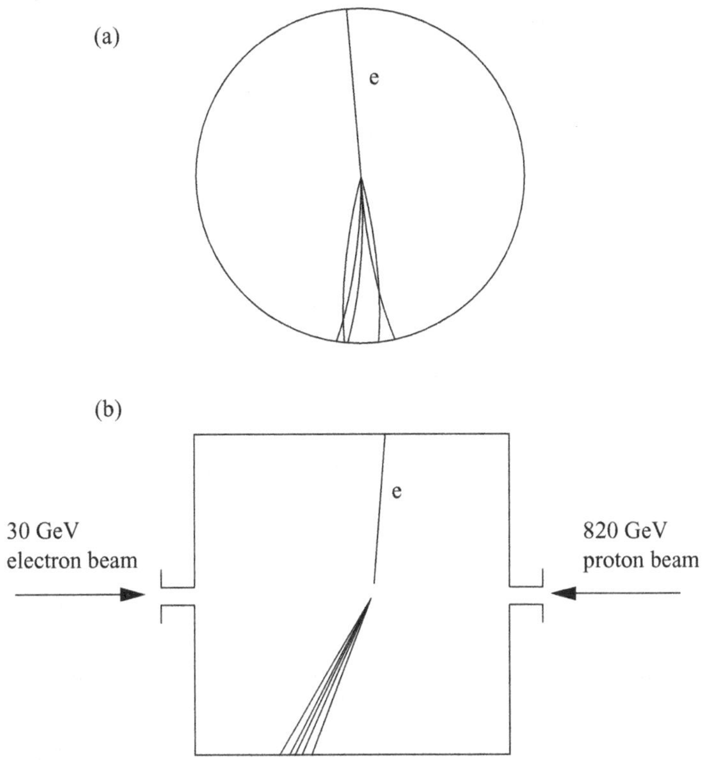

Figure 1.10 This figure illustrating particle tracks is taken from an event in the ZEUS detector at HERA, DESY. Figure 1.10(a) is the event projected onto a plane perpendicular to the axis of the beams. Figure 1.10(b) is the event projected onto a plane passing through the axis of the beams.

A hadron jet has been ejected from the proton by an electron. The track of the recoiling electron is marked e. The initiating beams and the proton remnant jet are confined to the beam pipes and are not detected.

1.9 Particle accelerators

Progress in our understanding of Nature has come through the interplay between theory and experiment. In particle physics, experiment now depends primarily on the great particle accelerators and ingeneous and complex particle detectors, which have been built, beginning in the early 1930s with the Cockroft–Walton linear accelerator at Cambridge, UK, and Lawrence's cyclotron at Berkeley, USA. The Cambridge machine accelerated protons to 0.7 MeV; the first Berkeley cyclotron accelerated protons to 1.2 MeV. For a time after 1945 important results were obtained using cosmic radiation as a source of high energy particles, events being detected in photographic emulsion, but in the 1950s new accelerators

Table 1.5. *Some particle accelerators*

Machine	Particles collided	Start date–end date
TEVATRON (Fermilab, Batavia, Il)	p: 900 GeV $\bar{\text{p}}$: 900 GeV	1987
SLC (SLAC, Stanford)	e^+ : 50 GeV e^- : 50 GeV	1989–1998
HERA (DESY, Hamburg)	e: 30 GeV p: 820 GeV	1992
LEP2 (CERN, Geneva)	e^+ : 81GeV e^- : 81GeV	1996–2000
PEP-II (SLAC, Stanford)	e^- : 9 GeV e^+ : 3.1 GeV	1999–2008
LHC (CERN, Geneva)	p: 7 TeV p: 7 TeV	2008

provided beams of particles of increasingly high energies. Some of the machines, past, present and future, are listed in Table 1.5.. Detailed parameters of these machines, and of others, may be found in Particle Data Group (2005).

The TEVATRON at Fermilab is where the top quark was discovered. The physics of the top quark is as yet little explored. It makes only a brief appearance in our text, though it is an essential part of the pattern of the Standard Model. The upgraded LEP2 at CERN is able to create $W^+ W^-$ pairs, and will allow detailed studies of the weak interaction. At Stanford, PEP-II and the associated 'BaBar' ($B\bar{B}$) detector is designed to study charge conjugation, parity (*CP*) violation. The way in which *CP* violation appears in the Standard Model is discussed in Chapter 18.

The most ambitious machine likely to be built in the immediate future is the Large Hadron Collider (LHC) at CERN. It is expected that with this machine it will be possible to observe the Higgs boson, if such a particle exists. The Higgs boson is an essential component of the Standard Model; we introduce it in Chapter 10. It is also widely believed that the physics of Supersymmetry, which perhaps underlies the Standard Model, will become apparent at the energies, up to 14 TeV, which will be available at the LHC.

1.10 Units

In particle physics it is usual to simplify the appearance of equations by using units in which $\hbar = 1$ and $c = 1$. In electromagnetism we set $\varepsilon_0 = 1$ (so that the force between charges q_1 and q_2 is $q_1 q_2/4\pi r^2$), and $\mu_0 = 1$, to give $c^2 = (\mu_0\varepsilon_0)^{-1} = 1$.

We shall occasionally reinsert factors of $\hbar$ and c where it may be reassuring or illuminating, or for the purposes of calculation. It is useful to remember that

$$\hbar c \approx 197\,\text{MeV fm}, \quad e^2\,4\pi \approx 1.44\,\text{MeV fm},$$
$$\alpha = e^2/4\pi\hbar c \approx (1/137), \quad c \approx 3 \times 10^{23}\,\text{fm s}^{-1}.$$

Energies, masses and momenta are usually quoted in MeV or GeV, and we shall follow this convention.

2

Lorentz transformations

The equations of the Standard Model must be consistent with Einstein's principle of relativity, which states that the laws of Nature take the same form in every inertial frame of reference. An inertial frame is one in which a free body moves without acceleration. An earth-bound frame approximates to an inertial frame if the gravitational field of the earth is introduced as an external field. We shall assume that the reader is familiar with rotations, and with proper Lorentz transformations and the relativistic mechanics of particle collisions. This chapter is very largely about notation, which may make for dry reading; however an appropriate notation is crucial to the exposition of any theory, and particularly so to a relativistic theory, such as the Standard Model.

2.1 Rotations, boosts and proper Lorentz transformations

The time and space coordinates of an event measured in different inertial frames of reference are related by a Lorentz transformation. A rotation is a special case of a Lorentz transformation. Consider, for example, a frame K' that is rotated about the z-axis with respect to a frame K, by an angle θ. If $(t, \mathbf{r})$ are the time and space coordinates of an event observed in K, then in K' the event is observed at $(t', \mathbf{r}')$ and

$$\begin{aligned} t' &= t \\ x' &= x\cos\theta + y\sin\theta \\ y' &= -x\sin\theta + y\cos\theta \\ z' &= z. \end{aligned} \tag{2.1}$$

Lorentz transformations also relate events observed in frames of reference that are moving with constant velocity, one with respect to the other. Consider, for example, an inertial frame K' moving in the z-direction in a frame K with velocity v, the spatial axes of K and K' being coincident at $t = 0$. If $(t, \mathbf{r})$ are the time and space coordinates of an event observed in K, and $(t', \mathbf{r}')$ are the coordinates of the same event observed in K', the transformation takes the form

$$\begin{aligned} ct' &= \gamma(ct - \beta z) \\ x' &= x \\ y' &= y \\ z' &= \gamma(z - \beta ct), \end{aligned} \tag{2.2}$$

where c is the velocity of light, $\beta = v/c,\ \gamma = (1 - \beta^2)^{-1/2}$.

Putting $x^0 = ct,\ x^1 = x,\ x^2 = y,\ x^3 = z$, the x^μ are dimensionally homogeneous, and an event in K is specified by the set x^μ, where $\mu = 0, 1, 2, 3$. Greek indices in the text will in general take these values. With this more convenient notation, we may write the Lorentz transformation (2.2) as

$$\begin{aligned} x'^0 &= x^0 \cosh\theta - x^3 \sinh\theta \\ x'^1 &= x^1 \\ x'^2 &= x^2 \\ x'^3 &= -x^0 \sinh\theta + x^3 \cosh\theta, \end{aligned} \tag{2.3}$$

where we have put $\beta = v/c = \tanh\theta$; then $\gamma = \cosh\theta$.

Transformations to a frame with parallel axes but moving in an arbitrary direction are called *boosts*. A general Lorentz transformation between inertial frames K and K' whose origins coincide at $x^0 = x'^0 = 0$ is a combination of a rotation and a boost. It is specified by six parameters: three parameters to give the orientation of the K' axes relative to the K axes, and three parameters to give the components of the velocity of K' relative to K. Such a general transformation is of the form

$$x'^\mu = L^\mu{}_\nu x^\nu, \tag{2.4}$$

where the elements $L^\mu{}_\nu$ of the transformation matrix are real and dimensionless. We use here, and subsequently, the *Einstein summation convention*: a repeated 'dummy' index is understood to be summed over, so that in (2.4) the notation $\sum_{\nu=0}^{3}$ has been omitted on the right-hand side. The matrices $L^\mu{}_\nu$ form a group, called the *proper Lorentz group* (Problem 2.6 and Appendix B). The significance of the placing of the superscript and the subscript will become evident shortly.

The *interval* $(\Delta s)^2$ between events x^μ and $x^\mu + \Delta x^\mu$ is defined to be

$$(\Delta s)^2 = (\Delta x^0)^2 - (\Delta x^1)^2 - (\Delta x^2)^2 - (\Delta x^3)^2. \tag{2.5}$$

It is a fundamental property of a Lorentz transformation that it leaves the interval between two events invariant:

$$(\Delta s')^2 = (\Delta s)^2. \tag{2.6}$$

We can express $(\Delta s)^2$ more compactly by introducing the *metric tensor* $(g_{\mu\nu})$:

$$(g_{\mu\nu}) = \begin{pmatrix} 1 & 0 & 0 & 0 \\ 0 & -1 & 0 & 0 \\ 0 & 0 & -1 & 0 \\ 0 & 0 & 0 & -1 \end{pmatrix}. \tag{2.7}$$

Then

$$(\Delta s)^2 = g_{\mu\nu}\Delta x^\mu \Delta x^\nu, \tag{2.8}$$

where the repeated upper and lower indices are summed over. Note that $g_{\mu\nu} = g_{\nu\mu}$; it is a symmetric tensor. It has the same elements in every frame of reference.

2.2 Scalars, contravariant and covariant four-vectors

Quantities, such as $(\Delta s)^2$, which are invariant under Lorentz transformations are called *scalars*. We define a *contravariant four-vector* to be a set a^μ which transforms like the set x^μ under a proper Lorentz transformation:

$$a'^\mu = L^\mu{}_\nu a^\nu. \tag{2.9}$$

A familiar example of a contravariant four-vector is the energy–momentum vector of a particle $(E/c, \mathbf{p})$.

We define the corresponding *covariant four-vector* a_μ, carrying a subscript, rather than a superscript, by

$$a_\mu = g_{\mu\nu} a^\nu. \tag{2.10}$$

Hence if $a^\mu = (a^0, \mathbf{a})$, then $a_\mu = (a^0, -\mathbf{a})$.

We can write the invariant Δs^2 as

$$\Delta s^2 = g_{\mu\nu}\Delta x^\mu \Delta x^\nu = \Delta x_\nu \Delta x^\nu.$$

More generally, if a^μ, b^μ are contravariant four-vectors, the *scalar product*

$$g_{\mu\nu} a^\mu b^\nu = a_\mu b^\mu = a^\mu b_\mu = a^0 b^0 - \mathbf{a}\cdot\mathbf{b} \tag{2.11}$$

is invariant under a Lorentz transformation.

We can define the contravariant metric tensor $g^{\mu\nu}$ so that

$$\alpha^\mu = g^{\mu\nu} a_\nu. \tag{2.12}$$

The elements of $g^{\mu\nu}$ are evidently identical to those of $g_{\mu\nu}$.

The transformation law for covariant vectors, which we write

$$a'_\mu = L_\mu{}^\nu a_\nu, \tag{2.13}$$

follows from that for contravariant vectors (Problem 2.1). Note that, in general, $L_\mu{}^\nu$ is not equal to $L_\nu{}^\mu$ (Problem 2.1). Using the invariance of the scalar product (2.11), we have

$$a'_\mu b'^\mu = L_\mu{}^\nu L^\mu{}_\rho a_\nu b^\rho = a_\nu b^\nu$$

and

$$a'^\mu b'_\mu = L^\mu{}_\nu L_\mu{}^\rho a^\nu b_\rho = a^\nu b_\nu.$$

Since the a_μ and b_μ are arbitrary, it follows that

$$L^\mu{}_\nu L_\mu{}^\rho = L_\mu{}^\nu L^\mu{}_\rho = \delta^\rho_\nu \tag{2.14}$$

where

$$\delta^\rho_\nu = \delta^\nu_\rho = \begin{cases} 1, & \rho = \nu \\ 0, & \rho \neq \nu. \end{cases}$$

2.3 Fields

The Standard Model is a theory of fields. We shall be concerned with fields that at each point x of space and time transform as scalars, or vectors, or tensors (defined later in this section). We use x to stand for the set (x^0, x^1, x^2, x^3). For example, we shall see that the electromagnetic potentials form a four-vector field, and the electromagnetic field is a tensor field. We shall also be concerned with scalar fields $\phi(x)$, which by definition transform simply as

$$\phi'(x') = \phi(x), \tag{2.15}$$

where x' and x refer to the same point in space-time.

We can construct a vector field from a scalar field. Consider the change of field $\mathrm{d}\phi$ in moving from x to a neighbouring point $x + \mathrm{d}x$, with $\mathrm{d}x$ infinitesimal. Then

$$\mathrm{d}\phi = \frac{\partial \phi}{\partial x^\mu}\mathrm{d}x^\mu$$

is invariant under a Lorentz transformation. Since the set $\mathrm{d}x^\mu$ make up an arbitrary contravariant infinitesimal vector, the set $\partial\phi/\partial x^\mu$ must make up a covariant vector (Problem 2.3). Following the subscript convention we write

$$\frac{\partial \phi}{\partial x^\mu} = \left(\frac{1}{c}\frac{\partial \phi}{\partial t}, \nabla\phi\right) = \partial_\mu \phi. \tag{2.16}$$

We can then also define the contravariant vector

$$\partial^\mu \phi = g^{\mu\nu}\partial_\nu \phi = \frac{\partial \phi}{\partial x_\mu} = \left(\frac{1}{c}\frac{\partial \phi}{\partial t}, -\nabla\phi\right). \tag{2.17}$$

It follows that

$$\partial_\mu \phi \partial^\mu \phi = \left(\frac{1}{c}\frac{\partial \phi}{\partial t}\right)^2 - (\nabla\phi)^2 \tag{2.18}$$

and

$$\partial_\mu \partial^\mu \phi = \frac{1}{c^2}\frac{\partial^2 \phi}{\partial t^2} - \nabla^2\phi \tag{2.19}$$

are invariant under Lorentz transformations.

We can define, and we shall need, tensor quantities. Tensors $T^{\mu\nu}$, $T_{\mu\nu}$, $T^{\mu}{}_{\nu}$, $T^{\mu\nu}{}_{\lambda}$, etc., are defined as quantities which transform under a Lorentz transformation in the same way as $a^\mu a^\nu$, $a_\mu a_\nu$, $a^\mu a_\nu$, $a^\mu a^\nu a_\lambda$, etc. For example,

$$T'^{\mu\nu} = L^{\mu}{}_{\rho} L^{\nu}{}_{\lambda} T^{\rho\lambda}.$$

The 'contraction' by summation of a repeated upper and lower index leaves the transformation properties determined by what remains. For example, $T^{\mu}{}_{\mu}$ is a scalar, $T^{\mu\nu}{}_{\mu}$ is a contravariant four-vector. The metric tensors $g_{\mu\nu}$, $g^{\mu\nu}$ conform with the definition, and this leads to the conditions on the matrix elements $L^{\mu}{}_{\nu}$:

$$g_{\mu\nu} = g_{\rho\lambda} L^{\rho}{}_{\mu} L^{\lambda}{}_{\nu}. \tag{2.20}$$

The conditions (2.20) and (2.14) are equivalent.

As well as scalars, vectors and tensors there are also very important objects called *spinors*, and *spinors fields*, which have well-defined rules of transformation under a Lorentz transformation of the coordinates. Their properties are discussed in Appendix B and Chapter 5.

2.4 The Levi–Civita tensor

The Levi–Civita tensor $\varepsilon_{\mu\nu\lambda\rho}$ is defined by

$$\varepsilon_{\mu\nu\lambda\rho} = \begin{cases} +1 & \text{if } \mu, \nu, \lambda, \rho \text{ is an even permutation of } 0, 1, 2, 3; \\ -1 & \text{if } \mu, \nu, \lambda, \rho \text{ is an odd permutation of } 0, 1, 2, 3; \\ 0 & \text{otherwise.} \end{cases} \tag{2.21}$$

For example, $\varepsilon_{1023} = -1$, $\varepsilon_{1203} = +1$, $\varepsilon_{0023} = 0$.

It is straightforward to verify that $\varepsilon_{\mu\nu\lambda\rho}$ satisfies

$$\begin{aligned} \varepsilon'_{\mu\nu\lambda\rho} &= L_{\mu}{}^{\alpha} L_{\nu}{}^{\beta} L_{\lambda}{}^{\gamma} L_{\rho}{}^{\delta} \varepsilon_{\alpha\beta\gamma\delta} \\ &= \varepsilon_{\mu\nu\lambda\mu} \det(L) = \varepsilon_{\mu\nu\lambda\mu}, \end{aligned}$$

using the definition of a determinant (Appendix A), and the result that the determinant of the transformation matrix is 1 (Problems 2.4 and 2.5).

The corresponding Levi–Civita symbol in three dimensions, ε_{ijk}, is defined similarly. It is useful in the construction of volumes, since

$$\varepsilon_{ijk}A^i B^j C^k = \mathbf{A}\cdot(\mathbf{B}\times\mathbf{C})$$

is the volume of the parallelepiped defined by the vectors **A, B, C**. The four-dimensional Levi–Civita tensor enables one to construct four-dimensional volumes $\varepsilon_{\mu\nu\lambda\rho}a^\mu b^\nu c^\lambda d^\rho$. The contraction of indices leaves this a Lorentz scalar. In particular, taking *a,b,c,d* to be infinitesimal elements parallel to the axes $0x^\mu$ so that $a = (\mathrm{d}x^0, 0, 0, 0)$, $b = (0, \mathrm{d}x^1, 0, 0)$, $c = (0, 0, \mathrm{d}x^2, 0)$, $d = (0, 0, 0, \mathrm{d}x^3)$, it follows that the 'volume' element of space-time

$$\mathrm{d}^4x = \mathrm{d}x^0\mathrm{d}x^1\mathrm{d}x^2\mathrm{d}x^3 = cd^3x\ \mathrm{d}t$$

is a Lorentz invariant scalar (see also Problem 2.9).

2.5 Time reversal and space inversion

The operations of time reversal:

$$x'^0 = -x^0,$$
$$x'^i = x^i, \quad i = 1, 2, 3,$$

and space inversion:

$$x'^0 = x^0$$
$$x'^i = -x^i, \quad i = 1, 2, 3,$$

also leave $(\Delta s)^2$ invariant, but these transformations are excluded from the proper Lorentz group. They are however of interest, and will arise in later chapters.

Problems

2.1 Show that $L_\mu{}^\nu = g_{\mu\rho}L^\rho{}_\lambda g^{\lambda\nu}$. Verify $L_0{}^1 = -L^1{}_0$.

2.2 Using (2.14), show that the inverse transformations to (2.9) and (2.13) are

$$a^\mu = a'^\nu L_\nu{}^\mu, \quad a_\mu = a'_\nu L^\nu{}_\mu.$$

Hence show

$$L_\nu{}^\mu L^\rho{}_\mu = \delta^\rho_\nu.$$

2.3 Prove that if $\phi(x)$ is a scalar field, the set $(\partial\phi/\partial x^\mu)$ makes up a covariant vector field.

2.4 Using Problem 2.1, show that $\det(L^{\mu}{}_{\nu}) = \det(L_{\mu}{}^{\nu})$ and hence show, using equation (2.14), that

$$\det(L^{\mu}{}_{\nu}) = \pm 1.$$

2.5 Show that $\det(L^{\mu}{}_{\nu})$ for both the rotation (2.1) and the boost (2.3) is equal to $+1$. This is a general property of proper Lorentz transformations that distinguishes them from space reflections and time reversal (Section 2.5), for which the determinant of the transformation equals -1.

2.6 Show that the matrices $L_{\mu}{}^{\nu}$ corresponding to proper Lorentz transformations form a group.

2.7 Show that δ^{μ}_{ν} is a tensor.

2.8 The frequency ω and wave vector $\mathbf{k}$ of an electromagnetic wave in free space make up a contravariant four-vector

$$k = (\omega/c, \mathbf{k}).$$

The invariant $k_{\mu}k^{\mu} = 0$; this corresponds to the dispersion relation $\omega^2 = c^2\mathbf{k}^2$. Show that a wave propagating with frequency ω in the z-direction, if viewed from a frame moving along the z-axis with velocity v, is seen to be Doppler shifted in frequency, with

$$\omega' = \mathrm{e}^{-\theta}\omega = \sqrt{\frac{1 - v/c}{1 + v/c}}\,\omega.$$

2.9 By considering the Jacobian of the Lorentz transformation, show that the four-dimensional volume element $\mathrm{d}^4x = \mathrm{d}x^0\mathrm{d}x^1\mathrm{d}x^2\mathrm{d}x^3$ is a Lorentz invariant.

2.10 Show that $\varepsilon_{\mu\nu\lambda\rho}$ is a *pseudo-tensor*, i.e. it changes sign under the operation of space inversion.

3

The Lagrangian formulation of mechanics

In most introductory texts on quantum mechanics you will find 'Hamiltonian' in the index (see our equation (3.8)) but you are less likely to find 'Lagrangian'. However, quantum field theories are most conveniently described in a Lagrangian formalism, to which this chapter is an introduction.

3.1 Hamilton's principle

The classical dynamics of a mechanical (non-dissipative) system is most elegantly derived from *Hamilton's principle.* A closed mechanical system is completely characterised by its Lagrangian $L(q, \dot{q})$; the variables $q(t)$, which are functions of time, are a set of coordinates $q_1(t), q_2(t), \ldots, q_s(t)$ which determine the configuration of the system at time t. In particular, the q_i might be the Cartesian coordinates of a set of interacting particles. We restrict our discussion to the case where all the $q_i(t)$ are independent. In non-relativistic mechanics we take $L = T - V$, where $T(q, \dot{q})$ is the kinetic energy of the system and $V(q)$ is its potential energy.

Given L, *the action* S is defined by

$$S = \int_{t_1}^{t_2} L(q, \dot{q})\, \mathrm{d}t. \tag{3.1}$$

The value of S depends on the path of integration in q-space. The end-points of the path are fixed at times t_1 and t_2, but the path is otherwise unrestricted. S is said to be a *functional* of $q(t)$. Hamilton's principle states that S is stationary for that particular path in q-space determined by the equations of motion, so that if we consider a variation to an arbitrary neighbouring path (Fig. 3.1), $\delta S = 0$, where

$$\begin{aligned}\delta S &= \delta \int_{t_1}^{t_2} L(q, \dot{q})\, \mathrm{d}t \\ &= \int_{t_1}^{t_2} \sum_i \left[\frac{\partial L}{\partial q_i}\delta q_i + \frac{\partial L}{\partial \dot{q}}\delta \dot{q}_i\right] \mathrm{d}t.\end{aligned}$$

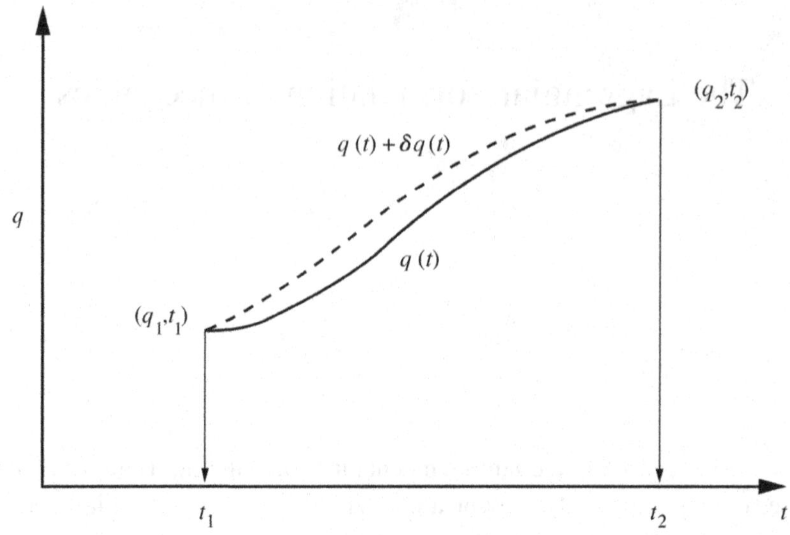

Figure 3.1 A schematic representation of the path in q-space determined by the equations of motion (full line) and a neighbouring path (dashed line).

Since $\delta\dot{q} = \mathrm{d}(\delta q)/\mathrm{d}t$, we can integrate the second term in this integral by parts, to give

$$\delta S = \int_{t_1}^{t_2} \sum_i \left[\frac{\partial L}{\partial q_i} - \frac{\mathrm{d}}{\mathrm{d}t}\left(\frac{\partial L}{\partial \dot{q}_i} \right) \right] \delta q_i \, \mathrm{d}t. \tag{3.2}$$

The 'end-point' contributions from the integration by parts are zero, since $\delta q(t_1) = \delta q(t_2) = 0$.

The variations $\delta q_i(t)$ are arbitrary. It follows from (3.2) that the condition $\delta S = 0$ requires

$$\frac{\mathrm{d}}{\mathrm{d}t}\left(\frac{\partial L}{\partial \dot{q}_i} \right) - \frac{\partial L}{\partial q_i} = 0, \quad i = 1, ..., s. \tag{3.3}$$

These are the *Euler–Lagrange equations of motion.* In classical non-relativistic mechanics they are equivalent to Newton's equations of motion. As a simple example, consider a particle of mass m moving in one dimension in a potential $V(x)$. Then $L = T - V = (m\dot{x}^2/2) - V(x)$. From (3.3) we have immediately $m\ddot{x} = -\partial V/\partial x$, which is Newton's equation of motion for the particle.

An external, and possibly time-dependent, field can be included in the Lagrangian formalism through a time-dependent potential. In our one-dimensional example above, $V(x)$ may be replaced by $V(x,t)$. Making the Lagrangian L depend explicitly on t does not affect the derivation of the field equations.

It is important to note that the Lagrangian of a given system is *not unique:* we can add to L any function of the form $\mathrm{d}f(q,t)/\mathrm{d}t$ where $f(q,t)$ is an arbitrary function of q and t. Such a term gives a contribution $[f(q_2, t_2) - f(q_1, t_1)]$ to S, independent of the path, and hence leaves the equations of motion unchanged.

3.2 Conservation of energy

In the case of a closed system of particles, interacting only among themselves, the equations of motion of the system do not depend explicitly on the time t, since the physics of a closed system does not depend on our choice of the origin of time. There is no reason to doubt that the laws of physics at the time of Archimedes, or the time of Newton, were the same as they are for us. Hence for a closed system we must be able to construct a Lagrangian $L(q, \dot{q})$ that does not depend explicitly on t. For such a Lagrangian,

$$\frac{\mathrm{d}L}{\mathrm{d}t} = \sum_i \left[\frac{\partial L}{\partial q_i}\dot{q}_i + \frac{\partial L}{\partial \dot{q}_i}\ddot{q}_i \right].$$

Taking the $q_i(t)$ to obey the equations of motion and substituting for $\partial L/\mathrm{d}q_i$ from (3.3) we obtain

$$\frac{\mathrm{d}L}{\mathrm{d}t} = \sum_i \left[\left(\frac{\mathrm{d}}{\mathrm{d}t}\frac{\partial L}{\partial \dot{q}_i} \right)\dot{q}_i + \frac{\partial L}{\partial \dot{q}_i}\ddot{q}_i \right] = \sum_i \frac{\mathrm{d}}{\mathrm{d}t}\left(\frac{\partial L}{\partial \dot{q}_i}\dot{q}_i \right)$$

or

$$\frac{\mathrm{d}}{\mathrm{d}t}\left[\sum_i \frac{\partial L}{\partial \dot{q}_i}\dot{q}_i - L \right] = 0. \tag{3.4}$$

Thus

$$E = \left[\sum_i \frac{\partial L}{\partial \dot{q}_i}\dot{q}_i - L \right] \tag{3.5}$$

remains constant during the motion, and is called the *energy* of the system. This result exemplifies Noether's theorem (Section 1.2): we have here a conservation law stemming from the symmetry of the Lagrangian under a translation in time.

For a closed system of non-relativistic particles, with a potential function $V(q_i)$, $\partial L/\partial \dot{q}_i = \partial T/\partial \dot{q}_i$. Since the kinetic energy T is a quadratic function of the $\dot{q}_i$ (Problem 3.1), $(\partial T/\partial \dot{q}_i)\dot{q}_i = 2T$. Hence

$$E = 2T - (T - V) = T + V.$$

We recover the result of elementary mechanics.

The *generalised momenta*, p_i, are defined by

$$p_i = \frac{\partial L}{\partial \dot{q}_i}. \tag{3.6}$$

The *Hamiltonian* of a system is defined by

$$H(p, q) = \sum_i p_i \dot{q}_i - L. \tag{3.7}$$

In terms of p and q, the energy equation (3.5) for a closed system becomes

$$H(p, q) = E. \tag{3.8}$$

This equation, which is a consequence of the homogeneity of time, is a foundation stone for making the transition from classical to quantum mechanics.

3.3 Continuous systems

To see how Hamilton's principle may be extended to continuous systems, we consider a flexible string, of mass ρ per unit length, stretched under tension F between two fixed points at $x = 0$ and $x = l$, say, but subject to small transverse displacements in a plane. Gravity is neglected. If $\phi(x, t)$ is the transverse displacement from equilibrium of an element $\mathrm{d}x$ of the string at x, at time t, then the length of the string is

$$\int_0^l (\mathrm{d}x^2 + \mathrm{d}\phi^2)^{1/2} = \int_0^l [1 + (\partial\phi/\partial x)^2]^{1/2}\mathrm{d}x.$$

To leading order in $\partial\phi/\partial x$, which we take to be small for small displacements, the extension of the string is $\int_0^l \frac{1}{2}(\partial\phi/\partial x)^2\mathrm{d}x$, and the potential energy of stretching under the tension F is $\int_0^1 \frac{1}{2}F(\partial\phi/\partial x)^2\mathrm{d}x$. The kinetic energy of the string is $\int_0^1 \frac{1}{2}\rho(\partial\phi/\partial t)^2\mathrm{d}x$. Hence

$$L = T - V = \int_0^1 \mathcal{L}\,\mathrm{d}x, \tag{3.9}$$

where

$$\mathcal{L} = \frac{1}{2}\rho\left(\frac{\partial\phi}{\partial t}\right)^2 - \frac{1}{2}F\left(\frac{\partial\phi}{\partial x}\right)^2 \tag{3.10}$$

is called the *Lagrangian density*.

The corresponding action is

$$S = \int_0^1 \mathrm{d}x \int_{t_1}^{t_2} \mathrm{d}t\,\mathcal{L}(\dot{\phi}, \phi'),$$

writing $\partial\phi/\partial t = \dot{\phi}$ and $\partial\phi/\partial x = \phi'$.

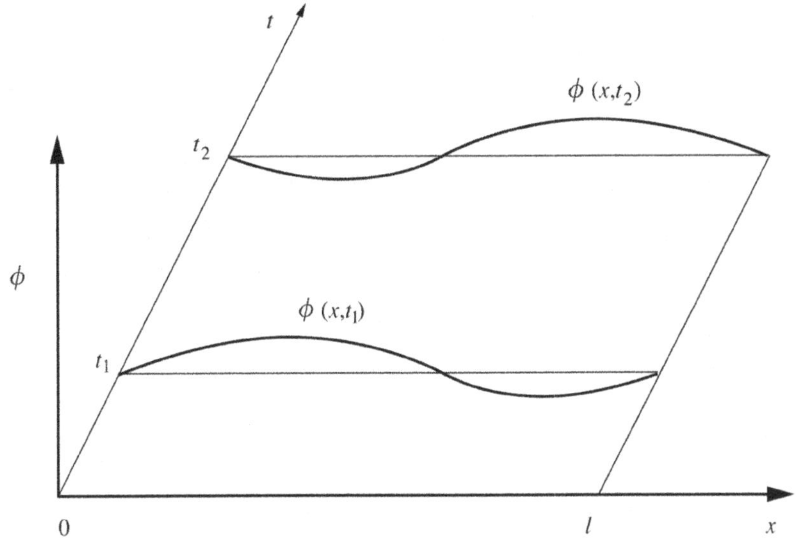

Figure 3.2 The actual motion of the string between an initial displacement $\phi(x, t_1)$ and a final displacement $\phi(x, t_2)$ generates a surface in space-time.

Hamilton's principle states that the action is stationary for that surface that describes the actual motion of the string between its initial displacement $\phi(x, t_1)$ and its final displacement $\phi(x, t_2)$ (Fig. 3.2). We have

$$\delta S = \int_0^1 \mathrm{d}x \int_{t_1}^{t_2} \mathrm{d}t \left[\frac{\partial \pounds}{\partial \dot{\phi}} \delta(\dot{\phi}) + \frac{\partial \pounds}{\partial \phi'} \delta(\phi') \right].$$

Using $\delta(\dot{\phi}) = \partial(\delta\phi)/\partial t$ and $\delta(\phi') = \partial(\delta\phi)/\mathrm{d}x$ we integrate each term by parts. Again, the boundary contributions are zero since

$$\begin{aligned} \delta\phi(x, t_1) = \delta\phi(x, t_2) = 0 \quad &\text{for all } x, \\ \delta\phi(0, t) = \delta\phi(l, t) = 0 \quad &\text{for all } t. \end{aligned}$$

We are left with

$$\delta S = -\int_0^1 \mathrm{d}x \int_{t_1}^{t_2} \mathrm{d}t \left[\frac{\partial}{\partial t} \left(\frac{\partial \pounds}{\partial \dot{\phi}} \right) + \frac{\partial}{\partial x} \left(\frac{\partial \pounds}{\partial \phi'} \right) \right] \delta\phi. \tag{3.11}$$

Since $\delta\phi(x, t)$ is arbitrary, the condition $\delta S = 0$ gives

$$\frac{\partial}{\partial t} \left(\frac{\partial \pounds}{\partial \dot{\phi}} \right) + \frac{\partial}{\partial x} \left(\frac{\partial \pounds}{\partial \phi'} \right) = 0. \tag{3.12}$$

Inserting the Lagrangian density (3.10), we obtain the familiar wave equation for small amplitude waves on a string:

$$\rho \frac{\partial^2 \phi}{\partial t^2} - F \frac{\partial^2 \phi}{\partial x^2} = 0.$$

Thus continuous systems can be described in a Lagrangian formalism by a suitable choice of Lagrangian density, and clearly the method can be extended to waves in any number of dimensions. By analogy with (3.6) and (3.7), we can define the momentum density

$$\Pi(\dot{\phi}) = \frac{\partial \mathcal{L}}{\partial \dot{\phi}}$$

and the Hamiltonian density

$$\mathcal{H} = \Pi \dot{\phi} - \mathcal{L}. \tag{3.13}$$

Since the Lagrangian density (3.10) does not depend explicitly on t, it follows that

$$E = \int \mathcal{H} \, \mathrm{d}x = \int \left(\frac{\partial \mathcal{L}}{\partial \dot{\phi}} \dot{\phi} - \mathcal{L} \right) \mathrm{d}x \tag{3.14}$$

remains constant during the motion (Problem 3.2). This result is the analogue of (3.5).

3.4 A Lorentz covariant field theory

In three spatial dimensions, the action is of the form

$$S = \int \mathcal{L} \, \mathrm{d}x \, \mathrm{d}y \, \mathrm{d}z \, \mathrm{d}t = \int \mathcal{L} \, \mathrm{d}x^0 \mathrm{d}x^1 \mathrm{d}x^2 \mathrm{d}x^3. \tag{3.15}$$

The 'volume element' $\mathrm{d}x^0 \mathrm{d}x^1 \mathrm{d}x^2 \mathrm{d}x^3 = \mathrm{d}^4 x$ is a Lorentz invariant (Section 2.4). Hence S is a Lorentz invariant if the Lagrangian density $\mathcal{L}$ transforms like a scalar field. The covariance of the field equations is then assured. Other symmetries required of a theory may be built into $\mathcal{L}$.

Consider a Lorentz invariant Lagrangian density of the form

$$\mathcal{L} = \mathcal{L}(\phi, \partial_\mu \phi), \tag{3.16}$$

where $\phi(x) = \phi(x^0, \boldsymbol{x})$ is a scalar field. At any point x in space-time, such a Lagrangian density depends only on the field and its first derivatives at that point. The field theory is said to be *local*: there is no 'action at a distance'. This will be an important feature of the Standard Model. The field equation is easily derived from the condition $\delta S = 0$, together with the condition that the field vanishes at large distances, and we find

$$\frac{\partial \mathcal{L}}{\partial \phi} - \partial_\mu \left(\frac{\partial \mathcal{L}}{\partial(\partial_\mu \phi)} \right) = 0. \tag{3.17}$$

3.5 The Klein–Gordon equation

The Lorentz invariant Lagrangian density

$$\mathcal{L} = \frac{1}{2}[g^{\mu\nu}\partial_\mu\phi\partial_\nu\phi - m^2\phi^2] = \frac{1}{2}[\partial_\mu\phi\partial^\mu\phi - m^2\phi^2], \tag{3.18}$$

where $\phi(x)$ is a real scalar field, is a particular case of (3.16). The field equation (3.17) becomes

$$-\partial_\mu\partial^\mu\phi - m^2\phi = 0,$$

or

$$\left(-\frac{\partial^2}{\partial t^2} + \nabla^2 - m^2\right)\phi = 0. \tag{3.19}$$

This equation is known as the *Klein–Gordon equation.*

The equation has wave-like solutions

$$\phi(\mathbf{r}, t) = a\cos(\mathbf{k}\cdot\mathbf{r} - \omega_{\mathbf{k}}t + \theta_{\mathbf{k}})$$

where the frequency $\omega_{\mathbf{k}}$ is related to the wave vector $\mathbf{k}$ by the dispersion relation

$$\omega_{\mathbf{k}}^2 = \mathbf{k}^2 + m^2, \tag{3.20}$$

and $\theta_{\mathbf{k}}$ is an arbitrary phase angle.

For mathematical simplicity we shall take the solutions $\phi(\mathbf{r}, t)$ to lie in a large cube of side l, volume $V = l^3$, and apply periodic boundary conditions, so that $\mathbf{k} = (2\pi n_1/l, 2\pi n_2/l, 2\pi n_3/l)$ where n_1, n_2, n_3 are any integers $0, \pm1, \pm2, \ldots$

The general solution of (3.19) is a superposition of such plane waves:

$$\phi(\mathbf{r}, t) = \frac{1}{\sqrt{V}}\sum_{\mathbf{k}}\left(\frac{a_k}{\sqrt{2\omega_k}}\mathrm{e}^{\mathrm{i}(\mathbf{k}\cdot\mathbf{r}-\omega t)} + \frac{a_k^*}{\sqrt{2\omega_k}}\mathrm{e}^{-\mathrm{i}(\mathbf{k}\cdot\mathbf{r}-\omega t)}\right). \tag{3.21}$$

The factors $\sqrt{2\omega_k}$ are introduced for later convenience, and the phase factors have been absorbed into the complex wave amplitudes a_k. The sum is over all allowed values of $\mathbf{k}$.

With the de Broglie identifications of $E = \omega_k$, $\mathbf{p} = \mathbf{k}$ (recall $\hbar = 1, c = 1$) the dispersion relation for ω_k is equivalent to the Einstein equation for a free particle,

$$E^2 = \mathbf{p}^2 + m^2.$$

We may conjecture that the Klein–Gordon equation for ϕ describes a scalar particle of mass m. There is no vector associated with a one-component scalar field, and the intrinsic angular momentum associated with such a particle is zero.

We shall see a Lagrangian density of the form (3.18) arising in the Standard Model to describe the Higgs particle. At a less fundamental level, the overall motion of the π^0 meson, which is an uncharged composite particle, is described by a similar Lagrangian density.

3.6 The energy–momentum tensor

The equations expressing both conservation of energy and conservation of linear momentum are obtained by considering the change in $\mathcal{L}$ corresponding to a uniform infinitesimal space-time displacement

$$x^\mu \to x^\mu + \delta a^\mu, \tag{3.22}$$

where δa^μ does not depend on x. The corresponding change in ϕ is

$$\delta\phi = (\partial_\nu \phi)\delta a^\nu. \tag{3.23}$$

Since $\mathcal{L}$ does not depend explicitly on the x^μ,

$$\delta\mathcal{L} = \frac{\partial \mathcal{L}}{\partial \phi}\delta\phi + \frac{\partial \mathcal{L}}{\partial(\partial_\mu \phi)}\delta(\partial_\mu \phi).$$

Using the field equation (3.17) for $\partial\mathcal{L}/\partial\phi$, and the fact that $\delta(\partial_\mu\phi) = \partial_\mu(\delta\phi)$, we can rewrite this as

$$\delta\mathcal{L} = \partial_\mu\left[\left(\frac{\partial \mathcal{L}}{\partial(\partial_\mu \phi)}\right)\delta\phi\right],$$

and then, from (3.23),

$$\delta\mathcal{L} = \partial_\mu\left[\frac{\partial \mathcal{L}}{\partial(\partial_\mu \phi)}\partial_\nu\phi\right]\delta a^\nu.$$

We have also

$$\delta\mathcal{L} = \frac{\partial \mathcal{L}}{\partial x^\mu}\delta a^\mu = \delta^\mu_\nu \frac{\partial \mathcal{L}}{\partial x^\mu}\delta a^\nu,$$

where, as in (2.14),

$$\delta^\mu_\nu = \begin{cases} 1, & \mu = \nu \\ 0, & \mu \neq \nu. \end{cases}$$

Since the δa^ν are arbitrary, it follows on comparing these expressions for $\delta\mathcal{L}$ that

$$\partial_\mu \left[\frac{\partial \mathcal{L}}{\partial(\partial_\mu \phi)} \partial_\nu \phi - \delta^\mu_\nu \mathcal{L} \right] = 0, \tag{3.24}$$

or

$$\partial_\mu T^\mu_\nu = 0, \quad \text{where } T^\mu_\nu = \left[\frac{\partial \mathcal{L}}{\partial(\partial_\mu \phi)} \partial_\nu \phi - \delta^\mu_\nu \mathcal{L} \right]. \tag{3.25}$$

T^μ_ν is the *energy–momentum* tensor. The component

$$T^0_0 = \frac{\partial \mathcal{L}}{\partial \dot{\phi}} \dot{\phi} - \mathcal{L}$$

corresponds to the Hamiltonian density defined in equation (3.13), and is interpreted as the energy density of the field; in a relativistic theory, the energy density transforms like a component of a tensor. The $\nu = 0$ component of (3.25) may be written

$$\frac{\partial}{\partial t}(T^0_0) + \boldsymbol{\nabla} \cdot \mathbf{T}_0 = 0, \tag{3.26}$$

and expresses local conservation of energy, with $\mathrm{T}_0 = (T^1_0, T^2_0, T^3_0)$ interpreted as the energy flux. Integrating (3.26) over all space and using the divergence theorem yields

$$\frac{\partial}{\partial t} \int T^0_0 \mathrm{d}^3\mathbf{x} = 0, \tag{3.27}$$

provided the field vanishes at large distances. This equation expresses the overall conservation of energy.

Similarly the $\nu = 1, 2, 3$ components of (3.24) correspond to local conservation of momentum, with the overall total momentum of the field given by

$$P_i = \int T^0_i \mathrm{d}^3\mathbf{x}. \tag{3.28}$$

As with the energy, the total momentum of the field is conserved if the field vanishes at large distances.

In the case of the Klein–Gordon Lagrangian density (3.19),

$$\frac{\partial \mathcal{L}}{\partial \dot{\phi}} = \dot{\phi},$$

and the energy density of the field is

$$T^0_0 = \frac{1}{2}[\dot{\phi}^2 + (\nabla\phi)^2 + m^2\phi^2]. \tag{3.29}$$

Expressing ϕ in terms of the field amplitudes $a_{\mathbf{k}}$ and $a^*_{\mathbf{k}}$, and integrating over all space, gives the total field energy

$$H = \int T^0_0 \mathrm{d}^3\mathbf{x} = \sum_{\mathbf{k}} a^*_{\mathbf{k}} a_{\mathbf{k}} \omega_{\mathbf{k}}. \tag{3.30}$$

In obtaining this expression we have used the orthogonality of the plane waves

$$\frac{1}{V}\int \mathrm{e}^{\mathrm{i}(\mathbf{k}-\mathbf{k}')\cdot\mathbf{r}} \mathrm{d}^3\mathbf{x} = \delta_{\mathbf{k}\mathbf{k}'}.$$

Similarly from (3.28) the total momentum of the field can be shown to be

$$\mathbf{P} = \sum_{\mathbf{k}} a^*_{\mathbf{k}} a_{\mathbf{k}} \mathrm{k}. \tag{3.31}$$

3.7 Complex scalar fields

It is instructive to consider also complex scalar fields $\Phi = (\phi_1 + \mathrm{i}\phi_2)/\sqrt{2}$ satisfying the Klein–Gordon equation. We shall see in Section 7.6 that if the field Φ carries charge q, then the field Φ^* carries charge $-q$. The Klein–Gordon equation for a complex field Φ is obtained from the (real) Lagrangian density

$$\mathcal{L} = \partial_\mu \Phi^* \partial^\mu \Phi - m^2 \Phi^* \Phi. \tag{3.32}$$

We introduce here a device that we shall often find useful. Instead of varying the real and imaginary parts of Φ to obtain the field equations, we may vary Φ and its complex conjugate Φ^* independently. These procedures are equivalent. Varying Φ^* in the action constructed from (3.32) yields, easily,

$$-\partial_\mu \partial^\mu \Phi - m^2 \Phi = 0. \tag{3.33}$$

(Varying Φ gives the complex conjugate of this equation.)

Note that the Lagrangian density (3.32) is the sum of contributions from the scalar fields ϕ_1 and ϕ_2:

$$\begin{aligned} \mathcal{L} = \partial_\mu \Phi^* \partial^\mu \Phi - m^2 \Phi^* \Phi = & \frac{1}{2}\left[\partial_\mu \phi_1 \partial^\mu \phi_1 - m^2 \phi_1^2\right] \\ & + \frac{1}{2}\left[\partial_\mu \phi_2 \partial^\mu \phi_2 - m^2 \phi_2^2\right]. \end{aligned} \tag{3.34}$$

The general solution of (3.33) is a superposition of plane waves of the form

$$\Phi = \frac{1}{\sqrt{V}} \sum_{\mathbf{k}} \left(\frac{a_{\mathbf{k}}}{\sqrt{2\omega_{\mathbf{k}}}} \mathrm{e}^{\mathrm{i}(\mathbf{k}\cdot\mathbf{r}-\omega t)} + \frac{b^*_{\mathbf{k}}}{\sqrt{2\omega_{\mathbf{k}}}} \mathrm{e}^{-\mathrm{i}(\mathbf{k}\cdot\mathbf{r}-\omega t)} \right) \tag{3.35}$$

where $a_{\mathbf{k}}$ and $b_{\mathbf{k}}$ are now independent complex numbers. The field energy becomes

$$H = \sum_{\mathbf{k}} (a^*_{\mathbf{k}} a_{\mathbf{k}} + b^*_{\mathbf{k}} b_{\mathbf{k}}) \omega_{\mathbf{k}}. \tag{3.36}$$

We shall see that we can interpret this expression as being made up of the distinct contributions of positively and negatively charged fields. (The π^+ and π^- mesons are composite particles whose overall motion is described by complex scalar fields.)

Problems

3.1 Show that the kinetic energy of a system of particles, whose positions are determined by $q(t)$, is a quadratic function of the $\dot{q}_i$.

3.2 Show that $\mathrm{d}E/\mathrm{d}t = 0$, where E is given by equation (3.14).

3.3 For the stretched string of Section 3.3, show that the Hamiltonian density is

$$\mathcal{H} = \frac{1}{2}\rho\left(\frac{\partial\phi}{\partial t}\right)^2 + \frac{1}{2}F\left(\frac{\partial\phi}{\partial x}\right)^2.$$

The nth normal mode of oscillation, with wave amplitude A_n, is given by

$$\phi_n(x,t) = A_n \sin(k_n x)\sin(\omega_n t)$$

where $k_n = n\pi/l$, $\omega_n = (F/\rho)^{1/2}k_n$. Show that the total energy is ${A_n}^2{\omega_n}^2\rho l/4$ and oscillates harmonically between potential energy and kinetic energy.

3.4 Verify the expressions (3.30) and (3.31) for the energy and momentum of the scalar field given by equation (3.21).

3.5 Show that the Schrödinger equation for the wave function $\psi(\mathbf{r},t)$ of a particle of mass m moving in a potential $V(\mathbf{r})$ may be obtained from the Lagrangian density

$$\mathcal{L} = -(1/2\mathrm{i})\left(\psi^*\frac{\partial\psi}{\partial t} - \frac{\partial\psi^*}{\partial t}\psi\right) - (1/2m)\nabla\psi^*\cdot\nabla\psi - \psi^* V\psi.$$

(Note that $\mathcal{L}$ is real, but not Lorentz invariant.)

4
Classical electromagnetism

Maxwell's theory of electromagnetism is, along with Einstein's theory of gravitation, one of the most beautiful of classical field theories. In this chapter we exhibit the Lorentz covariance of Maxwell's equations and show how they may be obtained from Hamilton's principle. The important idea of a gauge transformation is introduced, and related to the conservation of electric charge. We analyse some properties of solutions of the field equations. Finally, we generalise the Lagrangian to describe massive vector fields, which will figure in later chapters.

4.1 Maxwell's equations

In common with much of the literature, we shall use units in which the force between charges q_1 and q_2 is $q_1 q_2/4\pi r^2$, and the velocity of light $c = 1$. (Thus in these units $\mu_0 = 1,\ \varepsilon_0 = 1$.) Maxwell's equations then take the form

$$\begin{aligned} &\nabla \cdot \mathbf{E} = \rho \quad \text{(a)}, \quad \nabla \times \mathbf{B} - \frac{\partial \mathrm{E}}{\partial t} = \mathbf{J} \quad \text{(b)}, \\ &\nabla \cdot \mathbf{B} = 0 \quad \text{(c)}, \quad \nabla \times \mathbf{E} + \frac{\partial \mathrm{B}}{\partial t} = 0 \quad \text{(d)}. \end{aligned} \tag{4.1}$$

E and **B** are the electric and magnetic fields, ρ and **J** are the electric charge and current densities. In this chapter we do not consider the dynamics of ρ and **J**, but take them to be 'external' fields that we are free to manipulate. The inhomogeneous equations (a) and (b) are consistent with the observed fact of charge conservation, which is expressed by the continuity equation:

$$\frac{\partial \rho}{\partial t} + \nabla \cdot \mathrm{J} = 0.$$

This equation takes the Lorentz invariant form

$$\partial_\mu J^\mu = 0 \tag{4.2}$$

if we postulate that the charge-current densities

$$J^\mu = (\rho, \mathbf{J}) \tag{4.3}$$

make up a contravariant four-vector field.

Introducing a scalar potential ϕ and a vector potential $\mathbf{A}$, the homogeneous equations (c) and (d) of the set (4.1) are satisfied identically by

$$\mathbf{B} = \nabla \times \mathbf{A}, \quad \mathbf{E} = -\nabla\phi - \frac{\partial \mathbf{A}}{\partial t}. \tag{4.4}$$

We postulate that the potentials

$$A^\mu = (\phi, \mathbf{A}) \tag{4.5}$$

make up a contravariant four-vector field also.

Maxwell's equations may be written in terms of the antisymmetric tensor $F^{\mu\nu}$, defined by

$$F^{\mu\nu} = \partial^\mu A^\nu - \partial^\nu A^\mu = \begin{pmatrix} 0 & -E_x & -E_y & -E_z \\ E_x & 0 & -B_z & B_y \\ E_y & B_z & 0 & -B_x \\ E_z & -B_y & B_x & 0 \end{pmatrix}. \tag{4.6}$$

It is apparent that the electromagnetic field is a tensor field. For example,

$$F^{01} = \partial A^1/\partial x_0 - \partial A^0/\partial x_1 = \partial A_x/\partial t + \partial\phi/\partial x = -E_x.$$

Thus the components of the electromagnetic field transform under a Lorentz transformation like the elements of a tensor.

The homogeneous Maxwell equations correspond to the identitities

$$\partial^\lambda F^{\mu\nu} + \partial^\nu F^{\lambda\mu} + \partial^\mu F^{\nu\lambda} \equiv 0, \tag{4.7}$$

where λ, μ, ν are any three of 0, 1, 2, 3, as the reader may easily verify. The inhomogeneous equations take the manifestly covariant form

$$\partial_\mu F^{\mu\nu} = J^\nu. \tag{4.8}$$

For example, with $\nu = 0$, looking at the first column of $F^{\mu\nu}$, and noting $\partial_\mu = (\partial/\partial t, \nabla)$, gives

$$\nabla \cdot \mathbf{E} = \rho.$$

4.2 A Lagrangian density for electromagnetism

We now seek a Lagrangian density $\mathcal{L}$ that will yield Maxwell's equations from Hamilton's principle. If $\mathcal{L}$ is Lorentz invariant, the action

$$S = \int \mathcal{L}\, \mathrm{d}^4 x = \int \mathcal{L}\, \mathrm{d}x^0 \mathrm{d}x^1 \mathrm{d}x^2 \mathrm{d}x^3 \tag{4.9}$$

is also Lorentz invariant, since d^4x is invariant (Section 2.4 and Section 3.4), and the field equations which follow from the condition $\delta S = 0$ will take the same form in every inertial frame of reference.

Although Maxwell's equations do not refer explicitly to the potentials A^μ, to derive the equations from Hamilton's principle requires the potentials to be taken as the basic fields which are to be varied. The 'stretched string' example of Section 3.4 suggests that $\mathcal{L}$ should be quadratic in the first derivatives of the field. A suitable Lorentz invariant choice is found to be

$$\mathcal{L} = -\frac{1}{4}F_{\mu\nu}F^{\mu\nu} - J^\mu A_\mu. \tag{4.10}$$

Varying the fields A^μ, while keeping the charge and current densities J_μ fixed, yields Maxwell's equations, as we shall show in some detail. (Subsequent arguments will be more terse!)

We may write

$$S = \int \left[-\frac{1}{4}g_{\mu\lambda}g_{\nu\rho}F^{\lambda\rho}F^{\mu\nu} - J^\mu A_\mu\right]d^4x. \tag{4.11}$$

Then

$$\begin{aligned}\delta S &= \int \left[-\frac{1}{2}g_{\mu\lambda}g_{\nu\rho}F^{\lambda\rho}\delta F^{\mu\nu} - J^\mu \delta A_\mu\right]d^4x \\ &= \int \left[-\frac{1}{2}F^{\lambda\rho}(\partial_\lambda \delta A_\rho - \partial_\rho \delta A_\lambda) - J^\mu \delta A_\mu\right]d^4x \\ &= \int [-F^{\lambda\rho}\partial_\lambda \delta A_\rho - J^\mu \delta A_\mu]d^4x, \quad \text{since } F^{\lambda\rho} = -F^{\rho\lambda}.\end{aligned}$$

The first term we integrate by parts. The boundary terms vanish for suitable conditions on the fields, so that we are left with

$$\delta S = \int [\partial_\lambda F^{\lambda\rho} - J^\rho]\delta A_\rho \, d^4x.$$

Setting $\delta S = 0$ for arbitrary δA_ρ gives the inhomogeneous Maxwell equations (4.8). (The homogeneous equations (4.7) are no more than identities.)

4.3 Gauge transformations

The four-potential $A^\mu = (\phi, \mathbf{A})$ is *not unique*: the same electromagnetic field tensor $F^{\mu\nu}$ is obtained from the potential

$$A^\mu + \partial^\mu \chi = (\phi + \partial\chi/\partial t, \mathbf{A} - \nabla\chi), \tag{4.12}$$

where $\chi(x)$ is an arbitrary scalar field, since the additional terms which appear in $F^{\mu\nu}$ are identically zero:

$$\partial^\mu \partial^\nu \chi - \partial^\nu \partial^\mu \chi = 0.$$

The transformation $A^\mu \rightarrow A'^\mu = A^\mu + \partial^\mu \chi$ is called a *gauge transformation*.

Under a gauge transformation, the action (4.11) acquires an additional term ΔS, where

$$\begin{aligned}\Delta S &= -\int J_\mu \partial^\mu \chi \, \mathrm{d}^4 x \\ &= \int (\partial^\mu J_\mu) \chi \, \mathrm{d}^4 x.\end{aligned}$$

We have integrated by parts to obtain the second line and again assumed that the boundary terms vanish. ΔS is zero for arbitrary χ if, and only if,

$$\partial^\mu J_\mu = \partial_\mu J^\mu = 0,$$

which is just equation (4.2). Thus the gauge invariance of the action requires, and follows from, the conservation of electric charge.

4.4 Solutions of Maxwell's equations

In terms of the potentials, the field equations (4.8) are

$$(\partial_\mu \partial^\mu) A^\nu - \partial^\nu (\partial_\mu A^\mu) = J^\nu. \tag{4.13}$$

We stress again that there is much arbitrariness in the solutions to these equations. Equivalent solutions differ by gauge transformations. It is usual to impose a gauge-fixing condition. For example in the 'radiation gauge' we set $\boldsymbol{\nabla} \cdot \mathbf{A} = 0$, everywhere and at all times (Problem 4.2). This has the disadvantage of not being a Lorentz invariant condition – it will not be true in another, moving, frame – but it does display important features of the theory. In the radiation gauge the field equation for A^0 becomes

$$(\partial_i \partial^i) A^0 = -\boldsymbol{\nabla}^2 A^0 = J^0$$

(setting $\nu = 0$ in (4.13), and noting $\partial_\mu A^\mu = \partial_0 A^0$ since in the radiation gauge $\partial_i A^i = 0$). This equation has the solution

$$A^0(\mathbf{r}, t) = \frac{1}{4\pi} \int \frac{\rho(\mathbf{r}', t)}{|\mathbf{r} - \mathbf{r}'|} \mathrm{d}^3 \mathbf{r}'.$$

Hence, in the radiation gauge, A^0 is determined entirely by the charge density to which it is rigidly attached! There are no wave-like solutions. The vector components A^i $(i = 1, 2, 3)$ satisfy the inhomogeneous wave equation

$$\frac{\partial^2 \mathbf{A}}{\partial t^2} - \boldsymbol{\nabla}^2 \mathbf{A} = \mathbf{J} - \frac{\partial}{\partial t} \nabla A^0. \tag{4.14}$$

Charges and currents act as a source (and sink) of the field $\mathbf{A}$.

In free space $\mathbf{J} = 0$, $\rho = 0$, $A^0 = 0$, and there are plane wave solutions with wave vector $\mathbf{k}$, frequency $\omega_{\mathbf{k}} = |\mathbf{k}|$, of the form

$$\mathbf{A}(\mathbf{r}, t) = a \boldsymbol{\varepsilon} \cos(\mathbf{k} \cdot \mathbf{r} - \omega_{\mathbf{k}} t).$$

Here $\boldsymbol{\varepsilon}$ is a unit vector and a is the wave amplitude. The gauge condition requires $\mathbf{k} \cdot \boldsymbol{\varepsilon} = 0$. Thus for a given $\mathbf{k}$ there are only two independent states of polarisation, $\boldsymbol{\varepsilon}_1(\mathbf{k})$ and $\boldsymbol{\varepsilon}_2(\mathbf{k})$ say, perpendicular to $\mathbf{k}$. The general solution in free space is

$$\mathbf{A}(\mathbf{r}, t) = \frac{1}{\sqrt{V}} \sum_{\mathbf{k}} \sum_{\alpha=1,2} \frac{\boldsymbol{\varepsilon}_\alpha(\mathbf{k})}{\sqrt{2\omega_{\mathbf{k}}}} [a_{\mathbf{k}\alpha} \mathrm{e}^{\mathrm{i}(\mathbf{k} \cdot \mathbf{r} - \omega t)} + a^*_{\mathbf{k}\alpha} \mathrm{e}^{-\mathrm{i}(\mathbf{k} \cdot \mathbf{r} - \omega t)}]. \tag{4.15}$$

The complex number $a_{\mathbf{k}\alpha}$ represents an amplitude and a phase, and the plane waves are normalised in a volume V, with periodic boundary conditions. The factor $\sqrt{2\omega_{\mathbf{k}}}$ is put in for convenience later.

An important point apparent in the radiation gauge is that although the vector potential has four components A^μ, one of these, A^0, has no independent dynamics and another is a gauge artifact, which is eliminated by fixing the gauge. There are only *two* physically significant dynamical fields.

The fields in any other gauge are related to the fields in the radiation gauge by a gauge transformation; the physics is the same but the mathematics is different. For some purposes it is better to work in the relativistically invariant 'Lorentz gauge'. In the Lorentz gauge

$$\partial_\mu A^\mu = 0 \tag{4.16}$$

and the field equations become

$$\left(\frac{\partial^2}{\partial t^2} - \boldsymbol{\nabla}^2 \right) A^\mu = J^\mu. \tag{4.17}$$

4.5 Space inversion

We now consider the operation of space inversion of the coordinate axes in the origin: $\mathbf{r} \to \mathbf{r}' = -\mathbf{r}$, $\boldsymbol{\nabla} \to \boldsymbol{\nabla}' = -\boldsymbol{\nabla}$ (Fig. 4.1), which was excluded from the group of proper Lorentz transformations. We shall also refer to this as the *parity* operation. The transformed coordinate axes are left-handed. By convention the

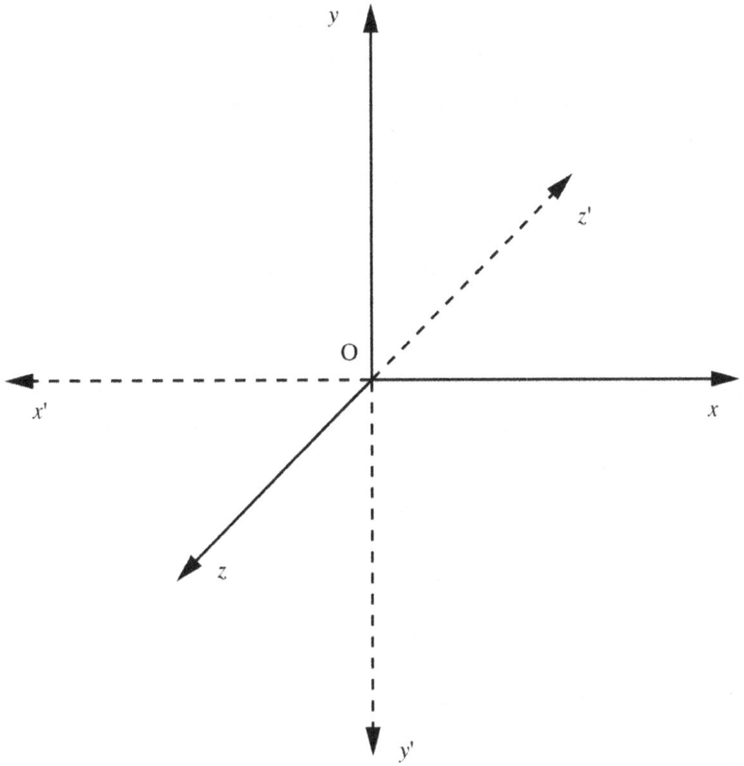

Figure 4.1 A normal right-handed set of axes (solid lines) and a space-inverted set (dashed lines). The space-inverted set is said to be left-handed. (Oz is out of the plane of the page.)

charge density is taken to be invariant under this transformation: if at some instant of time $\rho^P(\mathbf{r}')$ is the charge density referred to the inverted coordinate axes, then $\rho^P(\mathbf{r}') = \rho(\mathbf{r})$ when $\mathbf{r}' = -\mathbf{r}$. The current density $\mathbf{J}(\mathbf{r}) = \rho(\mathbf{r})\,\mathbf{u}(\mathbf{r})$, where $\mathbf{u}(\mathbf{r})$ is a velocity, and therefore transforms like $\mathrm{d}\mathbf{r}/\mathrm{d}t$, an ordinary vector: $\mathbf{J}^P(\mathbf{r}') = -\mathbf{J}(\mathbf{r})$. Maxwell's equations (4.1) retain the same form in the primed coordinate system if $\mathbf{E}(\mathbf{r}')$ also transforms like a vector, $\mathbf{E}^P(\mathbf{r}') = -\mathbf{E}(\mathbf{r})$, and $\mathbf{B}(\mathbf{r})$ transforms like an axial vector, $\mathbf{B}^P(\mathbf{r}') = \mathbf{B}(\mathbf{r})$.

In terms of the potentials, equation (4.4) shows that we must take

$$\phi^P(\mathbf{r}') = \phi(\mathbf{r}), \qquad \mathbf{A}^P(\mathbf{r}') = -\mathbf{A}(\mathbf{r}). \tag{4.18}$$

The field equations in a left-handed frame then have the same form as in a right-handed frame. The Lagrangian density (4.10) is invariant under space inversion. Electromagnetism is indifferent to handedness.

4.6 Charge conjugation

It will also be of interest to note that Maxwell's equations can be made to take the same form if matter is replaced by antimatter. As a consequence of this replacement both the charge and current densities change sign so that

$$\rho(\mathbf{r}) \to \rho^C(\mathbf{r}) = -\rho(\mathbf{r}) \qquad \text{and} \qquad \mathbf{J}(\mathbf{r}) \to \mathbf{J}^C(\mathbf{r}) = -\mathbf{J}(\mathbf{r}).$$

Maxwell's equations take the same form if we define

$$\phi^C(\mathbf{r}) = -\phi(\mathbf{r}), \quad \mathbf{A}^C(\mathbf{r}) = -\mathbf{A}(\mathbf{r}). \tag{4.19}$$

This operation is called *charge conjugation*. As with Lorentz transformations and the parity transformation, the Lagrangian is invariant under the charge conjugation transformation.

4.7 Intrinsic angular momentum of the photon

Without embarking here on the full quantisation of the electromagnetic field, we can discuss the quantised intrinsic angular momentum, or spin, of the photons associated with plane waves of the form (4.15).

The spin $\mathbf{S}$ of a particle with mass is defined as its angular momentum in a frame of reference in which it is at rest. In such a frame its orbital angular momentum $\mathbf{L} = 0$, and its total angular momentum $\mathbf{J} = \mathbf{L} + \mathbf{S} = \mathbf{S}$. This definition is inapplicable to a massless particle, which moves with the velocity of light in every frame of reference. However, for a massless particle moving in, say, the z-direction, it is possible to define the z-component S_z of its spin, since the z-component of the orbital angular momentum is $L_z = xp_y - yp_x$, and $p_x = p_y = 0$ for a particle moving in the z-direction, hence $L_z = 0$, and $J_z = S_z$.

In quantum mechanics, the component J_z of the total angular momentum operator of a system is given by

$$J_z = \mathrm{i}\hbar r_z = \mathrm{i}\hbar \lim_{\phi \to 0} [R_z(\phi) - 1]\,\phi, \tag{4.20}$$

where $R_z(\phi)$ is the operator that rotates the system through an angle ϕ about Oz in a positive sense.

Consider a term from (4.15) with $\mathbf{k} = (0, 0, k)$ along Oz:

$$\mathbf{A}(\mathbf{r}, t) = \frac{1}{\sqrt{2\omega V}}[(a_1 \boldsymbol{\varepsilon}_x + a_2 \boldsymbol{\varepsilon}_y)\mathrm{e}^{\mathrm{i}(kz - \omega t)} + \text{complex conjugate}]. \tag{4.21}$$

The wave amplitudes a_1 and a_2 are complex numbers, and we have taken the polarisation vectors $\boldsymbol{\varepsilon}_x$ and $\boldsymbol{\varepsilon}_y$ to be unit vectors aligned with the x- and y-axes. A rotation of **A** through an angle ϕ about Oz makes a change in the amplitudes that can be expressed by the rotation matrix equation

$$R_z(\phi)\begin{pmatrix} a_1 \\ a_2 \end{pmatrix} = \begin{pmatrix} a_1' \\ a_2' \end{pmatrix} = \begin{pmatrix} \cos\phi & -\sin\phi \\ \sin\phi & \cos\phi \end{pmatrix}\begin{pmatrix} a_1 \\ a_2 \end{pmatrix}.$$

In the limit $\phi \to 0$, we have

$$\lim[R_z(\phi) - 1]/\phi = \begin{pmatrix} 0 & -1 \\ 1 & 0 \end{pmatrix}$$

and

$$J_z = \hbar \begin{pmatrix} 0 & -\mathrm{i} \\ \mathrm{i} & 0 \end{pmatrix}.$$

The eigenvectors of $J_z/\hbar$ are

$$\begin{pmatrix} a_1 \\ a_2 \end{pmatrix} = \begin{pmatrix} 1 \\ \mathrm{i} \end{pmatrix}$$

with eigenvalue $+1$,

$$\begin{pmatrix} a_1 \\ a_2 \end{pmatrix} = \begin{pmatrix} 1 \\ -\mathrm{i} \end{pmatrix}$$

with eigenvalue -1.

Thus we may say that a photon represented by the plane wave (4.21) has ‘spin one’, with just two spin states aligned and anti-aligned with its direction of motion. No meaning can be given to spin components perpendicular to the direction of motion. Classically these waves are right circularly polarised and left circularly polarised, respectively (Problem 4.4).

A plane wave of any polarisation can be constructed by a suitable superposition of right-handed and left-handed circularly polarised waves.

4.8 The energy density of the electromagnetic field

The analysis of the energy density of the electromagnetic field in free space is a generalisation of the analysis for a scalar field set out in Section 3.6. Equation (3.25) becomes

$$T^\mu_\nu = \frac{\partial \mathcal{L}}{\partial(\partial_\mu A^\lambda)}\partial_\nu A^\lambda - \delta^\mu_\nu \mathcal{L}, \tag{4.22}$$

and using this formula gives

$$T^0_0 = -F_{0\mu}F^{0\mu} + \frac{1}{4}F_{\mu\nu}F^{\mu\nu} \tag{4.23}$$

(Problem 4.5). In terms of the physical fields **E** and **B**, (4.23) is the familiar expression

$$\text{energy density} = \frac{1}{2}(\mathbf{E}^2 + \mathbf{B}^2). \tag{4.24}$$

We can also express the fields in terms of the field amplitudes $a_{\mathbf{k}\alpha}$ introduced in equation (4.15) and obtain for the total energy of the field

$$H = \int T^0_0 \mathrm{d}^3\mathbf{x} = \sum_{\mathbf{k},\alpha} a^*_{\mathbf{k}\alpha} a_{\mathbf{k}\alpha} \omega_{\mathbf{k}}. \tag{4.25}$$

Similarly the total momentum of the field is

$$\mathbf{P} = \sum_{\mathbf{k},\alpha} a^*_{\mathbf{k}\alpha} a_{\mathbf{k}\alpha} \mathbf{k}. \tag{4.26}$$

4.9 Massive vector fields

Let us modify the Lagrangian density (4.10) by adding an additional Lorentz invariant term, and consider

$$\mathcal{L} = -\frac{1}{4} F_{\mu\nu} F^{\mu\nu} + \frac{1}{2} m^2 A_\mu A^\mu - J^\mu A_\mu \tag{4.27}$$

where J^μ is an external current. The additional term in the action is easily seen to modify the field equations to

$$\partial_\mu F^{\mu\nu} + m^2 A^\nu = J^\nu. \tag{4.28}$$

Since $\partial_\nu \partial_\mu F^{\mu\nu} \equiv 0$, it follows from (4.28) that

$$m^2 \partial_\nu A^\nu = \partial_\nu J^\nu. \tag{4.29}$$

This equation is a necessary consequence of the field equations: it is not a Lorentz gauge-fixing condition like equation (4.16), but it does imply that the A^ν are not independent. Using this equation, the field equations simplify to

$$\partial_\mu \partial^\mu A^\nu + m^2 A^\nu = J^\nu + \partial^\nu(\partial_\mu J^\mu)/m^2. \tag{4.30}$$

Hence in free space each component of A^ν of the field satisfies

$$\frac{\partial^2 A^\nu}{\partial t^2} - \boldsymbol{\nabla}^2 A^\nu + m^2 A^\nu = 0. \tag{4.31}$$

This wave equation is related by the quantisation rules $E \to \mathrm{i}\partial/\partial t$, $\mathbf{p} \to -\mathrm{i}\boldsymbol{\nabla}$, to the Einstein equation for a free particle,

$$E^2 = p^2 + m^2.$$

We may conclude that our modified Lagrangian, when quantised, describes particles of mass m associated with a four-component field, of which three components are independent.

Plane wave solutions of (4.31) are of the form

$$A^\nu = a\varepsilon^\nu \cos(\mathbf{k}\cdot\mathbf{r} - \omega_k t) = a\varepsilon^\nu \cos(k_\mu x^\mu),$$

where $\omega_k = k^0 = \sqrt{m^2 + \mathbf{k}^2}$. To satisfy the condition $\partial_\nu A^\nu = 0$ we need

$$k_\nu \varepsilon^\nu = 0. \tag{4.32}$$

For example, if we consider a plane wave in the z-direction with $k^\nu = (k^0, 0, 0, k)$ there are three independent polarisations, labelled 1, 2, 3, which we may take as the contravariant four-vectors

$$\begin{aligned} \varepsilon_1^\nu &= (0, 1, 0, 0), \\ \varepsilon_2^\nu &= (0, 0, 1, 0), \\ \varepsilon_3^\nu &= (k, 0, 0, k^0)/m. \end{aligned}$$

The intrinsic spin of a particle is its angular momentum in a frame of reference in which it is at rest (Section 4.7). In such a frame $\mathbf{k} = 0$, and $\varepsilon_1 = (0, \varepsilon_x)$, $\varepsilon_2 = (0, \varepsilon_y)$, $\varepsilon_3 = (0, \varepsilon_z)$. As in Section 4.7, the states with polarisation $\varepsilon_x \pm i\varepsilon_y$ correspond to $J_z = \pm 1$, but we now have also the state with polarisation ε_z, which corresponds to $J_z = 0$, since the operator r_z acting on ε_z gives $r_z\varepsilon_z = 0$.

Thus our modified Lagrangian describes massive particles having intrinsic spin $\mathbf{S}$ with $S = 1$ and $S_z = 1, 0, -1$. That such particles are important in the Standard Model will become evident in later chapters.

Problems

4.1 Show that the Lagrangian density of equation (4.10) can also be written

$$\mathcal{L} = \frac{1}{2}(\mathbf{E}^2 - \mathbf{B}^2) - J^\mu A_\mu.$$

4.2 Suppose that in a certain gauge $\boldsymbol{\nabla}\cdot\mathbf{A} = f(\mathbf{r}, t) \neq 0$. Find an expression for a gauge transforming function $\chi(\mathbf{r}, t)$ such that the new potentials given by equation (4.12) satisfy the radiation gauge condition.

4.3 Show that the tensor field $\tilde{F}_{\mu\nu} = \frac{1}{2}\varepsilon_{\mu\nu\alpha\beta}F^{\alpha\beta}$ has the same form as $F^{\mu\nu}$ but with the electric and magnetic fields interchanged. Show that

$$\frac{1}{4}\tilde{F}_{\mu\nu}F^{\mu\nu} = \mathbf{E}\cdot\mathbf{B}$$

and that it is a scalar field under Lorentz transformations but a pseudoscalar under the parity operation.

4.4 Show that the electric field of the wave of equation (4.21) with $a_1 = 1$, $a_2 = \mathrm{i}$, is

$$(E_x, E_y, E_z) = -\sqrt{\frac{2\omega}{V}}[\sin(kz - \omega t), \cos(kz - \omega t), 0].$$

Show that as a function of time, at a fixed z, $\mathbf{E}$ rotates in a positive sense about the z-axis. This is the definition of right circular polarisation.

4.5 Show that equation (4.22) gives immediately

$$T^0_0 = -F^{0\mu}\partial_0 A_\mu + \frac{1}{4}F_{\mu\nu}F^{\mu\nu}.$$

Show that the term $\partial_\mu(A_0 F^{0\mu}) = \partial_i(A_0 F^{0i})$ can be added to this without changing the total energy. Hence arrive at the form for T^0_0 given in equation (4.23).

4.6 A particle of mass m, charge q, is moving in a fixed external electromagnetic field described by the four-potential $(\phi, \mathbf{A})$. Show that the Lagrangian

$$L = \frac{1}{2}m\dot{\mathbf{x}}^2 - q\phi + q\dot{\mathbf{x}}\cdot\mathbf{A}$$

gives the non-relativistic equation of motion

$$m\ddot{\mathbf{x}} = q(\mathbf{E} + \dot{\mathbf{x}}\times\mathbf{B}),$$

and the Hamiltonian is

$$H(\mathbf{p}, \mathbf{x}) = \frac{1}{2m}(\mathbf{p} - q\mathbf{A})^2 + q\phi,$$

where $\mathbf{p} = m\dot{\mathbf{x}} + q\mathbf{A}$.

4.7 Show that for a particle the action $S = \int L\,\mathrm{d}t$ is Lorentz invariant if γL is Lorentz invariant. Verify that this condition is satisfied by the Lagrangian

$$L = -m/\gamma - qA^\mu(\mathrm{d}x_\mu/\mathrm{d}t).$$

(This gives the relativistic version of Problem 4.6.)

5

The Dirac equation and the Dirac field

The Standard Model is a quantum field theory. In Chapter 4 we discussed the classical electromagnetic field. The transition to a quantum field will be made in Chapter 8. In this chapter we begin our discussion of the *Dirac equation*, which was invented by Dirac as an equation for the relativistic quantum wave function of a single electron. However, we shall regard the Dirac wave function as a field, which will subsequently be quantised along with the electromagnetic field. The Dirac equation will be regarded as a field equation. The transition to a quantum field theory is called *second quantisation*. The field, like the Dirac wave function, is complex. We shall show how the Dirac field transforms under a Lorentz transformation, and find a Lorentz invariant Lagrangian from which it may be derived.

On quantisation, the electromagnetic fields $A_\mu(x)$, $F_{\mu\nu}(x)$ become space- and time-dependent operators. The expectation values of these operators in the environment described by the quantum states are the classical fields. The Dirac fields $\psi(x)$ also become space- and time-dependent operators on quantisation. However, there are no corresponding measurable classical fields. This difference reflects the Pauli exclusion principle, which applies to fermions but not to bosons. In this chapter and in the following two chapters, the properties of the Dirac fields as operators are rarely invoked: for the most part the manipulations proceed as if the Dirac fields were ordinary complex functions, and the fields can be thought of as single-particle Dirac wave functions.

5.1 The Dirac equation

Dirac invented his equation in seeking to make Schrödinger's equation for an electron compatible with special relativity. The Schrödinger equation for an electron wave function ψ is

$$\mathrm{i}\frac{\partial \psi}{\partial t} = H\psi.$$

To secure a symmetry between space and time, Dirac postulated the Hamiltonian for a free electron to be of the form

$$H_D = \boldsymbol{\alpha} \cdot \mathbf{p} + \beta m = -\mathrm{i}\boldsymbol{\alpha} \cdot \boldsymbol{\nabla} + \beta m, \tag{5.1}$$

where m is the mass of the electron, $\mathbf{p}$ its momentum, $\boldsymbol{\alpha} = (\alpha_1, \alpha_2, \alpha_3)$, and $\alpha_1, \alpha_2, \alpha_3$ and β are matrices. ψ is a column vector, and the Schrödinger equation becomes the multicomponent Dirac equation:

$$(\mathrm{i}\partial/\partial t + \mathrm{i}\boldsymbol{\alpha} \cdot \boldsymbol{\nabla} - \beta m)\psi = 0. \tag{5.2}$$

If this equation is to describe a free electron of mass m, its solutions should also satisfy the Klein–Gordon equation of Section 3.5. Multiplying the Dirac equation on the left by the operator $(\mathrm{i}\partial/\partial t - \mathrm{i}\boldsymbol{\alpha} \cdot \boldsymbol{\nabla} + \beta m)$, we obtain

$$\Big[-\partial^2/\partial t^2 + \sum_i \alpha_i^2 \partial_i \partial_i + \sum_{i<j} (\alpha_i \alpha_j + \alpha_j \alpha_i)\partial_i \partial_j + \mathrm{i}m \sum_i (\alpha_i \beta + \beta \alpha_i)\partial_i - \beta^2 m^2\Big]\psi = 0,$$

where $\partial_i = \partial/\partial x^i$. This equation is identical to the Klein–Gordon equation if

$$\begin{aligned} &\beta^2 = 1, \quad \alpha_1^2 = \alpha_2^2 = \alpha_3^2 = 1, \\ &\alpha_i \alpha_j + \alpha_j \alpha_i = 0, \quad i \neq j; \quad \alpha_i \beta + \beta \alpha_i = 0, \quad i = 1, 2, 3. \end{aligned} \tag{5.3}$$

The reader may recall that similar equations are satisfied by the set of 2×2 Pauli spin matrices $\boldsymbol{\sigma} = (\sigma^1, \sigma^2, \sigma^3)$, where it is conventional to take

$$\sigma^1 = \begin{pmatrix} 0 & 1 \\ 1 & 0 \end{pmatrix}, \quad \sigma^2 = \begin{pmatrix} 0 & -\mathrm{i} \\ \mathrm{i} & 0 \end{pmatrix}, \quad \sigma^3 = \begin{pmatrix} 1 & 0 \\ 0 & -1 \end{pmatrix}. \tag{5.4}$$

We shall also find it useful to write

$$\sigma^0 = \begin{pmatrix} 1 & 0 \\ 0 & 1 \end{pmatrix}$$

for the 2×2 unit matrix.

However, here we have four anticommuting matrices, the α_i and β, to represent. It proves necessary to introduce a second set of Pauli matrices and represent the α_i and β by 4×4 matrices. The representation is not unique: different choices are appropriate for illuminating different properties of the Dirac equation. We shall use the so-called *chiral representation*, in which

$$\alpha^i = \begin{pmatrix} -\sigma^i & \mathbf{0} \\ \mathbf{0} & \sigma^i \end{pmatrix}, \quad \beta = \begin{pmatrix} \mathbf{0} & \sigma^0 \\ \sigma^0 & \mathbf{0} \end{pmatrix}, \tag{5.5}$$

writing the matrices in 2 × 2 'block' form. Here

$$\mathbf{0} = \begin{pmatrix} 0 & 0 \\ 0 & 0 \end{pmatrix}$$

and the 4 × 4 identity matrix may be written

$$\mathrm{I} = \begin{pmatrix} \sigma^0 & \mathbf{0} \\ \mathbf{0} & \sigma^0 \end{pmatrix}.$$

It can easily be checked that these matrices satisfy the conditions (5.3). (The block multiplication of matrices is described in Appendix A.)

Since the α_i and β are 4 × 4 matrices, the Dirac wave function ψ is a four-component column matrix. Regarded as a relativistic Schrödinger equation, the Dirac equation has, as we shall see, remarkable consequences: it describes a particle with intrinsic angular momentum $(\hbar/2)\boldsymbol{\sigma}$ and intrinsic magnetic moment $(q\hbar/2m)\boldsymbol{\sigma}$ if the particle carries charge q, and there exist 'negative energy' solutions, which Dirac interpreted as antiparticles.

A Lagrangian density that yields the Dirac equation from the action principle is

$$\begin{aligned} \mathcal{L} &= \psi^\dagger(\mathrm{i}\partial/\partial t + \mathrm{i}\boldsymbol{\alpha}\cdot\boldsymbol{\nabla} - \beta m)\psi \\ &= \psi_a^*(I_{ab}\mathrm{i}\partial/\partial t + \mathrm{i}\boldsymbol{\alpha}_{ab}\cdot\boldsymbol{\nabla}\beta_{ab}m)\psi_b, \end{aligned} \tag{5.6}$$

where we have written in the matrix indices. ψ_a^* is a row matrix, the Hermitian conjugate $\psi^\dagger = \psi^{\mathrm{T}*}$ of ψ. Instead of varying the real and imaginary parts of ψ_a independently, it is formally equivalent to treat ψ_a and its complex conjugate ψ_a^* as independent fields (cf. Section 3.7). The condition that $S = \int \mathcal{L}\,\mathrm{d}^4x$ be stationary for an arbitrary variation $\delta\psi_a^*$ then gives the Dirac equation immediately, since $\mathcal{L}$ does not depend on the derivatives of ψ_a^*.

5.2 Lorentz transformations and Lorentz invariance

The chiral representation (5.5) of the matrices α^i and β is particularly convenient for discussing the way in which the Dirac field must transform under a Lorentz transformation. We have written the Dirac matrices in blocks of 2 × 2 matrices, and it is natural to write similarly the four-component Dirac field as a pair of two-component fields

$$\psi = \begin{pmatrix} \psi_{\mathrm{L}} \\ \psi_{\mathrm{R}} \end{pmatrix} = \begin{pmatrix} \psi_{\mathrm{L}} \\ \mathbf{0} \end{pmatrix} + \begin{pmatrix} \mathbf{0} \\ \psi_{\mathrm{R}} \end{pmatrix}, \tag{5.7}$$

where $\psi_{\rm L}$ and $\psi_{\rm R}$ are, respectively, the top and bottom two components of the four-component Dirac field:

$$\psi_{\rm L} = \begin{pmatrix} \psi_1 \\ \psi_2 \end{pmatrix}, \quad \psi_{\rm R} = \begin{pmatrix} \psi_3 \\ \psi_4 \end{pmatrix}. \tag{5.8}$$

The Dirac equation (5.2) becomes

$$\mathrm{i}\begin{pmatrix} \sigma^0 & \mathbf{0} \\ \mathbf{0} & \sigma^0 \end{pmatrix}\begin{pmatrix} \partial_0\psi_{\rm L} \\ \partial_0\psi_{\rm R} \end{pmatrix} + \mathrm{i}\begin{pmatrix} -\sigma^i & \mathbf{0} \\ \mathbf{0} & \sigma^i \end{pmatrix}\begin{pmatrix} \partial_i\psi_{\rm L} \\ \partial_i\psi_{\rm R} \end{pmatrix} - m\begin{pmatrix} \mathbf{0} & \sigma^0 \\ \sigma^0 & \mathbf{0} \end{pmatrix}\begin{pmatrix} \psi_{\rm L} \\ \psi_{\rm R} \end{pmatrix} = 0. \tag{5.9}$$

Block multiplication then gives two coupled equations for $\psi_{\rm L}$ and $\psi_{\rm R}$:

$$\begin{aligned} \mathrm{i}\sigma^0\partial_0\psi_{\rm L} - \mathrm{i}\sigma^i\partial_i\psi_{\rm L} - m\psi_{\rm R} &= 0, \\ \mathrm{i}\sigma^0\partial_0\psi_{\rm R} + \mathrm{i}\sigma^i\partial_i\psi_{\rm R} - m\psi_{\rm L} &= 0. \end{aligned} \tag{5.10}$$

We shall find it highly convenient for displaying the Lorentz structure to define

$$\sigma^\mu = (\sigma^0, \sigma^1, \sigma^2, \sigma^3), \qquad \tilde{\sigma}^\mu = (\sigma^0, -\sigma^1, -\sigma^2, -\sigma^3).$$

With this notation, the equations (5.10) may be written

$$\begin{aligned} \mathrm{i}\tilde{\sigma}^\mu\partial_\mu\psi_{\rm L} - m\psi_{\rm R} &= 0, \\ \mathrm{i}\sigma^\mu\partial_\mu\psi_{\rm R} - m\psi_{\rm L} &= 0. \end{aligned} \tag{5.11}$$

To obtain the Lagrangian density (5.6) in terms of $\psi_{\rm L}$ and $\psi_{\rm R}$, we need to multiply the expression on the left-hand side of (5.9) by the row matrix $(\psi_{\rm L}^\dagger, \psi_{\rm R}^\dagger)$, where the Hermitian conjugate fields are $\psi_{\rm L}^\dagger = (\psi_1^*, \psi_2^*)$, $\psi_{\rm R}^\dagger = (\psi_3^*, \psi_4^*)$. Block multiplication gives

$$\mathcal{L} = \mathrm{i}\psi_{\rm L}^\dagger\tilde{\sigma}^\mu\partial_\mu\psi_{\rm L} + \mathrm{i}\psi_{\rm R}^\dagger\sigma^\mu\partial_\mu\psi_{\rm R} - m(\psi_{\rm L}^\dagger\psi_{\rm R} + \psi_{\rm R}^\dagger\psi_{\rm L}). \tag{5.12}$$

Variations $\delta\psi_{\rm L}^*$ and $\delta\psi_{\rm R}^*$ in the action give the field equations (5.11).

To show that the Lagrangian has the same form in every frame of reference, we must relate the field $\psi'(x')$ in the frame K' to $\psi(x)$ in the frame K, when x' and x refer to the same point in space-time, and are related by a proper Lorentz transformation

$$x'^\mu = L^\mu{}_\nu x^\nu. \tag{5.13}$$

The operator ∂_μ transforms like a covariant vector, so that

$$\partial'_\mu = L_\mu{}^\nu\partial_\nu,$$

which has the inverse

$$\partial_\mu = L^\nu{}_\mu\partial'_\nu. \tag{5.14}$$

(See Problem 2.2.)

It is shown in Appendix B (equations (B.17) and (B.18)) that with this Lorentz transformation we can associate 2×2 matrices $\mathbf{M}$ and $\mathbf{N}$ with determinant 1 and with the properties

$$\mathbf{M}^\dagger \tilde{\sigma}^\nu \mathbf{M} = L^\nu{}_\mu \tilde{\sigma}^\mu, \tag{5.15}$$

$$\mathbf{N}^\dagger \sigma^\nu \mathbf{N} = L^\nu{}_\mu \sigma^\mu. \tag{5.16}$$

The matrices $\mathbf{M}$ and $\mathbf{N}$ are related by (B.19):

$$\mathbf{M}^\dagger \mathbf{N} = \mathbf{N}^\dagger \mathbf{M} = 1. \tag{5.17}$$

In the frame K' the Lagrangian density (5.12) can be written

$$\mathcal{L} = \mathrm{i}\psi_{\mathrm{L}}^\dagger \mathbf{M}^\dagger \tilde{\sigma}^\nu \mathbf{M} \partial'_\nu \psi_{\mathrm{L}} + \mathrm{i}\psi_{\mathrm{R}}^\dagger \mathbf{N}^\dagger \sigma^\nu \mathbf{N} \partial'_\nu \psi_{\mathrm{R}} - m(\psi_{\mathrm{L}}^\dagger \psi_{\mathrm{R}} + \psi_{\mathrm{R}}^\dagger \psi_{\mathrm{L}}), \tag{5.18}$$

where we have used (5.14) along with (5.15) and (5.16) in the first two terms.

We must define

$$\psi'_{\mathrm{L}}(x') = \mathbf{M}\psi_{\mathrm{L}}(x), \tag{5.19}$$

$$\psi'_{\mathrm{R}}(x') = \mathbf{N}\psi_{\mathrm{R}}(x), \tag{5.20}$$

to give

$$\mathcal{L} = \mathrm{i}\psi_{\mathrm{L}}^{\prime\dagger} \tilde{\sigma}^\nu \partial'_\nu \psi'_{\mathrm{L}} + \mathrm{i}\psi_{\mathrm{R}}^{\prime\dagger} \sigma^\nu \partial'_\nu \psi'_{\mathrm{R}} - m(\psi_{\mathrm{L}}^{\prime\dagger} \psi'_{\mathrm{R}} + \psi_{\mathrm{R}}^{\prime\dagger} \psi'_{\mathrm{L}})$$

(noting that $\psi_{\mathrm{L}}^{\prime\dagger} \psi'_{\mathrm{R}} = \psi_{\mathrm{L}}^\dagger \mathbf{M}^\dagger \mathbf{N} \psi_{\mathrm{R}} = \psi_{\mathrm{L}}^\dagger \psi_{\mathrm{R}}$, since $\mathbf{M}^\dagger \mathbf{N} = \mathbf{I}$, and similarly $\psi_{\mathrm{R}}^{\prime\dagger} \psi'_{\mathrm{L}} = \psi_{\mathrm{R}}^\dagger \psi_{\mathrm{L}}$).

With the transformations (5.19) and (5.20) the Lagrangian, and hence the field equations, take the same form in every inertial frame. The way to construct an $\mathbf{M}$ and an $\mathbf{N}$ for any Lorentz transformation is given in Appendix B.

An example of a rotation is

$$L^\mu{}_\nu = \begin{pmatrix} 1 & 0 & 0 & 0 \\ 0 & \cos\theta & \sin\theta & 0 \\ 0 & -\sin\theta & \cos\theta & 0 \\ 0 & 0 & 0 & 1 \end{pmatrix}. \tag{5.21}$$

This is a rotation of the coordinate axes through an angle θ about the z-axis and is equivalent to equations (2.1). The corresponding matrix $\mathbf{M}$ is unitary:

$$\mathrm{M} = \begin{pmatrix} \mathrm{e}^{\mathrm{i}\theta/2} & 0 \\ 0 & \mathrm{e}^{-\mathrm{i}\theta/2} \end{pmatrix}. \tag{5.22}$$

Hence, from (5.17), $\mathbf{N} = (\mathbf{M}^\dagger)^{-1} = \mathrm{M}$, since $\mathbf{M}\mathbf{M}^\dagger = \mathbf{1}$. The reader may verify that (5.15) and (5.16) hold. $\mathbf{M}$ is unitary (and hence equal to $\mathbf{N}$) for all rotations.

An example of a Lorentz boost is

$$L^{\mu}{}_{\nu} = \begin{pmatrix} \cosh\theta & 0 & 0 & -\sinh\theta \\ 0 & 1 & 0 & 0 \\ 0 & 0 & 1 & 0 \\ -\sinh\theta & 0 & 0 & \cosh\theta \end{pmatrix}. \tag{5.23}$$

This is a boost with velocity $v/c = \tanh\theta$ along the z-axis and is equivalent to equations (2.3). The corresponding matrix $\mathbf{M}$ is

$$\mathbf{M} = \begin{pmatrix} e^{\theta/2} & 0 \\ 0 & e^{-\theta/2} \end{pmatrix}, \quad \text{and} \quad \mathbf{N} = (\mathbf{M}^{\dagger})^{-1} = \begin{pmatrix} e^{-\theta/2} & 0 \\ 0 & e^{\theta/2} \end{pmatrix} = \mathbf{M}^{-1}. \tag{5.24}$$

5.3 The parity transformation

The Lagrangian density (5.12) can also be made invariant under space inversion of the axes. Denoting by a prime the space coordinates of a point as seen from the inverted axes, we have

$$\mathbf{r}' = -\mathbf{r} \quad \text{and} \quad \boldsymbol{\nabla}' = -\boldsymbol{\nabla}. \tag{5.25}$$

Hence, from the definitions (5.10) of σ^{μ} and $\tilde{\sigma}^{\mu}$,

$$\tilde{\sigma}^{\mu}\partial'_{\mu} = \sigma^{\mu}\partial_{\mu}, \quad \sigma^{\mu}\partial'_{\mu} = \tilde{\sigma}^{\mu}\partial_{\mu}. \tag{5.26}$$

Our Lagrangian density (5.12) is evidently invariant if $\psi(\mathbf{r}) \to \psi^{P}(\mathbf{r}')$ where

$$\psi_{\mathrm{L}}^{P}(\mathbf{r}') = \psi_{\mathrm{R}}(\mathbf{r}), \quad \psi_{\mathrm{R}}^{P}(\mathbf{r}') = \psi_{\mathrm{L}}(\mathbf{r}). \tag{5.27}$$

Actually the Lagrangian density would also retain the same form if we were to take, for example,

$$\psi_{\mathrm{L}}^{P}(\mathbf{r}') = e^{i\alpha}\psi_{\mathrm{R}}(\mathbf{r}), \quad \psi_{\mathrm{R}}^{P}(\mathbf{r}') = e^{i\alpha}\psi_{\mathrm{L}}(\mathbf{r}),$$

for any real α. It is the standard convention to adopt the form (5.27) for the field transformation under space inversion.

5.4 Spinors

Two-component complex quantities that transform under a Lorentz transformation according to the rules (5.19) and (5.20) are called *left-handed spinors* and *right-handed spinors,* respectively. Our subscripts L and R anticipated this. The four-component Dirac field is often called a *Dirac spinor.*

Spinors have the remarkable property that they can be combined in pairs to make Lorentz scalars, pseudoscalars, four-vectors, pseudovectors and higher order tensors. For example, $(\psi_{\mathrm{L}}^{\dagger}\psi_{\mathrm{R}} + \psi_{\mathrm{R}}^{\dagger}\psi_{\mathrm{L}})$ is a Lorentz invariant real scalar and $i(\psi_{\mathrm{L}}^{\dagger}\psi_{\mathrm{R}} - \psi_{\mathrm{R}}^{\dagger}\psi_{\mathrm{L}})$ is a real pseudoscalar; it is invariant under proper Lorentz transformations but changes sign under space inversion. Using (5.15), (5.16) and (5.27), we can see that $(\psi_L^\dagger\tilde{\sigma}^\mu\psi_L + \psi_R^\dagger\sigma^\mu\psi_R)$ is a four-vector, the space-like components of which change sign under space inversion (since $\tilde{\sigma}^i = -\sigma^i$), and $(\psi_L^\dagger\tilde{\sigma}^\mu\psi_L - \psi_R^\dagger\sigma^\mu\psi_R)$ is an axial four-vector, the space-like components of which are unchanged under space inversion.

5.5 The matrices γ^μ

The separation of the Dirac spinor into left-handed and right-handed components will be particularly appropriate when we discuss the weak interaction. For describing the electromagnetic interactions of fermions it is convenient to introduce 4×4 matrices γ^μ defined by

$$\gamma^0 = \beta; \quad \gamma^i = \beta\alpha_i, \quad i = 1, 2, 3. \tag{5.28}$$

It follows from the properties of the β and α^i matrices that

$$\begin{aligned}(\gamma^0)^2 = \mathbf{I}; \quad (\gamma^i)^2 = -\mathbf{I}, \quad i = 1, 2, 3;\\ \gamma^\mu\gamma^\nu + \gamma^\nu\gamma^\mu = \mathbf{0}, \quad \mu \neq \nu.\end{aligned} \tag{5.29}$$

In the chiral representation,

$$\gamma^0 = \begin{pmatrix} \mathbf{0} & \sigma^0 \\ \sigma^0 & \mathbf{0} \end{pmatrix}, \quad \gamma^i = \begin{pmatrix} \mathbf{0} & \sigma^i \\ -\sigma^i & \mathbf{0} \end{pmatrix}. \tag{5.30}$$

Written with the γ^μ matrices, the Lagrangian density (5.6) becomes

$$\mathcal{L} = \bar{\psi}(\mathrm{i}\gamma^\mu\partial_\mu - m)\psi, \tag{5.31}$$

where $\bar{\psi}$ is the row matrix $\bar{\psi} = \psi^\dagger\gamma^0$, and the Dirac equation takes the symmetrical form

$$(\mathrm{i}\gamma^\mu\partial_\mu - m)\psi = 0. \tag{5.32}$$

Another useful matrix $\gamma^5 = \mathrm{i}\gamma^0\gamma^1\gamma^2\gamma^3$. In the chiral representation,

$$\gamma^5 = \begin{pmatrix} -\sigma^0 & \mathbf{0} \\ \mathbf{0} & \sigma^0 \end{pmatrix}.$$

The matrices $\frac{1}{2}(\mathbf{I} - \gamma^5)$, $\frac{1}{2}(\mathbf{I} + \gamma^5)$ are projection operators giving the left-handed and right-handed parts of a Dirac spinor:

$$\frac{1}{2}(\mathbf{I} - \gamma^5)\psi = \begin{pmatrix} \sigma^0 & \mathbf{0} \\ \mathbf{0} & \mathbf{0} \end{pmatrix}\begin{pmatrix} \psi_L \\ \psi_R \end{pmatrix} = \begin{pmatrix} \psi_L \\ \mathbf{0} \end{pmatrix}, \tag{5.33}$$

$$\frac{1}{2}(\mathbf{I} + \gamma^5)\psi = \begin{pmatrix} \mathbf{0} & \mathbf{0} \\ \mathbf{0} & \sigma^0 \end{pmatrix}\begin{pmatrix} \psi_L \\ \psi_R \end{pmatrix} = \begin{pmatrix} \mathbf{0} \\ \psi_R \end{pmatrix}. \tag{5.34}$$

It is straightforward to verify that the Lorentz scalars and vectors constructed in Section 5.4 from two-component spinors can be written:

$$\begin{aligned}
\psi_L^\dagger\psi_R + \psi_R^\dagger\psi_L &= \bar{\psi}\psi \text{ (scalar)}\\
i(\psi_L^\dagger\psi_R - \psi_R^\dagger\psi_L) &= i\bar{\psi}\gamma^5\psi \text{ (pseudoscalar)}\\
\psi_L^\dagger\tilde{\sigma}^\mu\psi_L + \psi_R^\dagger\sigma^\mu\psi_R &= \bar{\psi}\gamma^\mu\psi \text{ (contravariant four-vector)}\\
\psi_L^\dagger\tilde{\sigma}^\mu\psi_L - \psi_R^\dagger\sigma^\mu\psi_R &= \bar{\psi}\gamma^5\gamma^\mu\psi \text{ (contravariant axial vector).}
\end{aligned}$$

Note that these quantities are all real.

5.6 Making the Lagrangian density real

A potential problem with our Lagrangian density (5.6) or (5.12) is that it is not real. Regarding ψ as a wave function, $\mathcal{L}$ is a complex function; regarding ψ as an operator, $\mathcal{L}$ is not Hermitian. As a consequence, the energy–momentum tensor is complex. Indeed, to apply Hamilton's principle, the variation δS in the action must be real. The term $-m(\psi_L^\dagger\psi_R + \psi_R^\dagger\psi_L)$ in (5.12) is real, and the imaginary part of $\mathcal{L}$ may be written

$$\begin{aligned}
&(1/2i)[i\psi_L^\dagger\tilde{\sigma}^\mu\partial_\mu\psi_L + i\psi_R^\dagger\sigma^\mu\partial_\mu\psi_R - (i\psi_L^\dagger\tilde{\sigma}^\mu\partial_\mu\psi_L + i\psi_R^\dagger\sigma^\mu\partial_\mu\psi_R)^\dagger]\\
&\quad = (1/2i)[i\psi_L^\dagger\tilde{\sigma}^\mu\partial_\mu\psi_L + i\psi_R^\dagger\sigma^\mu\partial_\mu\psi_R + i(\partial_\mu\psi_L^\dagger)\tilde{\sigma}^\mu\psi_L + i(\partial_\mu\psi_R)^\dagger\sigma^\mu\psi_R],
\end{aligned}$$

(where we have used the Hermitian property of the matrices σ^μ and $\tilde{\sigma}^\mu$). The last expression is just

$$(1/2)\partial_\mu(\psi_L^\dagger\tilde{\sigma}^\mu\psi_L + \psi_R^\dagger\sigma^\mu\psi_R).$$

This is a sum of derivatives, which give only irrelevant end-point contributions to the action (cf. Section 3.1). Hence δS is real. The imaginary part of $\mathcal{L}$ can be discarded, and we can take

$$\mathcal{L} = \frac{1}{2}[(i\psi_L^\dagger\tilde{\sigma}^\mu\partial_\mu\psi_L + i\psi_R^\dagger\sigma^\mu\partial_\mu\psi_R) \tag{5.35}$$

$$+ \text{Hermitian conjugate}] - m(\psi_L^\dagger\psi_R + \psi_R^\dagger\psi_L). \tag{5.36}$$

For further interesting discussion of this question see Olive (1997).

Problems

5.1 Show that the matrix $\mathbf{M} = \mathbf{N}$ of equation (5.22) when inserted into equations (5.15) and (5.16) generates the rotation matrix (5.21).

5.2 Show that the matrices $\mathbf{M}$ and $\mathbf{N} = \mathbf{M}^{-1}$ given by equation (5.24) when inserted into equations (5.15) and (5.16) generate the Lorentz boost of equation (5.23).

5.3 Show that $\psi_R^\dagger \psi_L$ and $\psi_L^\dagger \psi_R$ are invariant under proper Lorentz transformations.

Show that $\psi_R^\dagger \sigma^\mu \psi_R$ and $\psi_L{}^\dagger \tilde{\sigma}^\mu \psi_L$ are contravariant four-vectors under proper Lorentz transformations.

Show that $\psi_R^\dagger \sigma^\mu \tilde{\sigma}^\nu \psi_L$ and $\psi_L^\dagger \tilde{\sigma}^\mu \sigma^\nu \psi_R$ are contravariant tensors under proper Lorentz transformations.

5.4 Demonstrate the equivalence of the expressions (5.6) and (5.31) for the Lagrangian density.

5.5 Show that γ^5 has the properties

$$(\gamma^5)^2 = \mathbf{I}; \quad \gamma^\mu \gamma^5 = -\gamma^5 \gamma^\mu; \quad \mu = 0, 1, 2, 3.$$

5.6 Show that $\mathrm{i}\bar{\psi}\gamma^5\psi$ is a pseudoscalar field and $\bar{\psi}\gamma^5\gamma^\mu\psi = -\bar{\psi}\gamma^\mu\gamma^5\psi$ is an axial vector field.

5.7 Show that $(\gamma^0)^\dagger = \gamma^0$, $(\gamma^i)^\dagger = -\gamma^i$.

6

Free space solutions of the Dirac equation

In this chapter we display the plane wave solutions of the Dirac equation. We show that a Dirac particle has intrinsic spin $\hbar/2$, and we shall see how the Dirac equation predicts the existence of antiparticles.

6.1 A Dirac particle at rest

In Chapter 5 we showed that the Dirac equation for a particle in free space is equivalent to the coupled two-component equations

$$\begin{aligned} \mathrm{i}\tilde{\sigma}^\mu \partial_\mu \psi_\mathrm{L} - m\psi_\mathrm{R} &= 0, \\ \mathrm{i}\sigma^\mu \partial_\mu \psi_\mathrm{R} - m\psi_\mathrm{L} &= 0. \end{aligned} \tag{6.1}$$

These equations have plane wave solutions of the form

$$\psi_\mathrm{L} = u_\mathrm{L} \mathrm{e}^{\mathrm{i}(\mathbf{p}\cdot\mathbf{r} - Et)}, \quad \psi_\mathrm{R} = u_\mathrm{R} e^{\mathrm{i}(\mathbf{p}\cdot\mathbf{r} - Et)}, \tag{6.2}$$

where u_L and u_R are two-component spinors. Since solutions of the Dirac equation also satisfy the Klein–Gordon equation (3.19), we must have

$$E^2 = p^2 + m^2. \tag{6.3}$$

It is simplest to find the solution in a frame K' in which the particle is at rest, and then obtain the solution in a frame in which the particle is moving with velocity v by making a Lorentz boost. Using primes to denote quantities in the frame K', the momentum $\mathbf{p}' = 0$, so that equations (6.1) and (6.3) become

$$\mathrm{i}\partial'_0 \psi'_\mathrm{L} = m\psi'_\mathrm{R}, \qquad \mathrm{i}\partial'_0 \psi_\mathrm{R} = m\psi'_\mathrm{L},$$

and

$$E'^2 = m^2, \quad E' = \pm m. \tag{6.4}$$

The solutions with positive energy $E' = m$ are

$$\psi'_L = u e^{-imt'}, \quad \psi'_R = u e^{-imt'}, \tag{6.5}$$

where

$$u = \begin{pmatrix} u_1 \\ u_2 \end{pmatrix} = u_1 \begin{pmatrix} 1 \\ 0 \end{pmatrix} + u_2 \begin{pmatrix} 0 \\ 1 \end{pmatrix}$$

is an arbitrary two-component spinor and we are adopting the standard convention of quantum mechanics that the time dependence of an energy eigenstate is given by the phase factor e^{-iEt}.

In the rest frame K', the left-handed and right-handed positive energy spinors are identical. As a consequence this solution is invariant under space inversion (see Section 5.3). It is said to have *positive parity.*

6.2 The intrinsic spin of a Dirac particle

The intrinsic spin operator **S** of a particle with mass is defined to be its angular momentum operator in a frame in which it is at rest. The component of **S** along the z-direction is given by

$$S_z = i\hbar \lim_{\phi \to 0} [R_z(\phi) - 1] / \phi,$$

where $R_z(\phi)$ is the operator that rotates the state of the particle through an angle ϕ about Oz (cf. Section 4.7). A rotation of the state through an angle ϕ, is equivalent to rotating the axes through an angle $-\phi$, and then $\psi_L \to M\psi_L$, $\psi_R \to N\psi_R$ where, from (5.22),

$$M = N = \begin{pmatrix} e^{-i\phi/2} & 0 \\ 0 & e^{i\phi/2} \end{pmatrix}.$$

Hence

$$S_z = i\hbar \lim_{\phi \to 0} \frac{1}{\phi} \begin{pmatrix} e^{-i\phi/2} - 1 & 0 \\ 0 & e^{i\phi/2} - 1 \end{pmatrix} = \frac{\hbar}{2} \begin{pmatrix} 1 & 0 \\ 0 & -1 \end{pmatrix} = \frac{\hbar}{2} \sigma_z.$$

In the state with $u_1 = 1, u_2 = 0$,

$$S_z \psi'_L = (\hbar/2) \psi'_L$$

and

$$S_z \psi'_R = (\hbar/2) \psi'_R.$$

Acting on the Dirac wave function, we have

$$S_z \begin{pmatrix} \psi'_L \\ \psi'_R \end{pmatrix} = (\hbar/2) \begin{pmatrix} \psi'_L \\ \psi'_R \end{pmatrix}. \tag{6.6}$$

Similarly, in the state with $u_1 = 0$, $u_2 = 1$,

$$S_z \begin{pmatrix} \psi'_L \\ \psi'_R \end{pmatrix} = -(\hbar/2) \begin{pmatrix} \psi'_L \\ \psi'_R \end{pmatrix}. \tag{6.7}$$

Thus in the rest frame of the particle there are two independent states which are eigenstates of S_z with eigenvalues $\pm(\hbar/2)$. The operator S_z on a Dirac wave function is represented by the matrix

$$\boldsymbol{\Sigma}_z = (\hbar/2) \begin{pmatrix} \sigma_z & \mathbf{0} \\ \mathbf{0} & \sigma_z \end{pmatrix}. \tag{6.8}$$

More generally, **S** is represented by

$$\boldsymbol{\Sigma} = (\hbar/2) \begin{pmatrix} \boldsymbol{\sigma} & \mathbf{0} \\ \mathbf{0} & \boldsymbol{\sigma} \end{pmatrix}. \tag{6.9}$$

Also, every Dirac wave function is an eigenstate of the square of the spin operator,

$$\boldsymbol{\Sigma}^2 = (3/4)\hbar^2 \mathbf{I},$$

with eigenvalue $(3/4)\hbar^2 = (1/2)((1/2) + 1)\hbar^2$. Recalling that the square J^2 of the angular momentum for a state with angular momentum j is $j(j+1)\hbar^2$; it is appropriate to say that a Dirac particle has intrinsic spin $\hbar/2$.

6.3 Plane waves and helicity

We now transform to a frame K in which the frame K', and the particle, are moving with velocity v. For simplicity we take $\mathbf{v} = (0, 0, v)$, along the z-axis with $v > 0$, and consider the state with $u_1 = 1$, $u_2 = 0$.

Transformations between K and K' are then given by (5.23), along with (5.24). Using (5.19) and (5.20),

$$\psi_L = M^{-1}\psi'_L = \begin{pmatrix} e^{-\theta/2} & 0 \\ 0 & e^{\theta/2} \end{pmatrix} e^{-imt'} \begin{pmatrix} 1 \\ 0 \end{pmatrix} = e^{-imt'} e^{-\theta/2} \begin{pmatrix} 1 \\ 0 \end{pmatrix},$$

$$\psi_R = N^{-1}\psi'_R = \begin{pmatrix} e^{\theta/2} & 0 \\ 0 & e^{-\theta/2} \end{pmatrix} e^{-imt'} \begin{pmatrix} 1 \\ 0 \end{pmatrix} = e^{-imt'} e^{\theta/2} \begin{pmatrix} 1 \\ 0 \end{pmatrix}.$$

Finally, substituting $t' = t\cosh\theta - z\sinh\theta$ (and noting that $m\cosh\theta = \gamma m = E$, $m\sinh\theta = \gamma m v = p$, where $\gamma = (1 - v^2/c^2)^{-1/2}$ we have

$$\psi_{\mathrm{L}} = \mathrm{e}^{\mathrm{i}(pz-Et)}\begin{pmatrix}\mathrm{e}^{-\theta/2}\\0\end{pmatrix}, \qquad \psi_{\mathrm{R}} = \mathrm{e}^{\mathrm{i}(pz-Et)}\begin{pmatrix}\mathrm{e}^{\theta/2}\\0\end{pmatrix}. \tag{6.10}$$

The *helicity* operator is useful in classifying plane wave states. It is defined by

$$\text{helicity} = \frac{\boldsymbol{\Sigma}\cdot\mathbf{p}}{|\mathbf{p}|}. \tag{6.11}$$

The expectation value of this operator in a given state is a measure of the alignment of a particle's intrinsic spin with its direction of motion in that state. For $\mathbf{p} = (0, 0, p)$, $p > 0$, the helicity operator $\boldsymbol{\Sigma}\cdot\mathbf{p}/|\mathbf{p}| = \Sigma_z$. Thus the state (6.10) is an eigenstate of the helicity operator with positive helicity 1/2, which we can write as a Dirac spinor

$$\psi_+ = \frac{1}{\sqrt{2}}\mathrm{e}^{\mathrm{i}(pz-Et)}\begin{pmatrix}\mathrm{e}^{-\theta/2}\\0\\\mathrm{e}^{\theta/2}\\0\end{pmatrix}, \quad p > 0. \tag{6.12}$$

We have inserted the normalisation factor $1/\sqrt{2}$ to conform with the standard normalisation of the Lorentz scalar $\bar{\psi}\psi$:

$$\bar{\psi}\psi = \psi^\dagger\gamma^0\psi = \psi_{\mathrm{L}}^\dagger\psi_{\mathrm{R}} + \psi_{\mathrm{R}}^\dagger\psi_{\mathrm{L}} = 1.$$

Similarly, taking $u_1 = 0$, $u_2 = 1$, we can construct an eigenstate of negative helicity $-1/2$:

$$\psi_- = \frac{1}{\sqrt{2}}\mathrm{e}^{\mathrm{i}(pz-Et)}\begin{pmatrix}0\\\mathrm{e}^{\theta/2}\\0\\\mathrm{e}^{-\theta/2}\end{pmatrix}, \quad p > 0. \tag{6.13}$$

All plane waves with positive energy can be generated by applying rotations to the states we have found. The helicity of a state is unchanged by a rotation, since it is defined by a scalar product. The evident generalisations of (6.12) and (6.13) to a wave with wave vector $\mathbf{p}$ are

$$\psi_+ = \mathrm{e}^{\mathrm{i}(\mathbf{p}\cdot\mathbf{r}-Et)}u_+(\mathbf{p}) \tag{6.14}$$

where

$$u_+(\mathbf{p}) = \frac{1}{\sqrt{2}}\begin{pmatrix}\mathrm{e}^{-\theta/2}\,|+\rangle\\\mathrm{e}^{\theta/2}\,|+\rangle\end{pmatrix},$$

and

$$\psi_- = e^{i(\mathbf{p}\cdot\mathbf{r}-Et)}u_-(\mathbf{p}) \tag{6.15}$$

where

$$u_-(\mathbf{p}) = \frac{1}{\sqrt{2}}\begin{pmatrix} e^{\theta/2}\,|-\rangle \\ e^{-\theta/2}\,|-\rangle \end{pmatrix}.$$

The Pauli spin states $|\pm\rangle$ are here the eigenstates of the operators $\boldsymbol{\sigma}\cdot\mathbf{p}/|\mathbf{p}|$ with eigenvalues ± 1 (Problem 6.6). A general state of positive energy can be constructed as a superposition of plane waves.

6.4 Negative energy solutions

In the frame K' in which the particle is at rest, there are also negative energy solutions of (6.4) with $E' = -m$:

$$\psi'_{\mathrm{L}} = v e^{imt'}, \quad \psi'_{\mathrm{R}} = -v e^{imt'}. \tag{6.16}$$

In this case the left-handed and right-handed spinors v differ in sign. Thus the negative energy solution changes sign under space inversion (see Section 5.3). It is said to have *negative parity*.

The same Lorentz boost we used above in Section 6.3 gives solutions ψ_+ and ψ_- with positive and negative helicity, respectively, which we can write as Dirac spinors

$$\psi_+ = \frac{1}{\sqrt{2}}e^{i(-pz+Et)}\begin{pmatrix} 0 \\ e^{\theta/2} \\ 0 \\ -e^{-\theta/2} \end{pmatrix}, \quad \psi_- = \frac{1}{\sqrt{2}}e^{i(-pz+Et)}\begin{pmatrix} -e^{-\theta/2} \\ 0 \\ e^{\theta/2} \\ 0 \end{pmatrix}, \quad p > 0. \tag{6.17}$$

These solutions generalise to

$$\psi_+ = e^{i(-\mathbf{p}\cdot\mathbf{r}+Et)}v_+(\mathbf{p}) \tag{6.18}$$

where

$$v_+(\mathbf{p}) = \frac{1}{\sqrt{2}}\begin{pmatrix} e^{\theta/2}\,|-\rangle \\ -e^{-\theta/2}\,|-\rangle \end{pmatrix},$$

and

$$\psi_- = e^{i(-\mathbf{p}\cdot\mathbf{r}+Et)}v_-(\mathbf{p}) \tag{6.19}$$

where

$$v_-(\mathbf{p}) = \frac{1}{\sqrt{2}}\begin{pmatrix} -e^{\theta/2}\,|+\rangle \\ e^{-\theta/2}\,|+\rangle \end{pmatrix}.$$

$|+\rangle$ and $|-\rangle$ remain eigenstates of $\boldsymbol{\sigma} \cdot \mathbf{p}/|\mathbf{p}|$ as defined below (6.15). Note that the Lorentz invariant $\bar{\psi}\psi$ acquires a minus sign; in the case of the negative energy solutions,

$$\bar{\psi}\psi = \psi_{\mathrm{L}}^{\dagger}\psi_{\mathrm{R}} + \psi_{\mathrm{R}}^{\dagger}\psi_{\mathrm{L}} = -1.$$

Negative energy solutions of the Dirac equation appear at first sight to be an embarrassment. In quantum theory a particle can make transitions between states. Hence all Dirac states would seem to be unstable to a transition to lower energy. Dirac's solution to the difficulty was to assume that nearly all negative energy states are occupied, so that the Pauli exclusion principle forbids transitions to them. An unoccupied negative energy state, or *hole*, will behave as a positive energy *antiparticle*, of the same mass but opposite momentum, spin, and electric charge. Left unfilled, the negative energy state ψ_+ of (6.17) corresponds to an antiparticle of positive energy E and positive momentum p, and positive helicity, since the spin of the hole is also opposite to that of the negative energy state.

A particle falling into an empty negative energy state will be seen as the simultaneous annihilation of a particle–antiparticle pair with the emission of electromagnetic energy $\geq 2mc^2$. Conversely, the excitation of a particle from a negative energy state to a positive energy state will be seen as pair production. The existence of the positron, the antiparticle of the electron, was established experimentally in 1932, and the observation of pair production soon followed.

The uniform background sea of occupied negative energy states, with its associated infinite electric charge, is assumed to be unobservable. In any case, it is clearly quite arbitrary whether, say, the electron is regarded as the particle and the positron as antiparticle, or vice versa. Evidently our starting interpretation of the Dirac equation as a single particle equation is not tenable. We are led, inevitably, to a quantum field theory in which particles and antiparticles appear as the quanta of the field, in somewhat the same way as photons appear as the quanta of the electromagnetic field. We shall take up this theme in Chapter 8.

6.5 The energy and momentum of the Dirac field

The Lagrangian density of the Dirac field is given by (5.31), which we display in more detail:

$$\begin{aligned}\mathcal{L} &= \bar{\psi}(\mathrm{i}\gamma^{\mu}\partial_{\mu} - m)\psi \\ &= \mathrm{i}\psi_a^*\partial_0\psi_a + \bar{\psi}_b\big(\mathrm{i}\gamma^i_{ba}\partial_i - m\delta_{ba}\big)\psi_a.\end{aligned} \tag{6.20}$$

As in Section 5.1 we may treat the fields ψ_a and $\psi_a{}^*$ as independent, and take the energy–momentum tensor to be

$$T^{\mu}_{\nu} = \frac{\partial\mathcal{L}}{\partial(\partial_{\mu}\psi_a)}\partial_{\nu}\psi_a - \mathcal{L}\delta^{\mu}_{\nu} \tag{6.21}$$

($\mathcal{L}$ does not depend on $\partial_{\mu}\psi_a{}^*$).

In particular, the energy density is

$$\begin{aligned} T_0^0 &= \mathrm{i}\psi_a^* \partial_0 \psi_a - \mathcal{L} \\ &= \bar{\psi}(-\mathrm{i}\gamma^i \partial_i + m)\psi \end{aligned} \tag{6.22}$$

and the momentum density is

$$T_i^0 = \mathrm{i}\psi_a^* \partial_i \psi_a = \mathrm{i}\psi^\dagger \partial_i \psi. \tag{6.23}$$

The general solution of the free space Dirac equation is a superposition of all possible plane waves, which we will write

$$\psi = \frac{1}{\sqrt{V}} \sum_{\mathbf{p},\varepsilon} \sqrt{\frac{m}{E_{\mathbf{p}}}} \left(b_{\mathbf{p}\varepsilon} u_\varepsilon(\mathbf{p}) \mathrm{e}^{\mathrm{i}(\mathbf{p}\cdot\mathbf{r} - E_p t)} + d_{\mathbf{p}\varepsilon}^* v_\varepsilon(\mathbf{p}) \mathrm{e}^{\mathrm{i}(-\mathbf{p}\cdot\mathbf{r} + E_p t)} \right). \tag{6.24}$$

ε is the helicity index, $\pm$, and $b_{\mathbf{p}\varepsilon}$ and $d_{\mathbf{p}\varepsilon}$ are arbitrary complex numbers. The factors $\sqrt{(m/E_p)}$ take the place of the factors $1/\sqrt{2\omega_k}$ we inserted in the boson field expansions of Chapter 3 and Chapter 4.

We can express the total energy and total momentum of the Dirac field in terms of the wave amplitudes, by inserting the field expansion into T_0^0 and T_i^0, and integrating over the normalisation volume V. The results are

$$H = \sum_{\mathbf{p},\varepsilon} \left(b_{\mathbf{p}\varepsilon}^* b_{\mathbf{p}\varepsilon} - d_{\mathbf{p}\varepsilon} d_{\mathbf{p}\varepsilon}^* \right) E_{\mathbf{p}}, \tag{6.25}$$

$$P = \sum_{\mathbf{p},\varepsilon} \left(b_{\mathbf{p}\varepsilon}^* b_{\mathbf{p}\varepsilon} - d_{\mathbf{p}\varepsilon} d_{\mathbf{p}\varepsilon}^* \right) \mathbf{p}. \tag{6.26}$$

$\varepsilon = \pm 1$ is the helicity index.

The (somewhat tedious) derivation of these results is left to the reader. Note that each plane wave is a solution of the Dirac equation (5.32), which implies

$$\begin{aligned} (\gamma^0 E_p - \gamma^i p^i) u_\varepsilon(\mathbf{p}) &= m u_\varepsilon(\mathbf{p}), \\ (\gamma^0 E_p - \gamma^i p^i) v_\varepsilon(\mathbf{p}) &= -m v_\varepsilon(\mathbf{p}). \end{aligned} \tag{6.27}$$

It is also necessary to use various orthogonality relations, which are set out in Problem 6.3.

For later convenience, we rewrite the Dirac field ψ (6.24) in terms of ψ_{L} and ψ_{R}. Using (6.14), (6.15), (6.18) and (6.19) gives

$$\begin{aligned} \psi_{\mathrm{L}} = \frac{1}{\sqrt{V}} \sum_{\mathbf{p}} \sqrt{\frac{m}{2E_p}} &\left[\left(b_{\mathbf{p}+} \mathrm{e}^{-\theta/2} \left|+\right\rangle + b_{\mathbf{p}-} \mathrm{e}^{\theta/2} \left|-\right\rangle \right) \mathrm{e}^{\mathrm{i}(\mathbf{p}\cdot\mathbf{r} - Et)} \right. \\ &\left. + \left(d_{\mathbf{p}+}^* \mathrm{e}^{\theta/2} \left|-\right\rangle - d_{\mathbf{p}-}^* \mathrm{e}^{-\theta/2} \left|+\right\rangle \right) \mathrm{e}^{\mathrm{i}(-\mathbf{p}\cdot\mathbf{r} + Et)} \right] \end{aligned} \tag{6.28}$$

$$\begin{aligned} \psi_{\mathrm{R}} = \frac{1}{\sqrt{V}} \sum_{\mathbf{p}} \sqrt{\frac{m}{2E_p}} &\left[\left(b_{\mathbf{p}+} \mathrm{e}^{\theta/2} \left|+\right\rangle + b_{\mathbf{p}-} \mathrm{e}^{-\theta/2} \left|-\right\rangle \right) \mathrm{e}^{\mathrm{i}(\mathbf{p}\cdot\mathbf{r} - Et)} \right. \\ &\left. + \left(-d_{\mathbf{p}+}^* \mathrm{e}^{-\theta/2} \left|-\right\rangle + d_{\mathbf{p}-}^* \mathrm{e}^{\theta/2} \left|+\right\rangle \right) \mathrm{e}^{\mathrm{i}(-\mathbf{p}\cdot\mathbf{r} + Et)} \right] \end{aligned} \tag{6.29}$$

6.6 Dirac and Majorana fields

The expansion (6.24) is the general solution of the free field Dirac equation. For every momentum **p** there are four independent complex coefficients: $b_{\mathbf{p}+}$, $b_{\mathbf{p}-}$, $d^*_{\mathbf{p}+}$ and $d^*_{\mathbf{p}-}$, which correspond to particles with helicities $+1/2$, $-1/2$ and antiparticles with helicities $+1/2$, $-1/2$, respectively.

It will be of interest, in Chapter 21, to consider solutions in which we impose the constraint that $d_{\mathbf{p}+} = b_{\mathbf{p}+}$, $d_{\mathbf{p}-} = b_{\mathbf{p}-}$, and hence $d^*_{\mathbf{p}+} = b^*_{\mathbf{p}+}$, $d^*_{\mathbf{p}-} = b^*_{\mathbf{p}-}$. These solutions are known as *Majorana fields*. On quantisation, we shall see that the Dirac fields create and annihilate particles, and antiparticles. For example, if ψ is an electron field it creates positrons and annihilates electrons, $\psi^\dagger$ creates electrons and annihilates positrons. With the Majorana constraint, particles and antiparticles are identical. Majorana fields are irrelevant for electrically charged particles, but it is possible that the electrically neutral neutrino fields have this property. It is still an open question whether neutrino fields are Dirac or Majorana.

6.7 The $E \gg m$ limit, neutrinos

The coefficients of the plane waves in the expansions (6.25) and (6.26) may be expressed as

$$\sqrt{(m/2E)}\mathrm{e}^{\pm\theta/2} = \{(1 \pm v/c)/2\}^{1/2}, \tag{6.30}$$

where v is the particle velocity (Problem 6.1). In the high energy limit, $E \gg m$, the velocity $v \to c$. The only significant terms in the field expansions which survive in this limit are

$$\psi_{\mathrm{L}} = \frac{1}{\sqrt{V}} \sum_{\mathbf{p}} \left(b_{\mathbf{p}-} \left|-\right\rangle \mathrm{e}^{\mathrm{i}(\mathbf{p}\cdot\mathbf{r}-Et)} + d^*_{\mathbf{p}+} \left|-\right\rangle \mathrm{e}^{\mathrm{i}(-\mathbf{p}\cdot\mathbf{r}+Et)} \right), \tag{6.31}$$

$$\psi_{\mathrm{R}} = \frac{1}{\sqrt{V}} \sum_{\mathbf{p}} \left(b_{\mathbf{p}+} \left|+\right\rangle \mathrm{e}^{\mathrm{i}(\mathbf{p}\cdot\mathbf{r}-Et)} + d^*_{\mathbf{p}-} \left|+\right\rangle \mathrm{e}^{\mathrm{i}(-\mathbf{p}\cdot\mathbf{r}+Et)} \right). \tag{6.32}$$

In the limit, ψ_{L} and ψ_{R} are completely independent: ψ_{L} involves only negative helicity particles and positive helicity antiparticles; ψ_{R} involves only positive helicity particles and negative helicity antiparticles.

Since neutrinos are electrically neutral, they are accessible to experimental investigation only through the weak interaction and we shall see in Chapter 9 that in the weak interaction Nature only employs ψ_{L}. In practice neutrino energies are usually many orders of magnitude greater than their mass, so that only negative helicity neutrinos and positive helicity antineutrinos are readily observed. It has not so far been established that the 'hard to see' positive helicity neutrino is different from the 'easy to see' positive helicity antineutrino.

Problems

6.1 With the normalisaion of ψ_+ determined by equation (6.14), show that

$$\psi_+^\dagger \psi_+ = \cosh\theta = E/m.$$

(Note that this is not the usual normalisation of particle quantum mechanics.)

Show that the probability of this positive helicity state being in the right-handed mode is

$$e^\theta/(2\cosh\theta) = (1 + v/c)/2$$

and the probability of its being in the left-handed mode is $(1 - v/c)/2$. What are the corresponding results for ψ_-?

6.2 Show that the negative energy positive helicity state of equation (6.18) has probability $(1 + v/c)/2$ of being in the left-handed mode.

6.3 Show that

$$\begin{aligned}
&u_\pm^\dagger(\mathbf{p})u_\pm(\mathbf{p}) = v_\pm^\dagger(\mathbf{p})v_\pm(\mathbf{p}) = E_p/m,\\
&u_\pm^\dagger(\mathbf{p})u_\mp(\mathbf{p}) = v_\pm^\dagger(\mathbf{p})v_\mp(\mathbf{p}) = 0,\\
&u_\pm^\dagger(\mathbf{p})v_\pm(-\mathbf{p}) = v_\pm^\dagger(-\mathbf{p})u_\pm(\mathbf{p}) = u_\pm^\dagger(\mathbf{p})v_\mp(-\mathbf{p}) = v_\mp^\dagger(-\mathbf{p})u_\pm(\mathbf{p}) = 0.
\end{aligned}$$

These results are useful in Problem 6.4.

6.4 Using the plane wave expansion (6.24) and the energy–momentum tensor components (6.22) and (6.23), show that the energy and momentum carried by the wave ψ are given by (6.25) and (6.26).

6.5 Consider a momentum $\mathbf{p}$ in the direction specified by the polar coordinates θ and ϕ.

$$\hat{\mathbf{p}} = (\sin\theta\cos\phi, \sin\theta\sin\phi, \cos\theta).$$

Show that

$$\boldsymbol{\sigma}\cdot\hat{\mathbf{p}} = \begin{pmatrix} \cos\theta & \sin\theta\, e^{-i\phi} \\ \sin\theta\, e^{i\phi} & -\cos\theta \end{pmatrix}$$

and the Pauli spin states

$$|+\rangle = \begin{pmatrix} \cos(\theta/2) \\ \sin(\theta/2)e^{i\phi} \end{pmatrix}, \quad |-\rangle = \begin{pmatrix} -\sin(\theta/2)e^{-i\phi} \\ \cos(\theta/2) \end{pmatrix}$$

are the helicity eigenstates appearing in (6.14) and (6.15). An overall phase is undetermined.

7
Electrodynamics

In this chapter we set up a Lagrangian for a field theory in which electrically charged Dirac particles and antiparticles, for example electrons and positrons, interact with and through the electromagnetic field. To facilitate reference to other texts, and for conciseness, we work with four-component Dirac spinors and the matrices γ^μ introduced in Section 5.5.

7.1 Probability density and probability current

We have seen in previous chapters how conservation laws are associated with symmetries of the Lagrangian. The Lagrangian density (5.31),

$$\mathcal{L} = \bar{\psi}(\mathrm{i}\gamma^\mu \partial_\mu - m)\psi,$$

is invariant under the transformation

$$\psi(x) \to \psi'(x) = \mathrm{e}^{-\mathrm{i}\alpha}\psi(x), \tag{7.1}$$

where α is a constant phase. These transformations form a group $U(1)$ (see Appendix B) and are said to be *global*: the same at every point in space and time.

If now we allow an arbitrary small space- and time-dependent variation in α, $\alpha \to \alpha'(x) = \alpha + \delta\alpha(x)$, and if the fields satisfy the field equations, the corresponding first-order variation δS in the action must be zero, since S is stationary for the actual fields. The variation comes from the operators ∂_μ acting on $\mathrm{e}^{-\mathrm{i}\delta\alpha(x)}$, so that

$$\begin{aligned} \partial S &= \int \bar{\psi}\gamma^\mu \psi \mathrm{i}\partial_\mu \mathrm{e}^{-\mathrm{i}\delta\alpha} \mathrm{d}^4x \\ &= \int \bar{\psi}\gamma^\mu \psi \partial_\mu (\delta\alpha)\, \mathrm{d}^4x, \text{ to first order.} \end{aligned}$$

Integrating by parts,

$$\delta S = -\int [\partial_\mu(\bar{\psi}\gamma^\mu\psi)]\delta\alpha\, \mathrm{d}^4x.$$

This is zero for any arbitrary function $\delta\alpha(x)$ only if

$$\partial_\mu(\bar{\psi}\gamma^\mu\psi) = 0. \tag{7.2}$$

At each point x of space and time, $\bar{\psi}(x)\gamma^\mu\psi(x)$ transforms like a contravariant four-vector (Section 5.5) and we may define the contravariant field

$$j^\mu(x) = \bar{\psi}\gamma^\mu\psi = (P(x), \mathbf{j}(x)) \tag{7.3}$$

where $P(x) = \bar{\psi}\gamma^0\psi = \psi^\dagger(y^0)^2\psi = \psi_a^*\psi_a = \sum_{a=1}^{4} |\psi_a|^2$. Then (7.2) takes the familiar form

$$\frac{\partial P}{\partial t} + \nabla \cdot \mathbf{j} = 0. \tag{7.4}$$

If $P(x)$ is interpreted as the particle probability density associated with the wave function $\psi(x)$ and $\mathbf{j}(x)$ as the probability current, (7.4) expresses local particle conservation. Integrating over all space, and using the divergence theorem, it follows that for fields that vanish at large distances

$$\frac{\mathrm{d}}{\mathrm{d}t}\int P\mathrm{d}^3\mathbf{x} = 0.$$

Hence

$$\int P(t, \mathbf{x})\,\mathrm{d}^3\mathbf{x} = \int \psi^\dagger\psi\,\mathrm{d}^3\mathbf{x}$$

is a constant independent of time. With $\psi(x)$ taken to be a normalised wave function for a particle, the constant is unity, and we see that a wave function once normalised stays normalised. In Chapter 8 we shall see that in a second quantised field theory, $\int P(t, \boldsymbol{x})\,\mathrm{d}^3\boldsymbol{x}$ is an operator that counts the number of particles minus the number of antiparticles, and thus this number is conserved.

We could have derived (7.2) from the field equation but the device introduced here, whereby the conservation law appears as a consequence of the $U(1)$ symmetry (7.1), is both elegant and economical.

7.2 The Dirac equation with an electromagnetic field

In classical mechanics, the Hamiltonian for a particle carrying charge q moving in an external electromagnetic field specified by the electromagnetic potentials $(\phi,\mathbf{A})$

is obtained from the free particle Hamiltonian by the substitution in (3.8)

$$E \rightarrow E - q\phi, \quad \mathbf{p} \rightarrow \mathbf{p} - q\mathbf{A},$$

or, equivalently

$$p^\mu \rightarrow p^\mu - qA^\mu, \tag{7.5}$$

where $p^\mu = (E,\mathbf{p})$ is the energy-momentum four-vector of the particle. (See Problems 4.6 and 4.7.) With the quantisation rule $p_\mu \rightarrow \mathrm{i}\partial_\mu$, (7.5) suggests that the Dirac equation in the presence of an electromagnetic field should be

$$[\gamma^\mu(\mathrm{i}\partial_\mu - qA_\mu) - m]\psi = 0, \tag{7.6}$$

and there should be a corresponding substitution in the Lagrangian density.

Using (4.10) and (5.31), we take the Lagrangian density for the Dirac field together with the electromagnetic field with external charge-current sources J^μ to be

$$\begin{aligned} \mathcal{L} &= \bar{\psi}[\gamma^\mu(\mathrm{i}\partial_\mu - qA_\mu) - m]\psi - \frac{1}{4}F_{\mu\nu}F^{\mu\nu} - J^\mu A_\mu \\ &= \bar{\psi}[\gamma^\mu \mathrm{i}\partial_\mu - m]\psi - \frac{1}{4}F_{\mu\nu}F^{\mu\nu} - \left(J^\mu + q\bar{\psi}\gamma^\mu\psi\right)A_\mu. \end{aligned} \tag{7.7}$$

The Lagrangian is still invariant under the transformation $\psi(x) \rightarrow \psi'(x) = \mathrm{e}^{-\mathrm{i}\alpha}\psi(x)$ with α constant, and this leads as before to particle conservation:

$$\partial_\mu j^\mu = 0, \; j^\mu = \bar{\psi}\gamma^\mu\psi. \tag{7.8}$$

Variation of the fields A_μ in the action, as in Section 4.2, yields the Maxwell equations, with charge-current density

$$J^\mu + q\bar{\psi}\gamma^\mu\psi = J^\mu + qj^\mu. \tag{7.9}$$

In (7.8) and (7.9), $j^\mu(x)$ is the conserved particle number density current (antiparticles being counted as negative), and $qj^\mu(x)$ is the conserved charge density current. Thus the Lagrangian density (7.7) includes the electromagnetic field produced by the charged particle current as well as the field produced by external sources.

Setting q = the electron charge $= -e$, and m to be the electron mass, the Lagrangian (7.7) is, after quantisation, the Lagrangian of quantum electrodynamics. With the external charge-current distribution $J^\mu(x)$ taken to be that of the atomic nuclei, and including the dynamics of the nuclei as an assembly of point particles, this is the basic Lagrangian that describes and explains most of chemistry and materials science. We shall review some of the astounding successes of quantum electrodynamics in the next chapter.

7.3 Gauge transformations and symmetry

In Chapter 4 we stressed that the four-potential A_μ is not unique: the same physical electric and magnetic fields are obtained after a gauge transformation

$$A_\mu(x) \to A'_\mu(x) = A_\mu(x) + \partial_\mu \chi(x)$$

where $\chi(x)$ is an arbitrary function of space and time.

If ψ is a solution of the Dirac equation with the four-potential A_μ, the corresponding solution in the gauge with four-potential A'_μ is given by

$$\psi \to \psi' = \mathrm{e}^{-\mathrm{i}q\chi}\psi.$$

This is easily verified:

$$\begin{aligned}(\mathrm{i}\partial_\mu - qA'_\mu)\psi' &= \mathrm{e}^{-\mathrm{i}q\chi}\left\{\mathrm{i}\partial_\mu + q\partial_\mu\chi - q(A_\mu + \partial_\mu\chi)\right\}\psi \\ &= \mathrm{e}^{-\mathrm{i}q\chi}(\mathrm{i}\partial_\mu - qA_\mu)\psi.\end{aligned}$$

Hence the Dirac equation (7.6) is equivalent to

$$\left[\gamma^\mu(\mathrm{i}\partial_\mu - qA'_\mu) - m\right]\psi' = 0.$$

The transformations:

$$A_\mu(x) \to A_\mu(x) + \partial_\mu\chi(x) \tag{7.10a}$$

$$\psi(x) \to \mathrm{e}^{-\mathrm{i}q\chi(x)}\psi(x) \tag{7.10b}$$

make up a general local gauge transformation.

The charge-current density $qj^\mu = q\bar{\psi}\gamma^\mu\psi$ is invariant under the transformation and so too is the action provided that (as in Section 4.3) $\partial_\mu J^\mu = 0$. It is also interesting to note that the phase of a charged Dirac field, for example that of an electron, is a gauge artefact without physical significance: this phase cannot be measured.

We can look at this transformation from a different point of view. The Lagrangian (7.7) is invariant under the global $U(1)$ transformation $\psi \to \psi' = \mathrm{e}^{-\mathrm{i}\alpha}\psi$ where α is constant. If we now ask for the Lagrangian to be invariant under a similar but *local* transformation, $\psi \to \psi'(x) = \mathrm{e}^{-\mathrm{i}q\chi}\psi(x)$, where $\chi(x)$ is an arbitrary function of space and time, we are forced into introducing the *gauge field* A_μ, with the transformation property $A_\mu \to A'_\mu = A_\mu + \partial_\mu\chi$, in order to cancel out the additional terms which arise.

From this point of view, the electromagnetic field appears as a consequence of the invariance of the Lagrangian under a local symmetry transformation. This idea will be generalised in later chapters.

7.4 Charge conjugation

Charge conjugation is the operation of replacing matter by antimatter so that, for example, an electron is interpreted as the antiparticle of the positron, which is then the particle. This would be the natural point of view if the Universe contained antimatter rather than matter. An interchange is achieved if we replace the Dirac field by its complex conjugate. Consider a positive energy solution of the field equation that has a phase factor e^{-iEt}. After complex conjugation it has a phase factor e^{iEt}, and with the standard phase convention is a negative energy solution. In the 'hole' interpretation, negative energy solutions are associated with antiparticles. However, the operation of complex conjugation does not leave $\mathcal{L}$ invariant: additional manipulations are needed to display the symmetry.

Taking the complex conjugate of the Dirac equation (7.6) gives

$$[(\gamma^\mu)^*(-\mathrm{i}\partial_\mu - qA_\mu) - m]\psi^* = 0.$$

Now in the chiral representation γ^0, γ^1 and γ^3 are real and $(\gamma^2)^* = -\gamma^2$. Multiplying the equation above by γ^2 and using the anticommuting properties of the γ matrices gives

$$[\gamma^\mu(\mathrm{i}\partial_\mu + qA_\mu) - m](\gamma^2\psi^*) = 0,$$

or

$$\left[\gamma^\mu\left(\mathrm{i}\partial_\mu - qA^c_\mu\right) - m\right](\gamma^2\psi^*) = 0.$$

Hence if ψ is a positive energy solution of the Dirac equation for a particle carrying charge q, $(\gamma^2\psi^*)$ is a negative energy solution in the charge conjugate field $A^c_\mu = -A_\mu$, which we introduced in Section 4.6.

There is some freedom of choice in the details of the transformation. We shall define the charge conjugate field ψ^c by

$$\psi^c = -\mathrm{i}\gamma^2\psi^* \tag{7.11a}$$

or, in terms of two-component spinors

$$\psi^c_{\mathrm{L}} = -\mathrm{i}\sigma^2\psi^*_{\mathrm{R}}, \qquad \psi^c_{\mathrm{R}} = \mathrm{i}\sigma^2\psi^*_{\mathrm{L}}. \tag{7.11b}$$

Using $\left(\gamma^2\right)^2 = -\mathbf{I}$, $\left(\gamma^2\right)^* = -\gamma^2$, we can invert the transformation (7.11a), obtaining

$$\psi = -\mathrm{i}\gamma^2\left(\psi^c\right)^* \tag{7.12a}$$

or

$$\psi_{\mathrm{L}} = -\mathrm{i}\sigma^2\left(\psi^c_{\mathrm{R}}\right)^*, \qquad \psi_{\mathrm{R}} = \mathrm{i}\sigma^2\left(\psi^c_{\mathrm{L}}\right)^*. \tag{7.12b}$$

Then (noting $(\gamma^2)^\dagger = -\gamma^2$) we have

$$\psi^\dagger = -\mathrm{i}(\psi^c)^{\mathrm{T}}\gamma^2 \tag{7.13a}$$

or

$$\psi_{\mathrm{L}}^\dagger = \mathrm{i}(\psi_{\mathrm{R}}^c)^{\mathrm{T}}\sigma^2, \qquad \psi_{\mathrm{R}}^\dagger = -\mathrm{i}(\psi_{\mathrm{L}}^c)^{\mathrm{T}}\sigma^2. \tag{7.13b}$$

Let us see how the various terms in the Lagrangian density (7.7) transform. Consider

$$\bar{\psi}\psi = \psi^\dagger\gamma^0\psi = -(\psi^c)^{\mathrm{T}}\gamma^2\gamma^0\gamma^2(\psi^c)^* = -(\psi^c)^{\mathrm{T}}\gamma^0(\psi^c)^*,$$

(using the properties of the γ-matrices).

To display the invariance of $\mathcal{L}$ we must anticipate Chapter 8. As operators, spinor fields anticommute: if a product of two fields is interchanged, a minus sign is introduced. For example, $\psi_a{}^*\psi_b = -\psi_b\psi_a{}^*$. Thus in transposing the last expression above we introduce a minus sign, and hence recover the form of the original term:

$$\bar{\psi}\psi = (\bar{\psi}^c)\psi^c$$

(since $(\gamma^0)^{\mathrm{T}} = \gamma^0$).

Other terms likewise acquire a minus sign:

$$\begin{aligned}\bar{\psi}\gamma^\mu\psi &= -(\psi^c)^{\mathrm{T}}\gamma^2\gamma^0\gamma^\mu\gamma^2(\psi^c)^* \\ &= (\psi^c)^\dagger(\gamma^2\gamma^0\gamma^\mu\gamma^2)^T(\psi^c).\end{aligned}$$

But, as the reader may verify,

$$(\gamma^2\gamma^0\gamma^\mu\gamma^2)^{\mathrm{T}} = -\gamma^0\gamma^\mu.$$

Hence

$$\bar{\psi}\gamma^\mu\psi = -(\bar{\psi}^c)\gamma^\mu(\psi^c).$$

Finally,

$$\begin{aligned}\bar{\psi}\gamma^\mu\mathrm{i}\partial_\mu\psi &= -(\psi^c)^{\mathrm{T}}\gamma^2\gamma^0\gamma^\mu\gamma^2\mathrm{i}\partial_\mu(\psi^c)^* \\ &= \mathrm{i}\partial_\mu(\psi^c)^\dagger(\gamma^2\gamma^0\gamma^\mu\gamma^2)^{\mathrm{T}}(\psi^c) \\ &= -\mathrm{i}\partial_\mu(\psi^c)^\dagger\gamma^0\gamma^\mu(\psi^c).\end{aligned}$$

Integration by parts in the action allows us to replace this last term by $(\bar{\psi}^c)\gamma^\mu\mathrm{i}\partial_\mu(\psi^c)$ in the Lagrangian density.

The Lagrangian can be seen to be of exactly the same form after charge conjugation, provided that the charge conjugate potentials A_μ^c are defined to be $A_\mu^c = -A_\mu$ (as in Section 4.6) and any external charge-current density J_μ also changes sign. In ordinary matter, where the Dirac particles are electrons, the external J_μ arise from the atomic nuclei, and these currents also change sign under charge conjugation.

7.5 The electrodynamics of a charged scalar field

In Section 3.5 we introduced the Klein–Gordon equation,

$$-\partial_\mu \partial^\mu \phi - m^2 \phi = 0,$$

which describes the motion of an uncharged scalar particle. The corresponding equation for a charged scalar particle is obtained from the Klein–Gordon equation by making the substitution (7.5), $\mathrm{i}\partial_\mu \to \mathrm{i}\partial_\mu - qA_\mu$, which gives

$$[(\mathrm{i}\partial_\mu - qA_\mu)(\mathrm{i}\partial^\mu - qA^\mu) - m^2]\Phi = 0. \tag{7.14}$$

A solution of (7.14) is necessarily complex. Thus a charged particle of zero spin in an electromagnetic field must be described by a complex, or two-component, wave function $\Phi = (\phi_1 + \mathrm{i}\phi_2)/\sqrt{2}$. We introduced complex scalar fields in Section 3.7. A real Lagrangian density that yields (7.14) and is Lorentz invariant is

$$\mathcal{L} = -\left[(\mathrm{i}\partial_\mu + qA_\mu)\Phi^*\right]\left[(\mathrm{i}\partial^\mu - qA^\mu)\,\Phi\right] - m^2\Phi^*\Phi. \tag{7.15}$$

$\mathcal{L}$ is invariant under a local gauge transformation, $\Phi \to \mathrm{e}^{-\mathrm{i}q\chi}\,\Phi$. Note that, since zero spin particles are bosons, the fields Φ and Φ^* commute.

Taking the complex conjugate of equation (7.14), we see that if $\Phi\,(x)$ is a solution for a particle carrying charge q in a given external field, then $\Phi^*(x)$ is a solution for a particle carrying a charge $-q$. We define the field $\Phi^c\,(x) = \Phi^*\,(x)$ to be the charge conjugate of Φ. The Lagrangian density (7.15) is invariant under charge conjugation, $\Phi \to \Phi^c$, if the charge conjugate potentials are again defined to be $A^c_\mu = -A_\mu$.

The charged π^+ and π^- mesons are composite, spin zero, particles whose overall motion is described by the generalised Klein–Gordon equation (7.14). We shall meet these particles and the fields Φ and Φ^* in the phenomenological discussions of Chapter 9.

7.6 Particles at low energies and the Dirac magnetic moment

In an electromagnetic field, the coupled Dirac equations (5.10) become

$$\begin{aligned}(\mathrm{i}\partial_0 - qA_0)\,\psi_\mathrm{L} - \sigma^i\,(\mathrm{i}\partial_i - qA_i)\,\psi_\mathrm{L} - m\psi_\mathrm{R} &= 0\\ (\mathrm{i}\partial_0 - qA_0)\,\psi_\mathrm{R} + \sigma^i\,(\mathrm{i}\partial_i - qA_i)\,\psi_\mathrm{R} - m\psi_\mathrm{L} &= 0\end{aligned} \tag{7.16}$$

where the σ^i are the Pauli spin matrices.

From Section 6.1, solutions of the Dirac equation that correspond to particles at low energies have $\psi_L \approx \psi_R$. We shall now show that at low energies the two-component wave function

$$\phi = e^{imt}(\psi_L + \psi_R) \tag{7.17a}$$

corresponds closely to the Schrödinger wave function for the particle. The factor e^{imt} has been inserted so that, as in the Schrödinger equation, the rest mass energy of the particle is omitted. If we define the orthogonal combination

$$\chi = e^{imt}(\psi_L - \psi_R), \tag{7.17b}$$

then by adding and subtracting the equations (7.16) we obtain an equivalent pair of equations:

$$\begin{aligned} (i\partial_0 - qA_0)\phi - \sigma^i(i\partial_i - qA_i)\chi &= 0, \\ (i\partial_0 - qA_0 + 2m)\chi - \sigma^i(i\partial_i - qA_i)\phi &= 0. \end{aligned} \tag{7.18}$$

The Schrödinger equation results if the term $(i\partial_0 - qA_0 + 2m)\chi$ is replaced by $2m\chi$. This approximation is reasonable if the Coulomb potential energy qA_0 and the kinetic energy are small compared with the rest mass of the particle. Then

$$\chi = (1/2m)\sigma^i(i\partial_i - qA_i)\phi,$$

and by substitution

$$i\frac{\partial\phi}{\partial t} = \left[\frac{1}{2m}\sigma^i(i\partial_i - qA_i)\sigma^j(i\partial_j - qA_j) + qA_0\right]\phi. \tag{7.19}$$

The Pauli spin matrices have the property

$$\sigma^i\sigma^j = i\varepsilon_{ijk}\sigma^k + \delta_{ij}\sigma^0,$$

and from the antisymmetry of ε_{ijk},

$$\varepsilon_{ijk}\partial_i\partial_j\phi = 0, \qquad \varepsilon_{ijk}A_iA_j = 0.$$

Also $\varepsilon_{ijk}[\partial_i(A_j\phi) + A_i\partial_j\phi] = \varepsilon_{ijk}[\partial_i(A_j\phi) - A_j\partial_i\phi] = \varepsilon_{ijk}(\partial_iA_j)\phi$, and recalling $A_\mu = (\phi, -\mathbf{A})$, $\varepsilon_{ijk}(\partial_iA_j) = B_k = -B^k$ gives the magnetic field $\mathbf{B}$. Using these results, we write (7.19) as

$$i\frac{\partial\phi}{\partial t} = \left[\frac{1}{2m}(-i\nabla - q\mathbf{A})^2 + qA_0 - \left(\frac{q\sigma}{2m}\right)\cdot\mathbf{B}\right]\phi. \tag{7.20}$$

Without the term $-(q\sigma/2m).\ \mathbf{B}$, this would be the Schrödinger equation for a charged particle in an electromagnetic field. The additional term we interpret as the energy in a magnetic field of an intrinsic magnetic moment associated with a Dirac particle. This is another remarkable consequence of the Dirac equation. For an electron, with $q = -e$, the magnetic moment is the Bohr magneton $\mu_B = e\hbar/2m$, anti-aligned with the electron spin. The observed magnetic moment agrees to better than 1% (cf. Section 8.5).

At the level of approximation of (7.20), the magnetic moment would play no role in a purely electrostatic field A_0. In better approximations, or indeed solving the Dirac equation directly, 'spin–orbit coupling' terms appear, which are of some importance in atomic physics and materials science.

Problems

7.1 Using the plane wave expansion (6.24), show that the conserved particle number can be written

$$\int P\left(x^0, \mathbf{x}\right) \mathrm{d}^3\mathbf{x} = \int \psi^\dagger \psi \mathrm{d}^3\mathbf{x} = \sum_{\mathbf{p},\varepsilon} (b^*_{\mathbf{p}\varepsilon} b_{\mathbf{p}\varepsilon} + d_{\mathbf{p}\varepsilon} d^*_{\mathbf{p}\varepsilon}).$$

7.2 Show that the charge conjugation operation acting on the positive energy solutions (7.12) and (7.13) yields the negative energy solutions (7.17).

7.3 Show that, taking the fields to be anticommuting and neglecting the neutrino mass, the neutrino Lagrangian density

$$\mathcal{L} = \mathrm{i}\psi^\dagger_{\mathrm{L}} \tilde{\sigma}^\mu \partial_\mu \psi_{\mathrm{L}}$$

is invariant under the combined operations of parity and charge conjugation. (Note equations (5.26) and (5.27).)

7.4 Show that $\mathrm{i}\,\sigma^2 \psi^*_{\mathrm{R}}$ transforms like a left-handed spinor under a Lorentz transformation.

7.5 Obtain the Klein–Gordon equation (7.14) from the Lagrangian density (7.15).

7.6 Using the method of Section 7.1, show that the global $U(1)$ symmetry $\Phi \to \mathrm{e}^{\mathrm{i}\alpha}\Phi$ of the Lagrangian density (7.15) leads to a conserved charge density current

$$qj^\mu = \mathrm{i}q[\Phi^*(\partial^\mu\Phi) - (\partial^\mu\Phi^*)\Phi] - 2q^2 A^\mu \Phi^*\Phi.$$

(Note that, in contrast to the result (7.9) for the Dirac Lagrangian, the current of a complex scalar field contains a term proportional to A^μ.)

7.7 Show that for the positive energy solutions (6.12) and (6.13) of the Dirac equation,

$$qj^\mu = -e\bar{\psi}\gamma^\mu\psi = -e\,(\cosh\theta, 0, 0, \sinh\theta) = -\,(eE/m)\,(1, 0, 0, v)$$

and also for the 'negative energy' solutions (6.17),

$$qj^\mu = -\,(eE/m)\,(1, 0, 0, v)\,.$$

With Dirac's interpretation, the hole that remains when this state is removed from the sea corresponds to a particle carrying charge e moving with velocity v along the z-axis.

7.8 Show that after the operation of charge conjugation a proton has negative charge and an electron has positive charge.

7.9 How do the electromagnetic potentials transform under the operation of time reversal, $t \to t' = -t$? Show that $\gamma^1\gamma^3\psi^*(t)$ is a solution of the time reversed Dirac equation, if $\psi(t)$ is a solution of the Dirac equation.

7.10 Show that, for a Dirac particle in a magnetic field **B** given by the vector potential **A**, both ψ_L and ψ_R satisfy the equation

$$\left[-\frac{\partial^2}{\partial t^2} - (-\mathrm{i}\nabla - q\mathbf{A})^2 - m^2 + q\sigma \cdot \mathbf{B}\right]\psi = 0.$$

Note that this differs from the Klein–Gordon equation for a charged scalar particle in a magnetic field, by the additional term $q\sigma \cdot \mathbf{B}$.

7.11 Using the parity transformations (4.18) and (5.27), show that the Lagrangian density (7.7) is invariant under space inversion.

8

Quantising fields: QED

We turn now to the quantisation of the electrodynamic fields introduced in Chapter 7. So far we have treated the electromagnetic field and the Dirac field as classical fields (though we were compelled in Chapter 7 to recognise that Dirac fields anticommute). On quantisation, these fields become operator fields, acting on the states of a system. The classical total field energy becomes the Hamiltonian operator, which determines the dynamics of the system. We shall use the formalism of annihilation and creation operators; this formalism is reviewed briefly in Appendix C for readers not already familiar with it.

Quantum electrodynamics, or QED, is an important component of the Standard Model. It is also the foundation of our understanding of the material world at the atomic level. However, we do not wish to enter into the technical complications of electrons in atoms or in material media. In this chapter we shall only consider more simple situations of a few interacting photons, electrons and positrons, at energies sufficiently high for bound systems of electrons and positrons to be ignored. In these situations, the free field approximation to QED provides a sound basis for understanding the interactions of particles as perturbations on their free behaviour.

This is not a text on quantum field theory, and our outline of perturbation theory in this chapter is necessarily sketchy. But our intention is to try to give some insight into how the results of calculations, presented in later chapters, are arrived at. We shall attempt to explain the necessity of *renormalisation*, which is an important concept in the formulation of the Standard Model.

8.1 Boson and fermion field quantisation

The simplest classical field we have introduced is that of a massive free scalar particle. It satisfies the Klein–Gordon equation (3.19). In the field expansion (3.21) we have so far regarded the classical wave amplitudes $a_{\mathbf{k}}$ and $a_{\mathbf{k}}^*$ as ordinary complex numbers. We now quantise the theory. We interpret $a_{\mathbf{k}}$ as an annihilation operator and $a_{\mathbf{k}}^*$ becomes the creation operator $a_{\mathbf{k}}^\dagger$, the Hermitian conjugate of $a_{\mathbf{k}}$. These operators are to obey the commutation relations

$$\left[a_{\mathbf{k}}, a_{\mathbf{k}'}^\dagger\right] = \delta_{\mathbf{k}\mathbf{k}'}, \qquad [a_{\mathbf{k}}, a_{\mathbf{k}'}] = 0, \qquad \left[a_{\mathbf{k}}^\dagger, a_{\mathbf{k}'}^\dagger\right] = 0. \tag{8.1}$$

The total field energy (3.30) becomes the Hamiltonian operator

$$H = \sum_{\mathbf{k}} a_{\mathbf{k}}^\dagger a_{\mathbf{k}} \omega_{\mathbf{k}} = \sum_{\mathbf{k}} N_{\mathbf{k}} \omega_{\mathbf{k}}, \tag{8.2}$$

where $\omega_{\mathbf{k}} = \sqrt{(\mathbf{k}^2 + m^2)}$ and it follows from the commutation relations that $N_{\mathbf{k}} = a_{\mathbf{k}}^\dagger a_{\mathbf{k}}$ is the number operator (Appendix C). As in Chapter 3, we shall in this chapter confine all particles to a cube of side l, volume $V = l^3$, and use periodic boundary conditions. By defining the Hamiltonian to be of the form (8.2), rather than the more symmetrical form

$$\frac{1}{2}\sum_{\mathbf{k}} \left(a_{\mathbf{k}}^\dagger a_{\mathbf{k}} + a_{\mathbf{k}} a_{\mathbf{k}}^\dagger\right) \omega_{\mathbf{k}} = \sum_{\mathbf{k}} \left(N_{\mathbf{k}} + \frac{1}{2}\right) \omega_{\mathbf{k}} \tag{8.3}$$

we discard 'zero-point energy' contributions and hence make the energy of the vacuum state $|0\rangle$ to be zero. The excited energy eigenstates of the Hamiltonian can then be interpreted as assemblies of particles (π^0 mesons, say, or Higgs particles) with an integer number $n_{\mathbf{k}}$ of particles in the state $\mathbf{k}$, where $n_{\mathbf{k}}$ is the eigenvalue of the number operator $N_{\mathbf{k}}$. The particles will obey Bose–Einstein statistics.

In the radiation gauge of Section 4.1, the electromagnetic field in free space is quantised in a very similar way to the Klein–Gordon field. The wave amplitudes $a_{\mathbf{k}\alpha}$ and $a_{\mathbf{k}\alpha}^*$ which appear in the expansion (4.15), become the annihilation and creation operators $a_{\mathbf{k}\alpha}$ and $a_{\mathbf{k}\alpha}^\dagger$, and the total field energy (4.25) becomes the Hamiltonian operator

$$H_{em} = \sum_{\mathbf{k},\alpha} a_{\mathbf{k}\alpha}^\dagger a_{\mathbf{k}\alpha} \omega_{\mathbf{k}} \tag{8.4}$$

where $\omega_{\mathbf{k}} = |\mathbf{k}|$. The operators $a_{\mathbf{k}\alpha}$ and $a_{\mathbf{k}\alpha}^\dagger$ annihilate and create *photons* of wave vector $\mathbf{k}$ and polarisation α, and satisfy commutation relations

$$\left[a_{\mathbf{k}\omega}, a_{\mathbf{k}'\alpha'}^\dagger\right] = \delta_{\mathbf{k}\mathbf{k}'} \delta_{\alpha\alpha'}, \qquad [a_{\mathbf{k}\alpha}, a_{\mathbf{k}'\alpha'}] = 0, \qquad \left[a_{\mathbf{k}\alpha}^\dagger, a_{\mathbf{k}'\alpha'}^\dagger\right] = 0. \tag{8.5}$$

$N(\mathbf{k}, \alpha) = a_{\mathbf{k}\alpha}^\dagger a_{\mathbf{k}\alpha}$ is the number operator. The energy eigenstates of the radiation field correspond to assemblies of photons. Photons, like scalar particles, obey Bose–Einstein statistics. (See Problem 8.1.)

On quantising the Dirac field of a free electron, the wave amplitudes appearing in the expansion (6.24), and their complex conjugates likewise become operators: $b_{\mathbf{p}\varepsilon}$ and $b_{\mathbf{p}\varepsilon}{}^\dagger$ annihilate and create electrons of momentum $\mathbf{p}$, helicity ε; $d_{\mathbf{p}\varepsilon}$ and $d_{\mathbf{p}\varepsilon}{}^\dagger$

annihilate and create positrons of momentum $\mathbf{p}$, helicity ε. Electrons and positrons are fermions, and these operators obey anticommutation relations, for example

$$b_{\mathbf{p}\varepsilon}b^{\dagger}_{\mathbf{p}'\varepsilon'} + b^{\dagger}_{\mathbf{p}'\varepsilon'}b_{\mathbf{p}\varepsilon} = \left\{b_{\mathbf{p}\varepsilon}, b^{\dagger}_{\mathbf{p}'\varepsilon'}\right\} = \delta_{\mathbf{pp}'}\delta_{\varepsilon\varepsilon'}, \quad \{b_{\mathbf{p}\varepsilon}, b_{\mathbf{p}'\varepsilon'}\} = 0, \quad \left\{b^{\dagger}_{\mathbf{p}\varepsilon}, b^{\dagger}_{\mathbf{p}'\varepsilon'}\right\} = 0 \tag{8.6}$$

$d_{\mathbf{p}\varepsilon}$ and $d^{\dagger}_{\mathbf{p}'\varepsilon'}$ obey similar rules. Also all electron operators anticommute with all positron operators. The electron number operator $N_{\mathbf{e}}(\mathbf{p}, \varepsilon) = b^{\dagger}_{\mathbf{p}\varepsilon}b_{\mathbf{p}\varepsilon}$ and the positron number operator $N_{\mathbf{p}}(\mathbf{p}, \varepsilon) = d^{\dagger}_{\mathbf{p}\varepsilon}d_{\mathbf{p}\varepsilon}$ have possible eigenvalues restricted to 0 and 1, in accord with the Pauli exclusion principle (Appendix C). Electrons and positrons obey Fermi–Dirac statistics. (See Problem 8.2.)

After second quantisation, the difficulties that were associated with the interpretation of the Dirac equation as a single particle wave equation disappear. Electrons and positrons are now on a similar footing and the 'sea' of filled negative energy states is no longer needed. The total field energy (6.25) becomes the Hamiltonian

$$H = \sum_{\mathbf{p},\varepsilon} \left(b^{\dagger}_{\mathbf{p}\varepsilon}b_{\mathbf{p}\varepsilon} - d_{\mathbf{p}\varepsilon}d^{\dagger}_{\mathbf{p}\varepsilon}\right)E_{\mathbf{p}}.$$

Using an anticommutation relation, we can replace this by

$$H = \sum_{\mathbf{p},\varepsilon} \left(b^{\dagger}_{\mathbf{p}\varepsilon}b_{\mathbf{p}\varepsilon} + d^{\dagger}_{\mathbf{p}\varepsilon}d_{\mathbf{p}\varepsilon} - 1\right)E_{\mathbf{p}}.$$

We shall discard the constant zero-point energy term (which we note is negative for fermions) and take

$$H = \sum_{\mathbf{p},\varepsilon} \left(b^{\dagger}_{\mathbf{p}\varepsilon}b_{\mathbf{p}\varepsilon} + d^{\dagger}_{\mathbf{p}\varepsilon}d_{\mathbf{p}\varepsilon}\right)E_{\mathbf{p}}. \tag{8.7}$$

The energy of the vacuum state is then zero, and the excited states correspond to assemblies of electrons and positrons.

Similarly, the field momentum (6.26) becomes the momentum operator

$$\mathbf{P} = \sum_{\mathbf{p},\varepsilon} \left(b^{\dagger}_{\mathbf{p}\varepsilon}b_{\mathbf{p}\varepsilon} + d^{\dagger}_{\mathbf{p}\varepsilon}d_{\mathbf{p}\varepsilon}\right)\mathbf{p}. \tag{8.8}$$

The conserved particle number (Problem 7.1) becomes the time independent operator

$$\int \mathrm{P}\left(x^0, \mathbf{x}\right)\mathrm{d}^3\mathbf{x} = \sum_{\mathbf{p},\varepsilon} \left(b^{\dagger}_{\mathbf{p}\varepsilon}b_{\mathbf{p}\varepsilon} + d_{\mathbf{p}\varepsilon}d^{\dagger}_{\mathbf{p}\varepsilon}\right). \tag{8.8}$$

which we replace by:

$$\text{conserved number operator} = \sum_{\mathbf{p},\varepsilon} \left(b^{\dagger}_{\mathbf{p}\varepsilon}b_{\mathbf{p}\varepsilon} + d^{\dagger}_{\mathbf{p}\varepsilon}d_{\mathbf{p}\varepsilon}\right). \tag{8.9}$$

This operator counts the number of electrons minus the number of positrons, a number which is therefore constant in quantum electrodynamics.

8.2 Time dependence

In the Schrödinger picture, a system described by a Hamiltonian H evolves in time from a state $|t_0\rangle$ at time t_0 to a state $|t\rangle$ at time t, where

$$|t\rangle = \mathrm{e}^{-\mathrm{i}H(t-t_0)}|t_0\rangle.$$

Thus time displacements are generated by the unitary operator $\mathrm{e}^{-\mathrm{i}Ht}$.

The expectation value of a time independent operator $\hat{O}$ at time t is

$$\begin{aligned}\langle t|\hat{O}|t\rangle &= \langle t_0|\mathrm{e}^{\mathrm{i}H(t-t_0)}\hat{O}\mathrm{e}^{-\mathrm{i}H(t-t_0)}|t_0\rangle \\ &= \langle t_0|\hat{O}_H(t-t_0)|t_0\rangle\end{aligned}$$

where

$$\hat{O}_H(t) = \mathrm{e}^{\mathrm{i}Ht}\hat{O}\mathrm{e}^{-\mathrm{i}Ht} \tag{8.10}$$

depends on t.

These last equations give the so-called Heisenberg picture, in which the states of a system remain fixed and the operators become time dependent. In the case of free fields, the time dependence of the annihilation and creation operators is very simple. For example, in the case of a scalar field (see (3.21)),

$$a_{\mathbf{k}}(t) = e^{-i\omega_{\mathbf{k}}t}a_{\mathbf{k}}, \quad a^{\dagger}_{\mathbf{k}}(t) = e^{\mathrm{i}\omega_{\mathbf{k}}t}a^{\dagger}_{\mathbf{k}}, \tag{8.11}$$

as may be seen by considering the effect of the operators on a state $|n_{\mathrm{k}}\rangle$ (Appendix C). It is usual in quantum field theory to work in the Heisenberg picture.

In the case of interacting fields, the basic free field states we have defined are no longer eigenstates of the total Hamiltonian. In QED we may write

$$H = H_0 + V, \tag{8.12}$$

where

$$H_0 = H\,(\text{photons}) + H\,(\text{electrons}) + H\,(\text{positrons})$$

is given by (8.4) and (8.7). The eigenstates of H_{o} are just collections of freely moving photons, electrons, and positrons.

V comes from the term $-q\left(\bar{\psi}\gamma^{\mu}\psi\right)A_{\mu}$ in the Lagrangian density, (7.7), which we constructed in Chapter 7. We are here excluding external fields. Since V does not depend on derivatives of the fields, its contribution to the energy density T^0_0 is just $q(\bar{\psi}\gamma^{\mu}\psi)A_{\mu}$, and setting $q = -e$ for electrons we obtain at $t = t_0$

$$V(t_0) = -e \int \bar{\psi}(\mathbf{r}, t_0)\gamma^{\mu}\psi(\mathbf{r}, t_0)A_{\mu}(\mathbf{r}, t_0)\mathrm{d}^3\mathbf{r}. \tag{8.13}$$

Note that the subsequent time development of the fields is not that of the free fields, since it is determined by the full Hamiltonian $H = H_0 + V$.

We can expand the fields A_μ and ψ at the initial time t_o using (4.15) and (6.24), replacing the wave amplitudes by appropriate operators. On expanding out V there will be several types of term. For example, setting $t_o = 0$ one can easily pick out a term

$$-\frac{em}{\sqrt{(2V\omega_{\mathbf{k}}E_{\mathbf{p}'}E_{\mathbf{p}''})}}[\bar{u}_{\varepsilon'}(\mathbf{p}')\gamma^{\mu}v_{\varepsilon''}(\mathbf{p}'')\varepsilon_{\mu}]d^{\dagger}_{\mathbf{p}'\varepsilon}d^{\dagger}_{\mathbf{p}''\varepsilon''}a_{\mathbf{k}\alpha}\delta_{(\mathbf{k}-\mathbf{p}'-\mathbf{p}''),0}. \tag{8.14}$$

This term annihilates a photon and creates an electron–positron pair. The condition $\mathbf{k} - \mathbf{p}' - \mathbf{p}'' = 0$ comes from the integration over space of the exponential factors, and explicitly conserves momentum.

Dynamical calculations in a quantum field theory can be viewed as the calculation of the unitary operator $\mathrm{e}^{-\mathrm{i}Ht}$ acting on some initial specified state. In QED, the coupling (8.13) between the radiation field and the Dirac field is determined by the charge on the electron e. It is natural to introduce the dimensionless parameter α, the *fine structure constant*:

$$\alpha = \frac{e^2}{4\pi\hbar c} \approx \frac{1}{137}.$$

α characterises the strength of the coupling, and is small. Much progress has been made in QED by the construction of the operator $\mathrm{e}^{-\mathrm{i}Ht}$ as an expansion of the form

$$\mathrm{e}^{-\mathrm{i}Ht} = \mathrm{e}^{-\mathrm{i}H_0t}[1 + e\hat{O}_1(t) + e^2\hat{O}_2(t) + \ldots] \tag{8.15}$$

where the $\hat{O}_n(t)$ are time-dependent operators.

8.3 Perturbation theory

To construct the perturbation expansion (8.15), one can start by considering

$$\mathrm{e}^{-\mathrm{i}Ht} = [\mathrm{e}^{-\mathrm{i}H\delta t}]^n \text{with } \delta t = t/n.$$

For large enough n (small enough δt), one can take

$$\mathrm{e}^{-\mathrm{i}H\delta t} = 1 - \mathrm{i}H\delta t$$

and discard higher order terms in the Taylor expansion. Then

$$\mathrm{e}^{-\mathrm{i}Ht} = [1 - \mathrm{i}(H_0 + V)\delta t]^n .$$

In the lowest order of perturbation theory only the terms linear in V are kept, so that

$$\begin{aligned} e^{-iH_0t} e\hat{O}_1(t) &= -i \sum_{r=0}^{n-1} [1 - iH_0\delta t]^{n-1-r} \, V\delta t \, [1 - iH_0\delta t]^r \\ &= -i \sum_{r=0}^{n-1} e^{-iH_0(t-t')} V\delta t e^{-iH_0t'} \end{aligned}$$

with $t' = r\delta t$ and n large.

In the limit of $\delta t \to 0$, we can replace the sum by an integral, so that

$$e\hat{O}_1(t) = -i \int_0^t dt' e^{iH^0t'} V e^{-iH_0t'}. \tag{8.16}$$

The operator $e^{-iH_0t'}$ is the simple free field time evolution operator. If we take V to be given at $t = 0$ by (8.13), we can write

$$\hat{O}_1(t) = i \int_0^t \bar{\psi}(\mathbf{r}', t')\gamma^\mu \psi(\mathbf{r}', t') A_\mu(\mathbf{r}', t') \, dt' d^3\mathbf{r}' \tag{8.17}$$

where the fields have the time dependence of free unperturbed fields. A term like (8.14), for example, will have time dependence (see equation (8.11)).

$$e^{-i(\omega_\mathbf{k} - E_{\mathbf{p}'} - E_{\mathbf{p}''})t'} \tag{8.18}$$

The evolution of a state from time $-t/2$ in the past to time $t/2$ in the future corresponds to taking the integral in (8.17) from $-t/2$ to $t/2$. This more symmetrical form is appropriate to the description of particle scattering processes. For example, if the initial state at time $-t/2$ consists of a photon in the state $(\mathbf{k}, \alpha)$, the operators in (8.14) annihilate this photon and create an electron in a state $(\mathbf{p}', \varepsilon')$ and a positron in the state $(\mathbf{p}'', \varepsilon'')$. Taking the limit $t \to \infty$ in the time factor (8.18) gives

$$\int_{-\infty}^{\infty} e^{-i(\omega_\mathbf{k} - E_{p'} - E_{p''})t'} dt' = 2\pi\,\delta(\omega_\mathbf{k} - E_{\mathbf{p}'} - E_{\mathbf{p}''}).$$

Thus energy conservation, as well as momentum conservation, is explicit. In free space it is impossible to satisfy both these conservation laws in the case of pair production from a photon (Problem 8.3), so that first-order perturbation theory contributes nothing. (In the presence of an external electromagnetic field, for example the Coulomb field of a nucleus, momentum conservation between electrons and photons is lost, and pair production is possible if $\omega_\mathbf{k} > 2m$.)

When the first-order transition amplitude at time t does not vanish, we have, using (8.16),

$$\langle\text{final state}|e\hat{O}_1(t)|\text{initial state}\rangle = \langle \text{f}|V(0)|i\rangle \int_{-t/2}^{t/2} \text{e}^{-\text{i}\Delta E t'} \text{d}t',$$

where $\Delta E = E_\text{i} - E_\text{f}$ and E_i and E_f are the energies of the initial state $|i\rangle$ and final state $|f\rangle$. It is shown in textbooks on quantum mechanics that the time dependence can be interpreted as a transition probability per unit time, from the initial state i to the final state f, given by

transition probability $= 2\pi\,|\langle \text{f}|V(0)|i\rangle|^2 \rho(E_\text{f})$,
where$\rho(E_\text{f})$is the density of final energy states at $E_\text{f} = E_\text{i}$.

It is straightforward to extract higher order terms of the perturbation expansion. For example

$$\hat{O}_2(t) = \int_{-t/2}^{t/2} \text{d}^4x_2\, \bar{\psi}(x_2)\gamma^\mu\psi(x_2)\,A_\mu(x_2) \int_{-t/2}^{t_2} \text{d}^4x_1\, \bar{\psi}(x_1)\,\gamma^\mu\psi(x_1)\,A_\mu(x_1) \tag{8.19}$$

where $x_1 = (t_1, \mathbf{r}_1)$, $x_2 = (t_2, \mathbf{r}_2)$ and $-t/2 < t_2 < t/2$.

8.4 Renormalisation and renormalisable field theories

In second-order perturbation theory, we can pick out terms corresponding to the creation of an electron–positron pair at a point x_1 in space-time and its destruction at a point x_2. They may be characterised by the diagrams of Fig. 8.1. In these diagrams time runs from left to right. Momentum is conserved at x_1 and x_2. Overall there is also conservation of energy and angular momentum, so that the 'unperturbed' photon that emerges at time t_2 is in the same state as the initial unperturbed photon.

We pointed out that in free space it is not possible to create a real e^-e^+ pair from a photon. The e^-e^+ pair of the diagram is a virtual pair, corresponding to a term in a mathematical expansion. The transition amplitude

$$\langle\mathbf{k}|\text{e}^{-\text{i}H_0 t}\,\hat{O}_2(t)\,|\mathbf{k}\rangle = \text{e}^{-i\omega_\mathbf{k} t}\langle\mathbf{k}|\hat{O}_2(t)\,|\mathbf{k}\rangle$$

is non-vanishing. The 'real' photon is evidently a complex object. Calculations show that the effect of virtual e^-e^+ pairs is to make the vacuum behave like an electrically polarisable medium. In particular, the Coulomb interaction between two 'bare' electrons is screened. We can envisage this effect as resulting from a screening cloud of virtual positrons around each bare electron, the corresponding

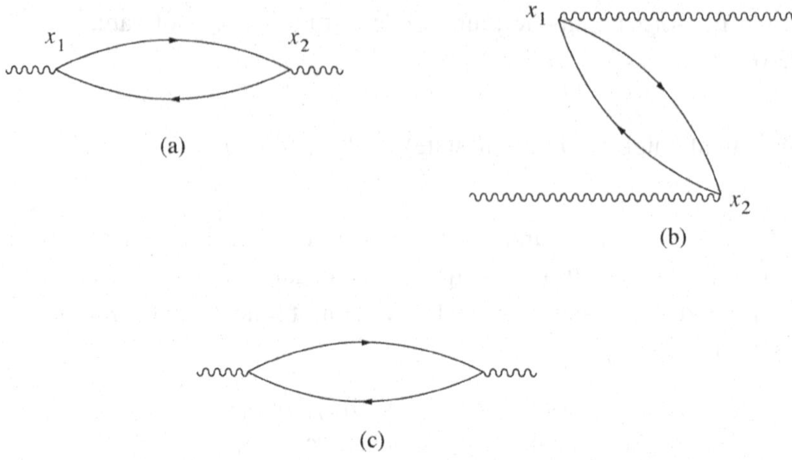

Figure 8.1 In these diagrams an unperturbed electron–positron pair is created at a point x_1 in space-time and destroyed at a point x_2. In (a) the initial unperturbed photon is destroyed at x_1 and recreated at x_2; vice versa in (b). In (a) and (b) time runs from left to right. As shown by Feynman it is convenient to characterise both processes by the single Feynman diagram (c). In all of these diagrams the arrows on the fermion lines follow the direction of electron number. (The arrows on positrons then run backwards in time.)

negative charge of the virtual e^-e^+ pairs appearing as charge at the surface of the confining volume.

What is measured experimentally as the charge $-e$ on an electron is the screened charge. To compensate for this screening effect, the parameter e that appears in the Lagrangian must be replaced by a 'bare' charge $e_o = e + \Delta e$. This gives 'counter terms' in the Lagrangian. Δe is chosen to cancel the screening effect. To second order the calculation gives $\Delta e = \alpha A_1 e$ where A_1 is a dimensionless quantity. With this adjustment and to this order, the screened charge on the electron becomes $-e$. In higher orders of perturbation theory one obtains

$$\Delta e = e[\alpha A_1 + \alpha^2 A_2 + \cdots].$$

To any order of perturbation theory an account must be kept of the readjustment of e, in order to extract from a calculation the significant physical effects which are also determined by terms in the perturbation expansion. The charge $-e$ on the electron is said to be *renormalised.* Δe itself can never be measured. Physical effects in atomic physics arising in part from vacuum polarisation terms have been calculated and measured with high precision. (See also Section 16.3.)

The other parameter appearing in electrodynamics is the mass of the electron. The bare mass of the electron is modified in second-order perturbation

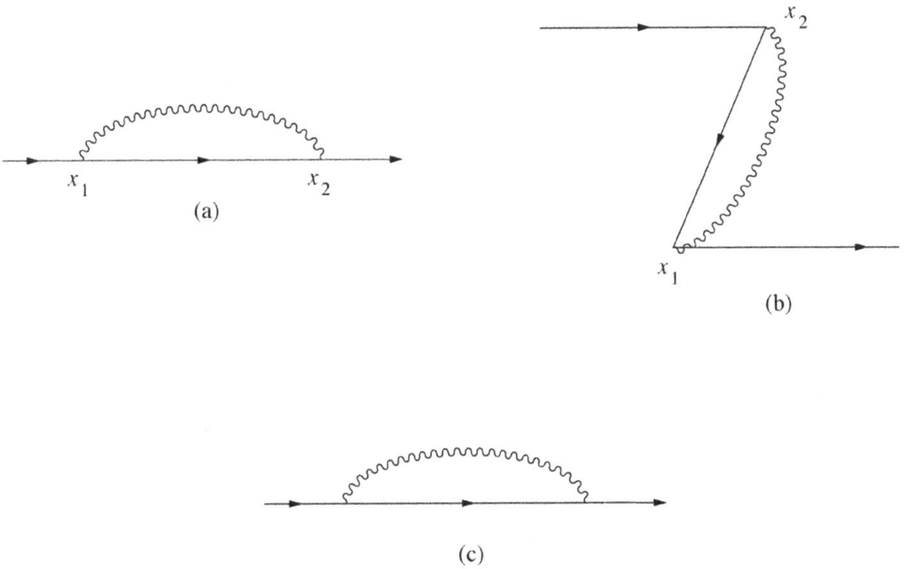

Figure 8.2 In these diagrams an unperturbed photon is created at a point x_1 in space-time and destroyed at a point x_2. In (a) the initial unperturbed electron is destroyed at x_1 and recreated at x_2; vice versa in (b). In (a) and (b) time runs from left to right. It is convenient to characterise both processes by the single Feynman diagram (c). In all of these diagrams the arrows on the fermion lines follow the direction of the electron number. (The arrows on positrons then run backwards in time.)

theory by the processes shown in Fig. 8.2. To compensate for these processes we must take $m_o = m - \Delta m$ in the Lagrangian where Δm is chosen to compensate for the shift in mass produced by the electron–photon interactions. We can think of the bare electron as 'dressed' by virtual photons. It is found that to second order $\Delta m = \alpha m B_1$, where B_1 is another dimensionless quantity, and more generally

$$\Delta m = m[\alpha B_1 + \alpha^2 B_2 + \cdots].$$

As with Δe, Δm has to be adjusted at each higher order of perturbation theory, and there is a systematic way of extracting physical answers from perturbation calculations. The physical mass m is the renormalised mass.

Diagrams like those of Fig. 8.3, in which virtual e^-e^+ pairs and virtual photons are created and annihilated together, give terms that modify the vacuum energy. Energy shifts in perturbation theory are to be expected, but since we have no unperturbed vacuum with which to compare, such shifts are not measurable. The cosmological constant of general relativity gives a measure of the vacuum energy

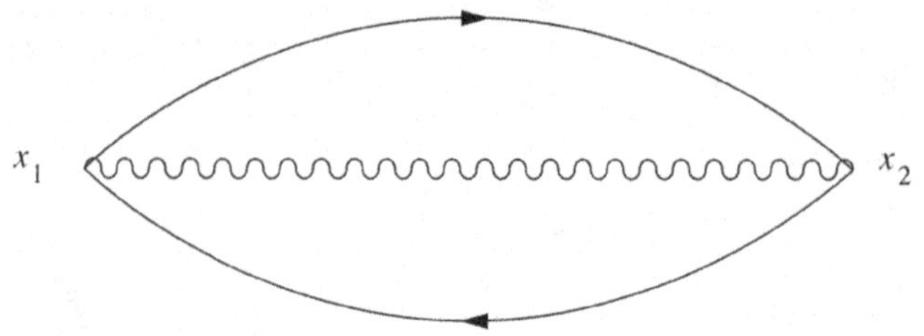

Figure 8.3 The vacuum state of quantum electrodynamics differs from the unperturbed vacuum by processes, one of which is illustrated in this figure.

density that is certainly very small, and is consistent with its being zero. We shall take the vacuum energy density, whatever its origin, to be zero.

It could have been anticipated without calculation that there would be perturbing effects of charge renormalisation and mass renormalisation. The unpalatable feature of quantum electrodynamics is that when the constants A_i, and B_i are calculated they all turn out to be infinite, as does the correction to the vacuum state energy. It is just as well that Δe and Δm have no physical significance. However, it is the case that an expansion in the small parameter α gives seemingly infinite corrections to quantities one cannot measure. An important feature of QED is that, leaving aside a scaling of the fields that is also part of the renormalisation scheme, infinities only appear in the renormalisation of the parameters of the theory, e, m and the vacuum energy. The only infinite counter terms that have to be added to the Lagrangian are contained in these parameters. Having made these adjustments, the remaining physical effects are calculable and finite.

QED is a local field theory, i.e. a theory in which the interaction terms involve a product of fields at the same point in space time. Infinities such as occur in QED are endemic in all local field theories. Field theories in which the infinities only appear in a finite number of parameters of the theory are said to be *renormalisable.*

The divergences in the coefficients A_i of Δe and B_i of Δm arise, for example, in the contribution from O_2 (see (8.19)), from the integration region where $x_2 \approx x_1$ and in particular where $\mathbf{r_2} \approx \mathbf{r_1}$. An important feature of QED is that the expansion parameter α and hence the coefficients, are dimensionless numbers. In Chapters 9 and 21 we will encounter theories in which the coupling constants and therefore the expansion parameters have the dimensions of inverse powers of mass. All the terms in perturbation expansions must have the same dimension, therefore the coefficients have a dimension to compensate those of the coupling constant. In the integration regions the integrands diverge with large inverse powers of $|\mathbf{r}_2 - \mathbf{r}_1|$ as $\mathbf{r}_2 \to \mathbf{r}_1$ to achieve the compensation, but they render the integrals infinite. Infinities occur for all multiparticle interactions, they can not be removed just by mass and coupling constant renormalisation. Such theories are unrenormalisable, they can not be taken seriously as quantum field theories.

8.5 The magnetic moment of the electron

We shall now illustrate the remarkable success of QED in calculating quantities of physical significance by giving an account of the calculation of the electron's magnetic moment. In Chapter 7 we showed that the Dirac equation before second quantisation implies that the electron carries a magnetic moment of magnitude $\mu_B = e\hbar/2m$ anti-aligned with its spin. The electron's magnetic moment has been measured with high precision: the experimental value μ_e is

$$\mu_e = \mu_B\,(1 + a)$$

where the 'anomaly' $a = 0.001\,159\,652\,188\,4(43)$ (Van Dyck *et al.*, 1987).

After second quantisation, the perturbative corrections to the Dirac value can be calculated. The Dirac value is contained in the operator $\hat{O}_1$ of equation (8.16), and is associated with diagram (a) of Fig. 8.4. This lowest order calculation reproduces the Dirac result $\mu_e = \mu_B$.

Since μ_B is the only combination of the parameters e, m_e and $\hbar$ which has the dimensions of magnetic moment, higher orders of perturbation theory will give terms of the form

$$\mu_e = \mu_B(1 + \alpha C_1 + \alpha^2 C_2 + \alpha^3 C_3 + \alpha^4 C_4 + \cdots),$$

where the C_i are dimensionless constants. To compare the theory with experiment we use the 1986 adjusted value of the fine structure constant,

$$\alpha^{-1} = 137.035\,9979\,(32)\,.$$

C_1 is associated with diagram (b) of Fig. 8.4; the calculation gives $C_1 = 1/(2\pi)$. Hence to this order

$$a = C_1\alpha = 0.001\,161\,409\,74,$$

which agrees with experiment to within five significant figures.

The next order correction, associated with diagrams (c) of Fig. 8.4, is

$$C_2 = \frac{1}{\pi^2}\left(\frac{197}{144} + \frac{3}{4}\zeta\,(3)\right) - \frac{1}{2}\ln 2 + \frac{1}{12}$$

where $\zeta(z)$ is the Riemann zeta function. To this order,

$$a = 0.001\,159\,637\,44,$$

in agreement to seven significant figures.

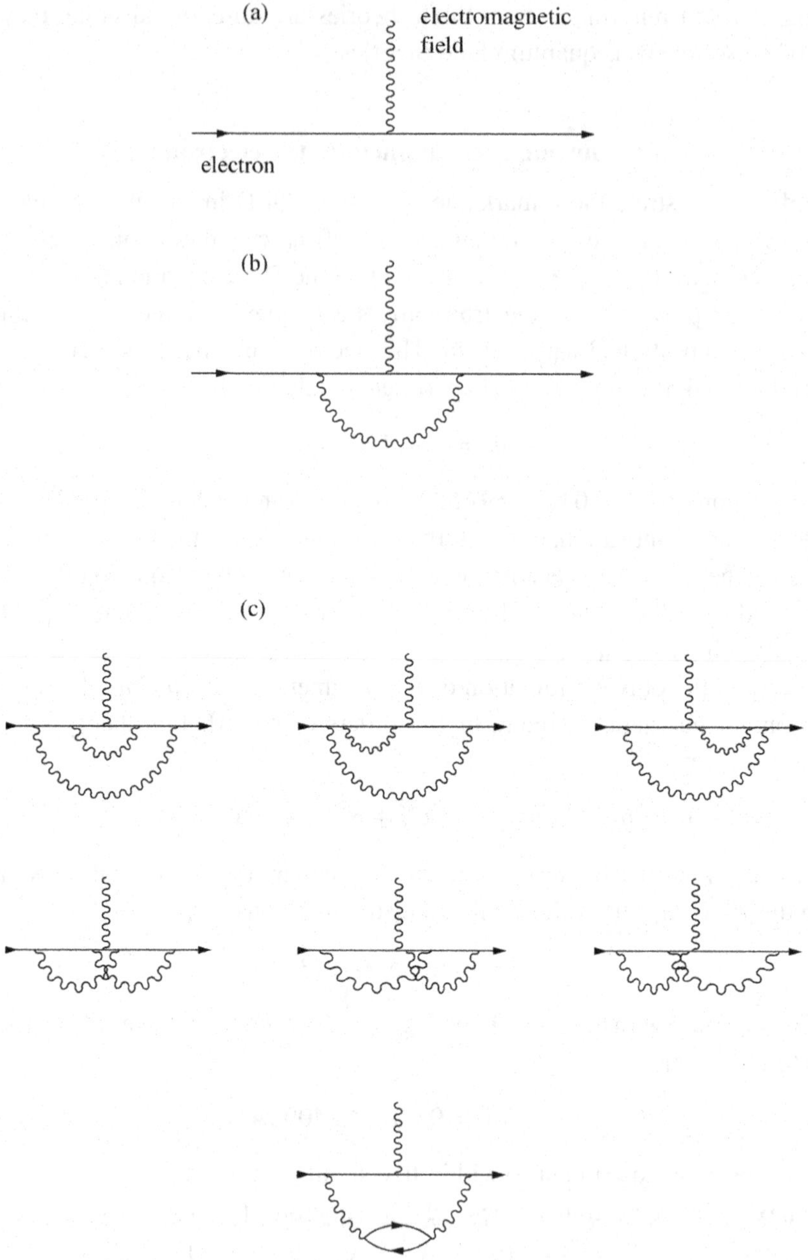

Figure 8.4 Perturbation theory Feynman diagrams that represent contnbutions to the electron magnetic moment. The anomalous moment, to order α^2, comes from calculations associated with diagrams (b) and (c).

Calculations of higher orders of perturbation theory become rapidly more intractable. Numerical estimates give $C_3 \approx 0.03792$, $C_4 \approx -0.014$. At this level of accuracy, corrections have to be made for processes that come from other parts of the Standard Model, in particular from the muon. The most recent comprehensive calculations (Kinoshita and Lindquist, 1990) give

$$a = 0.001\,159\,652\,140\,0\,(41 + 53 + 271),$$

in agreement with experiment to ten significant figures. The largest error in the theory is from the uncertainty in α^{-1}.

Within its range of applicability, quantum electrodynamics provides an astonishingly exact model of Nature. One may have some confidence that the techniques of renormalisation in perturbation theory are valid.

8.6 Quantisation in the Standard Model

In this chapter we have outlined the 'canonical quantisation' techniques that have been particularly successful in quantum electrodynamics. Many books have been written on this subject, for example Itzykson and Zuber (1980); some will have to be consulted if one is to be competent and confident in making detailed calculations. However, many of the decay rates and cross-sections given in the following chapters, which are needed to compare the predictions of the Standard Model with experiment, are quite well approximated by the so-called 'tree level' of perturbation theory. The tree-level diagrams have no closed loops (see Fig. 8.4(a)) and require no renormalisation. It is a fortunate circumstance that in low orders of perturbation theory these can be calculated quite easily.

The particles and forces of the weak and the strong interactions are also described by local gauge field theories, which will be exhibited at the classical level in the chapters that follow. The quantisation procedures used in these extensions of QED have been most successfully pursued by the path integral method of quantisation (see, for example, Cheng and Li (1984)). Both the theory of the weak interaction and the theory of the strong interaction pose their own special problems, but the principles of gauge symmetry and renormalisability have been essential in the construction of the Standard Model as it is today.

Problems

8.1 A general two-particle state of scalar bosons (Section 8.1) can be written

$$|\text{state}\rangle = \sum_{\mathbf{k}_1,\mathbf{k}_2} f(\mathbf{k}_1, \mathbf{k}_2)\, a^\dagger_{\mathbf{k}1}\, a^\dagger_{\mathbf{k}2}|0\rangle,$$

where, apart from normalisation, $f(\mathbf{k}_1, \mathbf{k}_2)$ is any function of $\mathbf{k}_1$ and $\mathbf{k}_2$. (f can be called the wave function of the state.)
Show that this state may be written

$$|\text{state}\rangle = \sum_{\mathbf{k}_1,\mathbf{k}_2} g\,(\mathbf{k}_1, \mathbf{k}_2)\, a^{\dagger}_{\mathbf{k}1}\, a^{\dagger}_{\mathbf{k}2}|0\rangle$$

with $g(\mathbf{k}_1, \mathbf{k}_2) = \{f(\mathbf{k}_1, \mathbf{k}_2) + f(\mathbf{k}_2, \mathbf{k}_1)\}/2$, symmetric under the interchange of labelling.

8.2 A general two-particle state of fermions can be written

$$|\text{state}\rangle = \sum_{\mathbf{p}_1,\varepsilon_1,\mathbf{p}_2,\varepsilon_2} f\,(\mathbf{p}_1, \varepsilon_1, \mathbf{p}_2, \varepsilon_2)\, b^{\dagger}_{\mathbf{p}_1\varepsilon_1}\, b^{\dagger}_{\mathbf{p}_2\varepsilon_2}|0\rangle$$

where apart from normalisation f is any function of $\mathbf{p}_1$, ε_1 and $\mathbf{p}_2$, ε_2.
Show that this state can also be written

$$|\text{state}\rangle = \sum_{\mathbf{p}_1,\varepsilon_1,\mathbf{p}_2,\varepsilon_2} g\,(\mathbf{p}_1, \varepsilon_1, \mathbf{p}_2, \varepsilon_2)\, b^{\dagger}_{\mathbf{p}_1\varepsilon_1}\, b^{\dagger}_{\mathbf{p}_2\varepsilon_2}|0\rangle$$

with $g(\mathbf{p}_1, \varepsilon_1; \mathbf{p}_2, \varepsilon_2) = \{f(\mathbf{p}_1, \varepsilon_1; \mathbf{p}_2, \varepsilon_2) - f(\mathbf{p}_2, \varepsilon_2; \mathbf{p}_1, \varepsilon_1\}/2$, antisymmetric under the interchange of labelling.

8.2 Use energy and momentum conservation to show that pair creation by a single photon, $\gamma \to e^+ + e^-$, is impossible in free space.

8.3 The energy density of an electromagnetic field is given by equation (4.24). Show that the total electric field energy of a point charge q outside a sphere of radius R centred on the particle is

$$\text{energy} = q^2/(8\pi R).$$

Note that this classical contribution to the particle rest energy is infinite in the limit $R \to 0$.

9

The weak interaction: low energy phenomenology

In this chapter we review some of the early phenomenology of the weak interaction that played an important guiding role in the construction of the Standard Model. The phenomenology discussed is insensitive to the very small effects of neutrino mass. These effects will be ignored.

9.1 Nuclear beta decay

In early investigations of nuclear physics, the existence of a 'weak interaction' responsible for nuclear β decay was discerned. It was regarded as weak since the mean lives of decays such as

$$
\begin{aligned}
{}^{17}_{9}\mathrm{F} &\to {}^{17}_{8}\mathrm{O} + \mathrm{e}^{+} + \nu_{\mathrm{e}}, \\
\mathrm{n} &\to \mathrm{p} + \mathrm{e}^{-} + \bar{\nu}_{\mathrm{e}},
\end{aligned}
$$

are very long, minutes in these examples, compared with typical nuclear electromagnetic decays, which have a mean life of $\sim 10^{-15}$s.

Nuclear physicists have by careful and ingenious experimentation established the principal features of the weak interaction and the properties of the electron neutrino ν_{e}. To conserve electric charge the neutrino must be electrically neutral, and angular momentum is conserved if it is a Dirac spin $\frac{1}{2}$ fermion. If the electron neutrino has a mass, it is certainly very small.

The surprising feature of the weak interaction, which was established experimentally in 1957 by Wu following a suggestion by Lee and Yang, is that it *does not conserve parity.* Nature is not ambidextrous. Indeed, parity is maximally violated, in that only the left-handed components of both the electron and neutrino fields participate in the interaction.

This phenomenon is clearly illustrated if one examines the longitudinal electron polarisation of electrons produced in 'allowed' β decays. An electron of negative helicity $-\frac{1}{2}$ and velocity v is in a left-handed state with probability

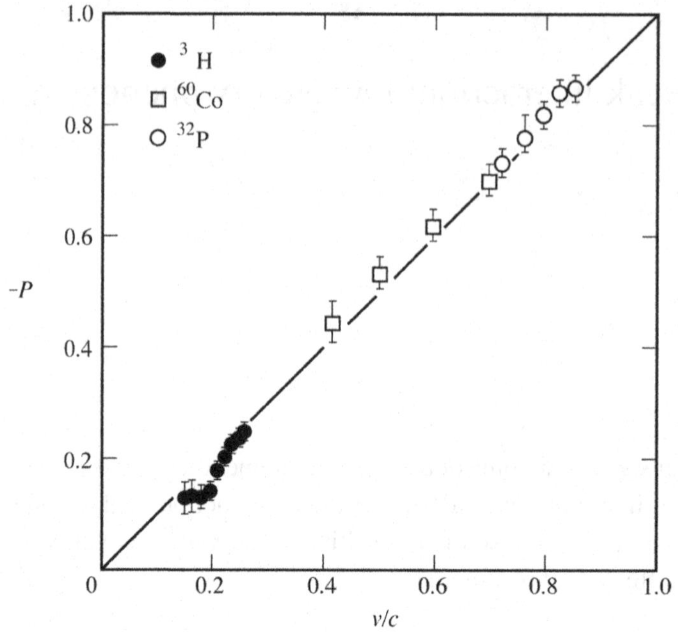

Figure 9.1 Measured degree of longitudinal polarisation P for allowed e^- decays. (Data from Koks and Van Klinken (1976).)

$\frac{1}{2}[1+(v/c)]$; an electron of positive helicity $+\frac{1}{2}$ is in a left-handed state with probability $+\frac{1}{2}[1-(v/c)]$ (Section 6.5). In allowed nuclear β decays there are no nuclear factors that favour one helicity state over another, so that if only the left-handed component of the electron field participates in the interaction, the degree of longitudinal polarisation of the emitted electron is

$$-\frac{1}{2}\left(1+\frac{v}{c}\right)+\frac{1}{2}\left(1-\frac{v}{c}\right)=-\frac{v}{c}.$$

For positrons, the probabilities are reversed (Section 6.5) and the longitudinal polarisation of a positron emitted in an allowed β decay is $+v/c$. Data from several such decays are shown in Fig. 9.1.

A direct measurement of the helicities of neutrinos emitted in β decay is almost impossible, but the helicities may be inferred from careful measurements of the angular momentum states of the participating nuclei. Within experimental error, only negative helicity neutrinos and positive helicity antineutrinos participate in the weak interaction.

Nuclear β decays do not release sufficient energy to produce either of the two other lepton families known to exist: muons and muon neutrinos, and tau leptons

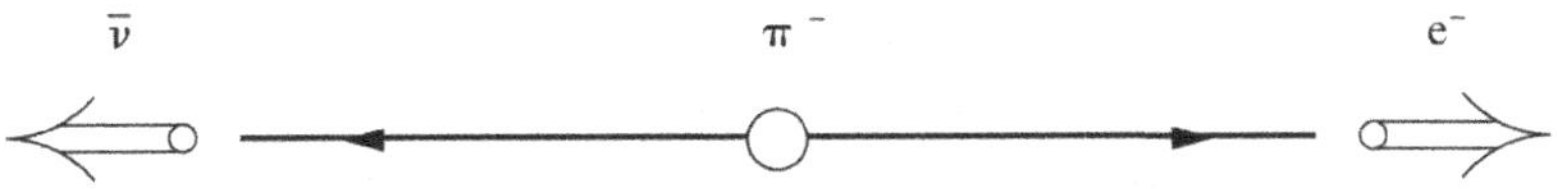

Figure 9.2 $\pi^- \to e^- + \bar{\nu}_e$. In this illustration the electron velocity is to the right, the antineutrino to the left, the spin directions are indicated Any orbital angular momentum is out of the plane of the page ($\mathbf{L} = \mathbf{r} \times \mathbf{p}$) and since the total angular momentum must be zero the spins have to be opposite.

and their partner neutrinos. We shall see in Chapter 13 that probably there are just these three, e, μ, τ, lepton families. Each family seems to play a similar role in Nature, an observation known as *lepton universality.* They differ only in the masses of the electrically charged leptons: $m_e \approx 0.511$ MeV, $m_\mu \approx 106$ MeV, $m_\tau = 1777$ MeV.

9.2 Pion decay

An important example that illustrates both the left-handedness of the lepton fields participating in β decay and lepton universality is provided by the decay of the charged pi mesons. These decays are common in the cosmic radiation and provide its principal component, muons, at ground level. Almost 100% of the pions decay through

$$\pi^- \to \mu^- + \bar{\nu}_\mu, \quad \pi^+ \to \mu^+ + \nu_\mu,$$

with a decay rate $1/\tau\ (\pi \to \mu\bar{\nu}_\mu) = 2.53 \times 10^{-14}$ MeV. The corresponding decays to electrons have much smaller decay rates: $1/\tau\ (\pi \to e\bar{\nu}_e) = 1.23 \times 10^{-4}(1/\tau\ (\pi \to \mu\bar{\nu}_\mu))$.

The decay rate to electrons is suppressed because only the left-handed fields of the electron and neutrino take part. Consider the π^- decay in a frame in which the pion is at rest (Fig. 9.2). The π^- has zero spin, the antineutrino has positive helicity. Hence to conserve angular momentum in this two-body decay the electron also must have positive helicity. The probability of its being in the left-handed state is $\frac{1}{2}[1 - (v_e/c)] = m_e^2/(m_\pi^2 + m_e^2) = 1.34 \times 10^{-5}$ (Problem 9.1). The μ^- decay is similarly inhibited, but the muon's much larger mass makes the factor less effective: $\frac{1}{2}[1 - (v_\mu/c)] = 0.36$.

An effective interaction Lagrangian density that incorporates these features is

$$\mathcal{L}_{int} = \alpha_\pi[j^\mu \partial_\mu \Phi_\pi + j^{\mu\dagger} \partial_\mu \Phi_\pi^\dagger], \tag{9.1}$$

where

$$j^{\mu} = e_{\mathrm{L}}^{\dagger}\tilde{\sigma}^{\mu}\upsilon_{\mathrm{eL}} + \mu_{\mathrm{L}}^{\dagger}\tilde{\sigma}^{\mu}\upsilon_{\mu\mathrm{L}} + \tau_{\mathrm{L}}^{\dagger}\tilde{\sigma}^{\mu}\upsilon_{\tau\mathrm{L}}, \tag{9.2}$$

and α_{π} is an effective (real) coupling constant.

Φ_{π} is a complex scalar field describing the charged $\pi^{\pm}$ mesons (Section 7.6). Φ_{π} destroys negative pions, and creates positive pions. It is not a fundamental field of the Standard Model, since it ignores the internal structure of the pions. The four-vector $e_{\mathrm{L}}^{\dagger}\tilde{\sigma}^{\mu}\upsilon_{\mathrm{eL}}$ is the simplest Lorentz structure we can construct from the two left-handed spinor fields, e_{L}, υ_{eL}, belonging to the electron and its neutrino (see Problem 5.3). Lepton universality is then incorporated in the model, the three lepton families contributing in a similar way to the 'current' j^{μ}; this structure survives in the Standard Model. A Lorentz invariant $\mathcal{L}_{\mathrm{int}}$ is obtained by taking the scalar product of j^{μ} with $\partial_{\mu}\Phi$, and, finally, we make $\mathcal{L}_{\mathrm{int}}$ real. Note that $\mathcal{L}_{\mathrm{int}}$ is a 'point' interaction: j^{μ} and $\partial_{\mu}\Phi$ are evaluated at the same point x in space-time. Since the pion is an extended object, this point interaction must be an approximation, not to be taken too seriously.

An effective interaction Lagrangian is to be used only in low orders of perturbation theory. It is not suitable for calculating high order corrections. One should not therefore demand high accuracy when comparing the results of a calculation with experiment.

Using our $\mathcal{L}_{\mathrm{int}}$ to lowest order, the partial decay rates for pions at rest are (Problem 9.4)

$$\frac{1}{\tau(\pi \to \mathrm{e}\bar{\nu}_{\mathrm{e}})} = \frac{\alpha_{\pi}^{2}}{4\pi}\left(1 - \frac{\upsilon_{\mathrm{e}}}{c}\right) p_{\mathrm{e}}^{2} E_{\mathrm{e}}, \qquad \frac{1}{\tau(\pi \to \mu\bar{\nu}_{\mu})} = \frac{\alpha_{\pi}^{2}}{4\pi}\left(1 - \frac{\upsilon_{\mu}}{c}\right) p_{\mu}^{2} E_{\mu}. \tag{9.3}$$

In these equations, E_{e}, E_{μ} and p_{e}, p_{μ} are the charged lepton's energy and momentum, and are determined by energy and momentum conservation. The factors $p_{\mathrm{e}}^{2}E_{\mathrm{e}}$, $p_{\mu}^{2}E_{\mu}$ come from the density of states factor in the expression for the transition probability (Problem 9.2). The factors $(1 - \upsilon_{\mathrm{e}}/c)$ and $(1 - \upsilon_{\mu}/c)$ are a consequence of the participation of left-handed fields only.

The ratio

$$\frac{\tau\,(\pi \to \mu\bar{\nu}_{\mu})}{\tau\,(\pi \to e\bar{\nu}_{\mathrm{e}})} = \frac{m_{\mathrm{e}}^{2}(m_{\pi}^{2} - m_{\mathrm{e}}^{2})^{2}}{m_{\mu}^{2}(m_{\pi}^{2} - m_{\mu}^{2})^{2}} = 1.28 \times 10^{-4} \tag{9.4}$$

(Problem 9.3). This lowest order calculation, which neglects the effects of nonlocality and electromagnetic corrections, agrees well with the experimental value of 1.23×10^{-4}, and gives strong support for lepton universality.

The observations give $1/\tau(\pi \to e\bar{\nu}_e) = 3.11 \times 10^{-18}$ MeV, $1/\tau(\pi \to \mu\bar{\nu}_\mu) = 2.53 \times 10^{-14}$ MeV, from which we may estimate

$$\alpha_\pi = 2.09 \times 10^{-9}\ \text{MeV}^{-1}.$$

The smallness of α_π reflects the weakness of the weak interaction.

Although the pion does not have enough mass to decay to tau leptons, the effective Lagrangian (9.1) also described the decays

$$\tau^+ \to \pi^+ + \bar{\nu}_\tau, \quad \tau^- \to \pi^- + \nu_\tau,$$

and in lowest order of perturbation theory, predicts

$$\frac{1}{\tau(\tau \to \pi\nu_\tau)} = \frac{\alpha_\pi^2}{32\pi} m_\tau^3 [1 - (m_\pi/m_\tau)^2]^2. \tag{9.5}$$

Using the estimate of α_π from $\pi^\pm$ decay to calculate $1/\tau\,(\tau \to \pi\nu_\tau)$ provides a further test of lepton universality: the predicted value 2.42×10^{-10} MeV compares quite well with the experimental value, $(2.6 \pm 0.1) \times 10^{-10}$ MeV.

9.3 Conservation of lepton number

In the model Lagrangian discussed so far, a single lepton can change only to another of the same family, and a lepton and antilepton of the same family can only be created or destroyed together. There is thus a conservation law, *the conservation of lepton number* (antileptons being counted negatively), for each separate family, exemplified in the decays we have so far considered.

We saw in Section 7.1 that particle conservation follows from a $U(1)$ symmetry of the Lagrangian, and it is interesting to see how this is accomplished with our model Lagrangian. We have

$$\mathcal{L} = \mathcal{L}_{\text{free}} + \mathcal{L}_{\text{int}}$$

where, using Dirac spinors for the lepton fields,

$$\begin{aligned}
\mathcal{L}_{\text{free}} &= \partial_\mu \Phi^\dagger \partial^\mu \Phi - m_\pi^2 \Phi^\dagger \Phi \\
&\quad + \bar{\psi}_e(\gamma^\mu \mathrm{i}\partial_\mu - m_e)\psi_e + \bar{\nu}_e \gamma^\mu \mathrm{i}\partial_\mu \nu_e \\
&\quad + \bar{\psi}_\mu(\gamma^\mu \mathrm{i}\partial_\mu - m_\mu)\psi_\mu + \bar{\nu}_\mu \gamma^\mu \mathrm{i}\partial_\mu \nu_\mu \\
&\quad + \bar{\psi}_\tau(\gamma^\mu \mathrm{i}\partial_\mu - m_\tau)\psi_\tau + \bar{\nu}_\tau \gamma^\mu \mathrm{i}\partial_\mu \nu_\tau, \\
\mathcal{L}_{\text{int}} &= \alpha_\pi [j^\mu \partial_\mu \Phi_\pi + j^{\mu\dagger} \partial_\mu \Phi_\pi^\dagger],
\end{aligned}$$

and, in terms of Dirac spinors, the current j_μ of equation (9.2) can be written

$$j^\mu = \bar{\psi}_e \gamma^\mu \frac{1}{2}(1 - \gamma^5)\nu_e + \bar{\psi}_\mu \gamma^\mu \frac{1}{2}(1 - \gamma^5)\nu_\mu + \bar{\psi}_\tau \gamma^\mu \frac{1}{2}(1 - \gamma^5)\nu_\tau. \tag{9.6}$$

By itself, $\mathcal{L}_{\text{free}}$ has seven $U(1)$ symmetries: seven independent phases on the seven free fields. Including $\mathcal{L}_{\text{int}}$ reduces these to four, which can be written

$$\begin{aligned} \psi_{\mathrm{e}} &\to \mathrm{e}^{\mathrm{i}\beta}\mathrm{e}^{\mathrm{i}\alpha_e}\Psi_{\mathrm{e}}, & \nu_{\mathrm{e}} &\to \mathrm{e}^{\mathrm{i}\alpha_e}\nu_{\mathrm{e}};\\ \psi_{\mu} &\to \mathrm{e}^{\mathrm{i}\beta}\mathrm{e}^{\mathrm{i}\alpha_\mu}\Psi_{\mu}, & \nu_{\mu} &\to \mathrm{e}^{\mathrm{i}\alpha_\mu}\nu_{\mu};\\ \psi_{\tau} &\to \mathrm{e}^{\mathrm{i}\beta}\mathrm{e}^{\mathrm{i}\alpha_\tau}\Psi_{\tau}, & \nu_{\tau} &\to \mathrm{e}^{\mathrm{i}\alpha_\tau}\nu_{\tau};\\ \Phi_{\pi} &\to \mathrm{e}^{\mathrm{i}\beta}\Phi_{\pi}. \end{aligned}$$

The phase factors $\alpha_e, \alpha_\mu, \alpha_\tau$ are associated with the conserved lepton currents (Problem 9.6). If we require $\mathcal{L}$ to be invariant under a *local* gauge symmetry, with $\beta = \beta(x)$ arbitrarily space and time dependent, we are led to the introduction of the electromagnetic field A^μ, as in Section 5.5. We shall see that not all these features of our effective Lagrangian survive the introduction of neutrino mass into the Standard Model.

9.4 Muon decay

The analysis of the muon decays

$$\mu^- \to \mathrm{e}^- + \bar{\nu}_{\mathrm{e}} + \nu_\mu, \quad \mu^+ \to \mathrm{e}^+ + \nu_{\mathrm{e}} + \bar{\nu}_\mu, \tag{9.7}$$

has played a very important role in establishing the Standard Model. The decays involve lepton fields only, so that the physics is not obscured by the phenomenology of strong interaction fields as was our example of pion decay.

An effective Lagrangian density that describes the decays again couples the participating particles into currents. In fact all decays seen so far that involve just leptons are well described by the effective interaction Lagrangian density

$$\mathcal{L}_{\text{lepton}} = -2\sqrt{2}G_F g_{\mu\nu} j^\mu j^{\nu\dagger}, \tag{9.8}$$

with j^μ again defined by (9.2) or (9.6). A similar form for nuclear β decay was introduced by Fermi, and G_{F} is called the Fermi constant. The $2\sqrt{2}$ is a related accident of history.

The term in (9.8) that describes μ^- decay is

$$\mathcal{L} = -2\sqrt{2}G_{\mathrm{F}} g_{\mu\nu}\left[\mathrm{e}_{\mathrm{L}}^{\dagger}\tilde{\sigma}^{\mu}\nu_{\mathrm{eL}}\nu_{\mu\mathrm{L}}^{\dagger}\tilde{\sigma}^{\nu}\mu_{\mathrm{L}}\right]. \tag{9.9}$$

The most ready supply of muons comes from pion decays and these, as we have seen, are almost 100% polarised. The interaction Lagrangian density (9.9) implies a strong correlation between the angle θ made by the direction of the electron with the direction of the muon spin, and the energy E_{e} of the electron. In the muon rest frame, to lowest order of perturbation theory, and neglecting terms in $(m_{\mathrm{e}}/m_\mu)^2$, the decay rate into an angular interval $\mathrm{d}\theta$ and energy interval $\mathrm{d}E_{\mathrm{e}}$ is (see Donoghue *et al.* 1992, p. 138)

$$R(\theta, E_e)\, d\theta\, dE_e = \frac{m_\mu G_F^2}{6\pi^3}\left[\left(\frac{3}{4}m_\mu - E_e\right) + \cos\theta\left(\frac{1}{4}m_\mu - E_e\right)\right] E_e^2\, dE_e \sin\theta d\theta. \qquad (9.10)$$

Integrating (9.10) over θ and E_e gives the total decay rate for this process

$$\frac{1}{\tau(\mu \to e\bar{\nu}_e\nu_\mu)} = \frac{m_\mu^5 G_F^2}{192\pi^3}. \qquad (9.11)$$

The total muon decay rate, which includes also decays with photons in the final state, for example the decays

$$\mu^- \to e^- + \gamma + \bar{\nu}_e + \nu_\mu,$$

has been very accurately measured, giving

$$\tau_\mu = (2.19703 \pm 0.00004) \times 10^{-6}\ \text{s}.$$

A corresponding accurate theoretical expression that corrects (9.11) by including terms in $(m_e/m_\mu)^2$ and electromagnetic effects, gives

$$G_F = 1.16639(2) \times 10^{-5}\ \text{GeV}^{-2}, \qquad (9.12)$$

which is the presently accepted value of this important constant.

Further tests of lepton universality are provided by the decays

$$\tau^- \to \mu^- + \bar{\nu}_\mu + \nu_\tau, \quad \tau^- \to e^- + \bar{\nu}_e + \nu_\tau,$$

and their charge conjugates. These, like muon decay, are described by appropriate terms in the interaction Lagrangian (9.8). Since both $(m_e/m_\tau)^2$ and $(m_\mu/m_\tau)^2$ are small, the first-order formula (9.11) with m_μ replaced by m_τ predicts these decay rates to be equal and $\approx 4 \times 10^{-10}$ MeV. They are indeed so within experimental error. Also from this formula

$$\frac{\tau\,(\tau \to e\bar{\nu}_e\nu_\tau)}{\tau\,(\mu \to e\bar{\nu}_e\nu_\mu)} \approx \left(\frac{m_\mu}{m_\tau}\right)^5.$$

The ratio of the decay rates is 7.36×10^{-7} and the ratio of the fifth power of the masses is 7.43×10^{-7}.

It should be noted that the coupling constant G_F has the dimension of $(\text{mass})^{-2}$. The effective interaction (9.8) cannot be elevated into a quantum field interaction; see Section 8.4.

9.5 The interactions of muon neutrinos with electrons

In the 1960s, intense muon neutrino beams were engineered at Brookhaven and at CERN. Muon neutrinos (or antineutrinos) were produced as secondary particles from the decay of π^+ (or π^-) mesons in flight. It was from the observation that these neutrino beams produced almost exclusively muons rather than electrons, when in interaction with a target, that the distinction between electron neutrinos and muon neutrinos was established.

The centre of mass energy $\sqrt{s}$ available in a collision of a neutrino with an electron at rest is relatively small, because of the smallness of the electron mass. If E_ν is the neutrino energy,

$$s = m_e(2E_\nu + m_e), \tag{9.13}$$

(Problem 9.8). For example, if $E_\nu = 30$ GeV then $s = (175 \text{ MeV})^2$, which will produce no more than a muon. Most neutrino interactions will be with the atomic nuclei in the target. However, here we consider only the interactions with electrons.

The interaction

$$\nu_\mu + e^- \to \mu^- + \nu_e$$

is included in the effective interaction Lagrangian density (9.8). In first-order perturbation theory and averaging over electron polarisations, this Lagrangian predicts an isotropic differential cross-section in the centre of mass system:

$$\frac{d\sigma}{d\Omega} = \frac{G_F^2}{4\pi^2}\frac{\left(s - m_\mu^2\right)^2}{s}, \qquad \sigma_{tot} = \frac{G_F^2}{\pi}\frac{\left(s - m_\mu^2\right)^2}{s} \tag{9.14}$$

with s the square of the centre of mass energy. (See Okun 1982, p. 134.)

At the low energies available experimentally, the cross-section appears to be consistent with the theoretical form. The high energy structure is not easily explored experimentally, because of (9.13), but clearly the theoretical formulae become inadequate at high energies: the expressions (9.14) increase without limit as s increases, and for a ‘point’ interaction this is inconsistent with unitarity. Nor is it possible to improve the expressions within this framework, since the effective Lagrangian does not give a renormalisable theory.

The most significant result to come from the experiments on neutrino–electron interactions was the observation of elastic scattering for both ν_μ and $\bar{\nu}_\mu$:

$$\nu_\mu + e^- \to \nu_\mu + e^-,$$
$$\bar{\nu}_\mu + e^- \to \bar{\nu}_\mu + e^-,$$

with cross-sections of a magnitude similar to those for muon production. Such elastic scattering is *not* included in our $\mathcal{L}_{int}$ (though there are terms corresponding to

$e\nu_e \to e\nu_e$ and $e\bar{\nu}_e \to e\bar{\nu}_e$). Thus another weak interaction must exist. The experimental investigation of this is difficult because of the smallness of the cross-sections at the available energies. We shall see from the Standard Model that the effective interaction Lagrangian required is again of current–current form,

$$\mathcal{L}_{\text{int}} = \frac{-G_F}{\sqrt{2}}(j_{\text{neutral}})_\mu (j_{\text{neutral}})^\mu, \tag{9.15}$$

where, in terms of Dirac spinors,

$$(j_{\text{neutral}})^\mu = \bar{\nu}_e \gamma^\mu \frac{1}{2}(1-\gamma^5)\nu_e + \bar{\psi}_e \gamma^\mu (c_V - c_A \gamma^5)\psi_e \tag{9.16}$$
$$+ \text{ similar terms for the } \mu \text{ and } \tau \text{ lepton families,}$$

and c_V and c_A are parameters. The current is called a *neutral current* because it does not induce a change of charge as do the currents (9.2). (Note that it will also contribute to the scattering $e\nu_e \to e\nu_e$.)

Rewriting (9.16) with two-component spinors,

$$(j_{\text{neutral}})^\mu = (\nu_{eL})^\dagger \tilde{\sigma}^\mu \nu_{eL} + (c_V + c_A) e_L^\dagger \tilde{\sigma}^\mu e_L$$
$$+ (c_V + c_A) e_R^\dagger \sigma^\mu e_R + \text{similar } \mu \text{ and } \tau \text{ terms.} \tag{9.17}$$

In this form it is evident that right-handed lepton fields as well as left-handed are involved in the neutral currents. The parameters c_V and c_A are related to the Weinberg angle θ_W, which appears in the Standard Model, as we shall see in Chapter 12 (equation (12.24)). The subscripts V and A refer, respectively, to the vector and axial vector nature of the terms in (9.16). (See Section 5.5.)

One might anticipate that neutral currents are also present in atomic physics, and indeed they are. However, their effects are hard to discern experimentally. For example, they induce parity violation in atoms, but at atomic energies the weak interaction gives a very small effect. Indeed the decay of an unstable nuclear or atomic system through the neutral current must always compete with faster electromagnetic decays, and for this reason neutral current decays in these systems have never been observed.

Problems

9.1 In the decay of the π^- at rest, $\pi^- \to e^- + \bar{\nu}_e$, show that

$$\frac{1}{2}\left(1 - \frac{v_e}{c}\right) = \frac{m_e^2}{m_\pi^2 + m_e^2}.$$

9.2 Show that the density of final states for the decay of Problem 9.1 is

$$\rho(E) = \frac{V}{(2\pi)^3} 4\pi p_e^2 \frac{dp_e}{dE}$$

where V is the normalisation volume and

$$\frac{dp_e}{dE} = \frac{E_e}{m_\pi}.$$

9.3 Obtain the ratio of decay rates given by equation (9.4).

9.4 The term in $\mathcal{L}_{int}$ describing the decay $\pi^- \to e^- + \bar{\nu}_e$ is

$$\mathcal{L}_{int} = \alpha_\pi e_L^\dagger \tilde{\sigma}^\mu \nu_{eL} \partial_\mu \Phi_\pi .$$

Assume that this gives a corresponding term $V(0)$ in the effective Hamiltonian,

$$V(0) = -\alpha_\pi \int e_L^\dagger \tilde{\sigma}^\mu \nu_{eL} \partial_\mu \Phi_\pi d^3\mathbf{x}.$$

(This assumption will be justified in Chapter 12.)

The transition probability per unit time for the decay is to lowest order

$$2\pi |\langle e_p, \bar{\nu}_{p'} | V(0) | \pi^-(\text{rest}) \rangle|^2 \rho(E)$$

where $\rho(E)$ is given by Problem 9.2.

Use the free field expansions given in equations (3.35) and (6.24), and Problem 6.5, to evaluate the matrix element above and hence verify equation (9.3).

9.5 Verify the equivalence of the expressions (9.2) and (9.6) for the current j^μ.

9.6 Taking the effective Lagrangian of Section 9.3, show that the conserved current associated with the $U(1)$ symmetry $\psi_e \to e^{i\alpha}\psi_e$, $\nu_e \to e^{i\alpha}\nu_e$, is the electron electron–neutrino current

$$j^\mu = \bar{\psi}_e \gamma^\mu \psi_e + \bar{\nu}_e \gamma^\mu \nu_e .$$

Show that the conserved current associated with $e^{i\beta}$ in the transformations (9.7) is

$$\bar{\psi}_e \gamma^\mu \psi_e + \bar{\psi}_\mu \gamma^\mu \psi_\mu + \bar{\psi}_\tau \gamma^\mu \psi_\tau + i[(\Phi^\dagger \partial^\mu \Phi - \Phi \partial^\mu \Phi^\dagger) + \alpha_\pi (j^{\mu\dagger} \Phi^\dagger - j^\mu \Phi)].$$

Construct the Lagrangian density that results, when the electromagnetic field is introduced by elevating the global $U(1)$ symmetry of the phase factor $e^{i\beta}$ into a local gauge symmetry.

9.7 Estimate G_F from the expression (9.11) and the experimental lifetime τ_μ.

9.8 Using a suitable Lorentz invariant, obtain equation (9.13).

9.9 Pick out the term in the effective Lagrangian density (9.8) that contributes to the scattering

$$e^- + \nu_e \to e^- + \nu_e,$$

and the term in (9.15) that contributes to the scattering

$$e^- + \nu_\mu \to e^- + \nu_\mu.$$

9.10 The K^- is like the π^-, but with an s quark replacing the d. An effective interaction with leptons is similar in form to equation (9.1), with Φ_K replacing Φ_π and α_K replacing α_π. Use the analogue of equation (9.4) to estimate the ratio $\tau\,(K \to \mu\bar{\nu}_\mu)/\tau\,(K \to e\bar{\nu}_e)$, and compare with the observed value $(2.44 \pm 0.1) \times 10^{-5}$ ($m_K = 493.68\,\text{MeV}$).

The mean life $\tau\,(K^- \to \mu^-\bar{\nu}_\mu)$ is measured to be 1.948×10^{-8} s. Estimate α_K/α_π.

9.11 Obtain the decay rate (9.5).

10

Symmetry breaking in model theories

In Chapter 9, 'effective' weak interaction Lagrangian densities were constructed. When used in low orders of perturbation theory, these account well for the observed phenomena at low energies. Difficulties arise in higher order perturbation theory, as they do in quantum electrodynamics. There is, however, an important difference: it has been proved that these effective Lagrangian theories cannot be renormalised and they are therefore unsatisfactory. Furthermore, at higher energies new phenomena appear, and it is now well established experimentally that the weak interaction is mediated by the W^+, W^- and Z bosons. How are these particles to be incorporated in a theory of the weak interaction that can be renormalised, and which has the same seeming inevitability as QED? The answer lies in the Weinberg–Salam unified theory of the electromagnetic and weak interactions. As an introduction to the Weinberg–Salam theory we shall in this chapter consider 'model' theories, the mathematics of which is fairly simple, but which contain the basic ideas we shall need.

10.1 Global symmetry breaking and Goldstone bosons

A possible Lagrangian density for a complex scalar field $\Phi = (\phi_1 + \mathrm{i}\phi_2)/\sqrt{2}$ is

$$\mathcal{L} = \partial_\mu \Phi^\dagger \partial^\mu \Phi - m^2 \Phi^\dagger \Phi \tag{10.1}$$

(cf. equation (3.32)).

In this expression $(\partial \Phi^\dagger/\partial t)(\partial \Phi/\partial t)$ can be regarded as the kinetic energy density and $\nabla \Phi^\dagger \cdot \nabla \Phi + m^2 \Phi^\dagger \Phi$ as the potential energy density (see Section 3.3). If Φ is constant, independent of space and time, the only contribution to the energy is $m^2 \Phi^\dagger \Phi$. Since m^2 is positive this will be a minimum when $\phi_1 = \phi_2 = 0$. Thus $\Phi = 0$ corresponds to the 'vacuum' state. Consider now the Lagrangian density obtained by changing the sign in front of m^2. This would be unstable: the potential

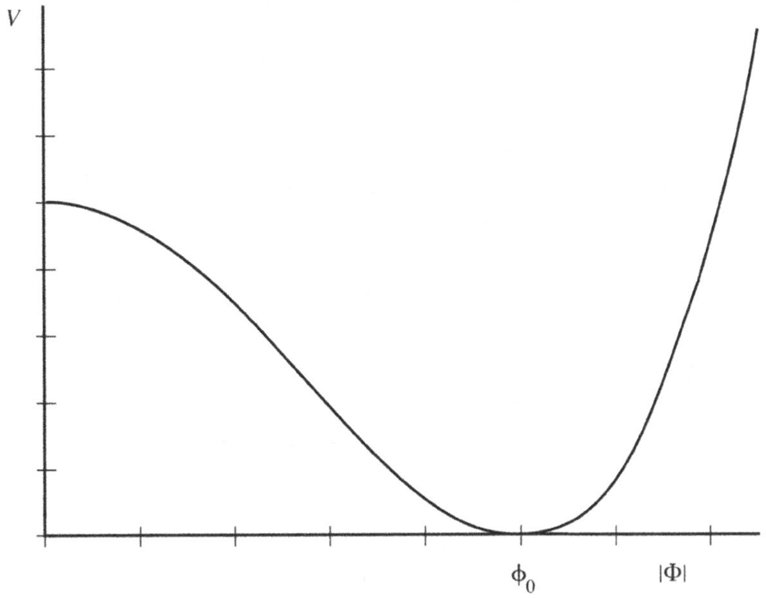

Figure 10.1 Plot of $V = (m^2/2\phi_0^2)[\Phi^*\Phi - \phi_0^2]^2$ as a function of $|\Phi|$; Φ is here a classical field.

energy density is then unbounded below. Stability can be restored by introducing a term $(m^2/2\phi_0^2)(\Phi^\dagger\Phi)^2$ where ϕ_0^2 is another (real) parameter. For convenience we add a constant term $m^2\phi_0^2/2$, and then

$$\mathcal{L} = \partial_\mu\Phi^\dagger\partial^\mu\Phi - V(\Phi^\dagger\Phi)$$

where

$$V(\Phi^\dagger\Phi) = \frac{m^2}{2\phi_0^2}\left[\Phi^\dagger\Phi - \phi_0^2\right]^2. \tag{10.2}$$

The form of V is shown in Fig. (10.1). The minimum field energy is now obtained with Φ constant independent of space and time, but such that $\Phi^\dagger\Phi = |\Phi|^2 = \phi_0^2$. Such a field is not unique but is defined by a point on the circle $|\Phi| = \phi_0$ in the state space (ϕ_1, ϕ_2), so that the number of possible vacuum states is infinite.

An analogy with magnetism is helpful. The Hamiltonian describing a Heisenberg ferromagnet has rotational symmetry: all directions in space are equivalent. However, in its ground state a ferromagnet is magnetised in some particular direction, which is not determined within the theory, and the rotational symmetry is lost. This is an example of *spontaneous symmetry breaking*.

The Lagrangian density (10.2) has a 'global' $U(1)$ symmetry: $\Phi \to \Phi' = \mathrm{e}^{-\mathrm{i}\alpha}\Phi$, $\mathcal{L} \to \mathcal{L}' = \mathcal{L}$, for any real α. Equivalently,

$$\begin{aligned}\phi_1' &= \phi_1 \cos\alpha + \phi_2 \sin\alpha,\\ \phi_2' &= -\phi_1 \sin\alpha + \phi_2 \cos\alpha.\end{aligned}$$

The transformation rotates the state round a circle $|\Phi|^2 =$ constant in the state space (ϕ_1, ϕ_2). If we pick out the particular direction in (ϕ_1, ϕ_2) space for which Φ is real, and take the vacuum state to be $(\phi_0, 0)$, we break the $U(1)$ symmetry.

Expanding about this ground state $(\phi_0, 0)$, we put $\Phi = \phi_0 + (1/\sqrt{2})(\chi + \mathrm{i}\psi)$. The Lagrangian density becomes

$$\mathcal{L} = \frac{1}{2}\partial_\mu\chi\partial^\mu\chi + \frac{1}{2}\partial_\mu\psi\partial^\mu\psi - \frac{m^2}{2\phi_0^2}\left[\sqrt{2}\phi_0\chi + \frac{\chi^2}{2} + \frac{\psi^2}{2}\right]^2. \tag{10.3}$$

After breaking the $U(1)$ symmetry we must interpret the new fields. (In much the same way, the excited states of a ferromagnet cannot be discussed until the spatial symmetry has been broken.) In place of the complex field Φ, we have two coupled scalar real fields χ and ψ. We write

$$\mathcal{L} = \mathcal{L}_{\text{free}} + \mathcal{L}_{\text{int}}$$

where

$$\mathcal{L}_{\text{free}} = \frac{1}{2}\partial_\mu\chi\partial^\mu\chi - m^2\chi^2 + \frac{1}{2}\partial^\mu\psi\partial^\mu\psi. \tag{10.4}$$

$\mathcal{L}_{\text{free}}$ represents free particle fields, and contains all the terms in $\mathcal{L}$ that are quadratic in the fields. For classical fields and small oscillations, these terms dominate. The rest of the Lagrangian density, $\mathcal{L}_{\text{int}}$, corresponds to interactions between the free particles and higher order corrections to their motion.

There is a quadratic term $-m^2\chi^2$ in (10.4), so that the χ field corresponds to a scalar spin-zero particle of mass $\sqrt{2}m$ (by comparison with (3.18)). In the case of the ψ field there is no such quadratic term: the corresponding scalar spin-zero particle is therefore massless. The massless particles that always arise as a result of global symmetry breaking are called *Goldstone bosons*.

10.2 Local symmetry breaking and the Higgs boson

We now generalise further, and construct a Lagrangian density that is invariant under a *local* $U(1)$ gauge transformation,

$$\Phi \to \Phi' = \mathrm{e}^{-\mathrm{i}q\theta}\Phi,$$

where $\theta = \theta(x)$ may be space and time dependent. This requires the introduction of a (massless) gauge field A_μ, as in Section 7.5, and we take

$$\mathcal{L} = [(\partial_\mu - iqA_\mu)\Phi^\dagger][(\partial^\mu + iqA^\mu)\Phi] - \frac{1}{4}F_{\mu\nu}F^{\mu\nu} - V(\Phi^\dagger\Phi), \tag{10.5}$$

where $F_{\mu\nu} = \partial_\mu A_\nu - \partial_\nu A_\mu$, and again

$$V(\Phi^\dagger\Phi) = \frac{m^2}{2\phi_0^2}\left[\Phi^\dagger\Phi - \phi_0^2\right]^2.$$

$\mathcal{L}$ is invariant under the local gauge transformation

$$\Phi(x) \to \Phi'(x) = e^{-iq\theta}\Phi(x), \quad A_\mu(x) \to A'_\mu(x) = A_\mu(x) + \partial_\mu\theta(x).$$

A minimum field energy is obtained when the fields A_μ vanish, and Φ is constant, defined by a point on the circle $|\Phi| = \phi_0$. Any gauge transformation on this field configuration is also a minimum. Again we have an infinity of vacuum states.

Given $\Phi(x)$, we can always choose $\theta(x)$ so that the field $\Phi'(x) = e^{-iq\theta}\Phi(x)$ is real. This breaks the symmetry, since we are no longer free to make further gauge transformations.

Putting $\Phi'(x) = \phi_0 + h(x)/\sqrt{2}$, where $h(x)$ is real, gives

$$\begin{aligned}\mathcal{L} = {} & [(\partial_\mu - iqA'_\mu)(\phi_0 + h/\sqrt{2})][(\partial^\mu + iqA'^\mu)(\phi_0 + h/\sqrt{2})] \\ & - \frac{1}{4}F'_{\mu\nu}F'^{\mu\nu} - \frac{m^2}{2\phi_0^2}\left[\sqrt{2}\phi_0 h + \frac{1}{2}h^2\right]^2.\end{aligned} \tag{10.6}$$

For clarity, we again separate this into

$$\mathcal{L} = \mathcal{L}_{\text{free}} + \mathcal{L}_{\text{int}}$$

where, dropping the primes on the gauge field,

$$\begin{aligned}\mathcal{L}_{\text{free}} &= \frac{1}{2}\partial_\mu h \partial^\mu h - m^2h^2 - \frac{1}{4}F_{\mu\nu}F^{\mu\nu} + q^2\phi_0^2 A_\mu A^\mu, \\ \mathcal{L}_{\text{int}} &= q^2 A_\mu A^\mu\left(\sqrt{2}\phi_0 h + \frac{1}{2}h^2\right) - \frac{m^2h^2}{2\phi_0^2}\left(\sqrt{2}\phi_0 h + \frac{1}{4}h^2\right).\end{aligned} \tag{10.7}$$

Before symmetry breaking, we had a complex scalar field $\Phi = (\phi_1 + i\phi_2)/\sqrt{2}$, and a massless vector field with two polarisation states (Section 4.4). In $\mathcal{L}_{\text{free}}$ we have a single scalar field $h(x)$ corresponding to a spinless boson of mass $\sqrt{2}m$, and a vector field A_μ, corresponding to a vector boson of mass $\sqrt{2}q\phi_0$, with three independent components (Section 4.9).

This mechanism for introducing mass into a theory was invented by Higgs (1964) and others (for example Anderson, 1963), and the particle corresponding to the field $h(x)$ is called a *Higgs boson*. As a consequence of local symmetry breaking the gauge field acquires a mass, and the massless spin-zero Goldstone boson that appeared in our example of global symmetry breaking in Section 10.1 is replaced by the longitudinal polarised state of this massive spin one boson.

In the Weinberg and Salam 'electroweak' theory, the masses of the $W^{\pm}$ and Z particles arise as a result of symmetry breaking. The resulting theory can be renormalised, whereas the phenomenological theory of Chapter 9 cannot be renormalised. The form of $V(\Phi^{\dagger}\Phi)$ that has been introduced in this chapter appears also in the electroweak theory. It may seem a somewhat arbitrary feature. However, it can be shown to be the most general form that can be renormalised.

Problems

10.1 What interaction term in the model Lagrangian density (10.3) allows the massive boson to decay into two Goldstone bosons? Show that the decay rate in lowest order perturbation theory is

$$\frac{1}{\tau(\chi \to \psi\psi)} = \frac{m_\chi}{128\pi}\left(\frac{m_\chi}{\phi_0}\right)^2.$$

10.2 Show that with the model Lagrangian density (10.7), the vector boson would be stable, but if the coupling constant $q < m/(2\phi_0)$ the scalar boson would decay into two vector bosons.

11

Massive gauge fields

In the preceding chapter (Section 10.2), we set up a simple Lorentz invariant Lagrangian density, which we required to be also invariant under a local $U(1)$ transformation. This requirement leads to the introduction of a 'gauge field' A_μ. The system has a degenerate ground state. Breaking the local symmetry results in the appearance of a vector field carrying mass, together with a scalar Higgs field also carrying mass. The motivation for introducing mass in this way is that the subsequent quantum theory can be renormalised. In this chapter we apply the same idea to a more complicated Lagrangian, which will turn out to have remarkable physical significance.

11.1 *SU*(2) symmetry

As a further generalisation, which is basic to the Standard Model, we shall construct a Lagrangian density that is invariant under a local $SU(2)$ transformation as well as a local $U(1)$ transformation. The idea was first explored by Yang and Mills (1954). We introduce a two-component field

$$\Phi = \begin{pmatrix} \Phi_A \\ \Phi_B \end{pmatrix}, \tag{11.1}$$

where now Φ_A and Φ_B are both complex scalar fields,

$$\Phi_A = \phi_1 + i\phi_2, \quad \Phi_B = \phi_3 + i\phi_4,$$

giving, in total, four real fields.

If $e^{-i\theta}$ is any element of the group $U(1)$ and $\mathbf{U}$ is any element of the group $SU(2)$ (discussed in Appendix B), so that $\mathbf{U}^\dagger\mathbf{U} = \mathbf{U}\mathbf{U}^\dagger = \mathbf{1}$, we require the Lagrangian density to be invariant under the $U(1) \times SU(2)$ transformation

$$\Phi \to \Phi' = e^{-i\theta}\mathbf{U}\Phi. \tag{11.2}$$

A simple Lagrangian density that has a *global* $U(1) \times SU(2)$ symmetry is

$$\mathcal{L}_\Phi = \partial_\mu \Phi^\dagger \partial^\mu \Phi - V(\Phi^\dagger \Phi). \tag{11.3}$$

In terms of the real fields,

$$\Phi^\dagger \Phi = \Phi_A^* \Phi_A + \Phi_B^* \Phi_B = \phi_1^2 + \phi_2^2 + \phi_3^2 + \phi_4^2,$$
$$\partial_\mu \Phi^\dagger \partial^\mu \Phi = \partial_\mu \phi_1 \partial^\mu \phi_1 + \partial_\mu \phi_2 \partial^\mu \phi_2 + \partial_\mu \phi_3 \partial^\mu \phi_3 + \partial_\mu \phi_4 \partial^\mu \phi_4.$$

If $V(\Phi^\dagger\Phi) = m^2 \Phi^\dagger \Phi$, this Lagrangian density corresponds to four independent free scalar fields, all with the same mass m (cf. (3.18)).

In the Standard Model, the $U(1)$ and $SU(2)$ global symmetries are promoted to local symmetries. The $U(1)$ transformation may be written

$$\Phi \to \Phi' = \mathrm{e}^{-\mathrm{i}\theta} \Phi = \exp(-\mathrm{i}\theta\tau^0)\Phi, \tag{11.4a}$$

where in this context we write τ^0 for the unit matrix

$$\tau^0 = \begin{pmatrix} 1 & 0 \\ 0 & 1 \end{pmatrix}.$$

For this to become a local symmetry, we must introduce a vector gauge field $B_\mu(x)\tau^0$ with the transformation law

$$B_\mu(x) \to B'_\mu(x) = B_\mu(x) + (2/g_1)\partial_\mu \theta, \tag{11.4b}$$

and make the replacement

$$\mathrm{i}\partial_\mu \to \mathrm{i}\partial_\mu - (g_1/2)B_\mu,$$

as in Chapter 7. Here the constant g_1 is a dimensionless parameter of the theory, and the factor 2 follows convention.

Any element of $SU(2)$ can be written in the form

$$\mathbf{U} = \exp(-\mathrm{i}\alpha^k \tau^k) \tag{11.5}$$

where the α^k are three real numbers and the τ^k are the three generators of the group *SU(2)*. The τ^k are identical to the Pauli spin matrices:

$$\tau^1 = \begin{pmatrix} 0 & 1 \\ 1 & 0 \end{pmatrix}, \tau^2 = \begin{pmatrix} 0 & -\mathrm{i} \\ \mathrm{i} & 0 \end{pmatrix}, \tau^3 = \begin{pmatrix} 1 & 0 \\ 0 & -1 \end{pmatrix}.$$

For the global $SU(2)$ symmetry to be made into a local $SU(2)$ symmetry, with $\mathbf{U} = \mathbf{U}(x)$ dependent on space and time coordinates, we must introduce a vector gauge field ${W_\mu}^k(x)$ for each generator τ^k. The transformation law for the matrices

$$\mathbf{W}_\mu(x) = {W_\mu}^k(x)\tau^k$$

is

$$\mathbf{W}_\mu(x) \to \mathbf{W}'_\mu(x) = \mathbf{U}(x)\mathbf{W}_\mu(x)\mathbf{U}^\dagger(x) + (2\mathrm{i}/g_2)(\partial_\mu \mathbf{U}(x))\mathbf{U}^\dagger(x), \tag{11.6}$$

which is a generalisation of (11.4). Here g_2 is another dimensionless parameter of the theory.

Note that the matrices

$$\mathbf{W}_\mu(x) = \begin{pmatrix} W_\mu^3 & W_\mu^1 - \mathrm{i}W_\mu^2 \\ W_\mu^1 + \mathrm{i}W_\mu^2 & -W_\mu^3 \end{pmatrix} \tag{11.7}$$

are Hermitian and have zero trace. These properties are preserved by the transformation (11.6) as is clearly necessary (Problem 11.1). A global $SU(2)$ transformation $\mathbf{W}'_\mu = \mathbf{U}\mathbf{W}_\mu\mathbf{U}^\dagger$ corresponds to a rotation of the vectors $W_\mu{}^k$ in the three-dimensional 'weak isospin' space defined by the generators τ^k. (See Appendix B.)

Finally we define

$$D_\mu \Phi = [\partial_\mu + (\mathrm{i}g_1/2)B_\mu + (\mathrm{i}g_2/2)\mathbf{W}_\mu]\Phi. \tag{11.8a}$$

It is straightforward to show

$$D'_\mu \Phi' = [\partial_\mu + (\mathrm{i}g_1/2)B'_\mu + (\mathrm{i}g_2/2)\mathbf{W}'_\mu]\Phi' = \mathrm{e}^{-\mathrm{i}\theta}\mathbf{U} D_\mu \Phi,$$

where

$$\Phi' = \mathrm{e}^{-\mathrm{i}\theta}\mathbf{U}\Phi. \tag{11.8b}$$

Hence the locally gauge invariant Lagrangian density corresponding to (11.3) is

$$\mathcal{L}_\Phi = (D_\mu \Phi)^\dagger D^\mu \Phi - V(\Phi^\dagger \Phi). \tag{11.9}$$

$\mathcal{L}_\Phi$ is also invariant under Lorentz transformations if we require B_μ and $\mathbf{W}_\mu$ to transform as covariant four-vectors.

11.2 The gauge fields

In the case of the gauge field B_μ, we define the field strength tensor $B_{\mu\nu}$ by

$$B_{\mu\nu} = \partial_\mu B_\nu - \partial_\nu B_\mu, \tag{11.10}$$

and take the dynamical contribution to the Lagrangian density to be $-(1/4)\, B_{\mu\nu}B^{\mu\nu}$, as in Section 4.2.

There are additional complications in introducing the field strength tensors for the gauge fields $\mathbf{W}_\mu$, stemming from the non-Abelian nature of the group $SU(2)$. The field strength tensor must be taken to be

$$\mathbf{W}_{\mu\nu} = [\partial_\mu + (ig_2/2)\mathbf{W}_\mu]\mathbf{W}_\nu - [\partial_\nu + (ig_2/2)\mathbf{W}_\nu]\mathbf{W}_\mu. \tag{11.11}$$

Under an $SU(2)$ transformation, $\mathbf{W}_\mu \to \mathbf{W}'_\mu$, given by (11.6), it is straightforward, if tedious, to show that

$$\mathbf{W}_{\mu\nu} \to \mathbf{W}'_{\mu\nu} = \mathbf{U}\mathbf{W}_{\mu\nu}\mathbf{U}^\dagger. \tag{11.12}$$

In verifying this result, note that, since $\mathbf{U}\mathbf{U}^\dagger = \mathbf{1}$,

$$\mathbf{U}(\partial_\mu \mathbf{U}^\dagger) + (\partial_\mu \mathbf{U})\mathbf{U}^\dagger = 0.$$

The complicated definition of $\mathbf{W}_{\mu\nu}$ given by (11.11) is necessary in order to achieve the simple transformation property (11.12).

We then take the total dynamical contribution to the Lagrangian density associated with the gauge fields to be

$$\mathcal{L}_{\text{dyn}} = -\frac{1}{4}B_{\mu\nu}B^{\mu\nu} - \frac{1}{8}\text{Tr}(\mathbf{W}_{\mu\nu}\mathbf{W}^{\mu\nu}). \tag{11.13}$$

Using (11.12) and the cyclic invariance of the trace, we can see that $\mathcal{L}_{\text{dyn}}$ is invariant under a local $SU(2)$ transformation.

Using the results $[\tau^2, \tau^3] = 2i\tau^1$, etc., the matrix $\mathbf{W}_{\mu\nu}$ may be written

$$\mathbf{W}_{\mu\nu} = W^i_{\mu\nu}\tau^i \tag{11.14}$$

where

$$W^1_{\mu\nu} = \partial_\mu W^1_\nu - \partial_\nu W^1_\mu - g_2\left(W^2_\mu W^3_\nu - W^2_\nu W^3_\mu\right), \tag{11.15a}$$
$$W^2_{\mu\nu} = \partial_\mu W^2_\nu - \partial_\nu W^2_\mu - g_2\left(W^3_\mu W^1_\nu - W^3_\nu W^1_\mu\right), \tag{11.15b}$$
$$W^3_{\mu\nu} = \partial_\mu W^3_\nu - \partial_\nu W^3_\mu - g_2\left(W^1_\mu W^2_\nu - W^1_\nu W^2_\mu\right). \tag{11.15c}$$

Since $\text{Tr}(\tau^i)^2 = 2$, and $\text{Tr}(\tau^i\tau^j) = 0, i \neq j$, we can use (11.14) to express the Lagrangian density in the more reassuring form:

$$\mathcal{L}_{dyn} = -\frac{1}{4}B_{\mu\nu}B^{\mu\nu} - \sum_{i=1}^{3}\frac{1}{4}W^i_{\mu\nu}W^{i\mu\nu}. \tag{11.16}$$

We shall see, later in this chapter, that the fields W^1_μ and W^2_μ are electrically charged, and it is convenient to define here the complex combinations

$$W^+_\mu = (W^1_\mu - iW^2_\mu)/\sqrt{2}, \quad W^-_\mu = (W^1_\mu + iW^2_\mu)/\sqrt{2}. \tag{11.17}$$

Note that the field W_μ^- is the complex conjugate of the field W_μ^+. We also define

$$\begin{aligned} W_{\mu\nu}^+ &= (W_{\mu\nu}^1 - \mathrm{i}W_{\mu\nu}^2)/\sqrt{2} \\ &= (\partial_\mu + \mathrm{i}g_2 W_\mu^3)W_\nu^+ - (\partial_\nu + \mathrm{i}g_2 W_\nu^3)W_\mu^+ \end{aligned} \tag{11.18}$$

using (11.15a) and (11.15b). $W_{\mu\nu}^-$ is defined similarly.

We can also write (11.15c) as

$$W_{\mu\nu}^3 = \partial_\mu W_\nu^3 - \partial_\nu W_\mu^3 - \mathrm{i}g_2(W_\mu^- W_\nu^+ - W_\nu^- W_\mu^+) \tag{11.19}$$

and (11.16) becomes

$$\mathcal{L}_{\text{dyn}} = -\frac{1}{4}B_{\mu\nu}B^{\mu\nu} - \frac{1}{4}W_{\mu\nu}^3 W^{3\mu\nu} - \frac{1}{2}W_{\mu\nu}^- W^{+\mu\nu}. \tag{11.20}$$

11.3 Breaking the $SU(2)$ symmetry

As in equation (10.2) we take $V(\Phi^\dagger\Phi)$ to be

$$\begin{aligned} V(\Phi^\dagger\Phi) &= \frac{m^2}{2\phi_0^2}\left[(\Phi^\dagger\Phi) - \phi_0^2\right]^2 \\ &= \frac{m^2}{2\phi_0^2}\left[\phi_1^2 + \phi_2^2 + \phi_3^2 + \phi_4^2 - \phi_0^2\right]^2 \end{aligned} \tag{11.21}$$

where ϕ_0 is a fixed parameter that is the analogue of (10.2). With this expression for V, the vacuum state of our system is degenerate in the four-dimensional space of the scalar fields. We now break the $SU(2)$ symmetry. At our disposal we have the three real parameters $\alpha^k(x)$ that specify an element of $SU(2)$. We use this freedom to adopt a gauge in which for any field configuration $\Phi_{\mathrm{A}} = 0$ (two conditions) and Φ_{B} is real (one condition). The ground state is then

$$\Phi_{\text{ground}} = \begin{pmatrix} 0 \\ \phi_0 \end{pmatrix}, \tag{11.22}$$

and excited states are of the form

$$\Phi = \begin{pmatrix} 0 \\ \phi_0 + h(x)/\sqrt{2} \end{pmatrix}, \tag{11.23}$$

where the field $h(x)$ is real.

A local $U(1)$ symmetry remains: the fields (11.23) are unchanged by a $U(1) \times SU(2)$ transformation of the form

$$\mathrm{e}^{-\mathrm{i}\theta/2}\begin{pmatrix} \mathrm{e}^{-\mathrm{i}\theta/2} & 0 \\ 0 & \mathrm{e}^{\mathrm{i}\theta/2} \end{pmatrix} = \begin{pmatrix} \mathrm{e}^{-\mathrm{i}\theta} & 0 \\ 0 & 1 \end{pmatrix}. \tag{11.24}$$

Such matrices give a 2×2 matrix representation of the group $U(1)$. This residual symmetry will turn out to be the $U(1)$ symmetry of electromagnetism.

We wish to express $\mathcal{L}_\Phi$ (equation (11.9)) in terms of the field $h(x)$. We have from (11.21)

$$V(\Phi^\dagger \Phi) = m^2 h^2 + \frac{m^2 h^3}{\sqrt{2}\phi_0} + \frac{m^2 h^4}{8\phi_0^2} = V(h),$$

and from (11.8a) and (11.7)

$$D^\mu \Phi = \begin{pmatrix} 0 \\ \partial^\mu h/\sqrt{2} \end{pmatrix} + \frac{\mathrm{i}g_1}{2}\begin{pmatrix} 0 \\ B^\mu(\phi_0 + h/\sqrt{2}) \end{pmatrix} + \frac{\mathrm{i}g_2}{2}\begin{pmatrix} \sqrt{2}W_\mu^+(\phi_0 + h/\sqrt{2}) \\ -W_\mu^3(\phi_0 + h/\sqrt{2}) \end{pmatrix}.$$

Multiplying $(D_\mu \Phi)^\dagger$ by $D^\mu \Phi$, we find

$$\begin{aligned}
\mathcal{L}_\Phi = {} & \frac{1}{2}\partial_\mu h \partial^\mu h + \frac{g_2^2}{2} W_\mu^- W^{+\mu}(\phi_0 + h/\sqrt{2})^2 \\
& + \left[\frac{g_2^2}{4} W_\mu^3 W^{3\mu} - \frac{g_1 g_2}{2} W_\mu^3 B^\mu + \frac{g_1^2}{4} B_\mu B^\mu\right](\phi_0 + h/\sqrt{2})^2 - V(h) \\
= {} & \frac{1}{2}\partial_\mu h \partial^\mu h + \frac{g_2^2}{2} W_\mu^- W^{+\mu}(\phi_0 + h/\sqrt{2})^2 \\
& + \frac{1}{4}(g_1^2 + g_2^2) Z_\mu Z^\mu (\phi_0 + h/\sqrt{2})^2 - V(h).
\end{aligned} \tag{11.25}$$

We have written

$$Z_\mu = W_\mu^3 \cos\theta_\mathrm{w} - B_\mu \sin\theta_\mathrm{w}, \tag{11.26}$$

where

$$\cos\theta_\mathrm{w} = \frac{g_2}{\left(g_1^2 + g_2^2\right)^{1/2}}, \quad \sin\theta_\mathrm{w} = \frac{g_1}{\left(g_1^2 + g_2^2\right)^{1/2}}. \tag{11.27}$$

θ_w is called the *Weinberg angle.*

Along with the field Z_μ, we define the orthogonal combination

$$A_\mu = W_\mu^3 \sin\theta_\mathrm{w} + B_\mu \cos\theta_\mathrm{w}. \tag{11.28}$$

Equations (11.26) and (11.28) correspond to a rotation of axes in (B_μ, W_μ^3) space. The rotation can be inverted to give

$$\begin{aligned}
B_\mu &= A_\mu \cos\theta_\mathrm{w} - Z_\mu \sin\theta_\mathrm{w}, \\
W_\mu^3 &= A_\mu \sin\theta_\mathrm{w} + Z_\mu \cos\theta_\mathrm{w}.
\end{aligned} \tag{11.29}$$

Substituting in (11.10) and (11.19) gives

$$\begin{aligned}
B_{\mu\nu} &= A_{\mu\nu} \cos\theta_\mathrm{w} - Z_{\mu\nu} \sin\theta_\mathrm{w}, \\
W_{\mu\nu}^3 &= A_{\mu\nu} \sin\theta_\mathrm{w} + Z_{\mu\nu} \cos\theta_\mathrm{w} - \mathrm{i}g_2\left(W_\mu^- W_\nu^+ - W_\nu^- W_\mu^+\right),
\end{aligned}$$

where

$$A_{\mu\nu} = \partial_\mu A_\nu - \partial_\nu A_\mu \qquad (A_{\mu\nu} \text{ is the } F_{\mu\nu} \text{ of Chapter 4})$$

and

$$Z_{\mu\nu} = \partial_\mu Z_\nu - \partial_\nu Z_\mu. \tag{11.30}$$

11.4 Identification of the fields

We are now in a position to rearrange the terms in the full Lagrangian density $\mathcal{L} = \mathcal{L}_\Phi + \mathcal{L}_{\text{dyn}}$ to reveal its physical content. In $\mathcal{L}_{\text{dyn}}$ (equation (11.20)) we use (11.29) and (11.30) to express the field B_μ and W_μ^3 in terms of the fields A_μ and Z_μ, and then we may write

$$\mathcal{L} = \mathcal{L}_1 + \mathcal{L}_2,$$

where

$$\begin{aligned}
\mathcal{L}_1 = {} & \frac{1}{2}\partial_\mu h \partial^\mu h - m^2 h^2 \\
& - \frac{1}{4} Z_{\mu\nu} Z^{\mu\nu} + \frac{1}{4}\phi_0^2 (g_1^2 + g_2^2) Z_\mu Z^\mu \\
& - \frac{1}{4} A_{\mu\nu} A^{\mu\nu} \\
& - \frac{1}{2}[(D_\mu W_\nu^+)^* - (D_\nu W_\mu^+)^*][D^\mu W^{+\nu} - D^\nu W^{+\mu}] + \frac{1}{2} g_2^2 \phi_0^2 W_\mu^- W^{+\mu},
\end{aligned} \tag{11.31}$$

and $D_\mu W_\nu^+ = (\partial_\mu + \mathrm{i} g_2 \sin\theta_w A_\mu) W_\nu^+$.

$\mathcal{L}_1$ is relatively simple: you will recognise it as the Lagrangian density for a free massive neutral scalar boson field *h(x)*, a free massive neutral vector boson field $Z_\mu(x)$, and a pair of massive charged vector boson fields $W_\mu^+(x)$ and $W_\mu^-(x)$, interacting with the electromagnetic field $A_\mu(x)$.

$\mathcal{L}_2$ is the sum of the remaining interaction terms. As the patient reader may verify,

$$\begin{aligned}
\mathcal{L}_2 = {} & \left(\frac{1}{4}h^2 + \frac{1}{\sqrt{2}} h\phi_0\right)\left(g_2^2 W_\mu^- W^{+\mu} + \frac{1}{2}(g_1^2 + g_2^2) Z_\mu Z^\mu\right) \\
& - \frac{m^2 h^3}{\sqrt{2}\phi_0} - \frac{m^2 h^4}{8\phi_0^2} + \frac{g_2^2}{4}(W_\mu^- W_\nu^+ - W_\nu^- W_\mu^+)(W^{-\mu} W^{+\nu} - W^{-\nu} W^{+\mu}) \\
& + \frac{\mathrm{i} g_2}{2}(A_{\mu\nu} \sin\theta_w + Z_{\mu\nu}\cos\theta_w)(W^{-\mu} W^{+\nu} - W^{-\nu} W^{+\mu}) \\
& - g_2^2 \cos^2\theta_w (Z_\mu Z^\mu W_\nu^- W^{+\nu} - Z_\mu Z^\nu W_\nu^- W^{+\mu})
\end{aligned}$$

$$+\frac{ig_2}{2}\cos\theta_w[(Z_\mu W_\nu^- - Z_\nu W_\mu^-)(D^\mu W^{+\nu} - D^\nu W^{+\mu})$$
$$-(Z_\mu W_\nu^+ - Z_\nu W_\mu^+)(D^\mu W^{+\nu})^* - (D^\nu W^{+\mu})^*)]. \quad (11.32)$$

Most of the $U(1) \times SU(2)$ symmetry with which we began has been lost on symmetry breaking. In particular, no trace of the original $SU(2)$ symmetry is to be seen in the interactions described by $\mathcal{L}_2$. Nevertheless it is precisely this complicated set of interactions that makes the theory renormalisable, as it would be if the symmetry were not broken.

We identify the three vector fields, W_μ^+, W_μ^-, Z_μ, with the mediators of the weak interaction, the W^+, W^-, Z particles, which, subsequent to the theory, were discovered experimentally. The masses are (Particle Data Group, 2004)

$$M_{\mathrm{w}} = 80.425 \pm 0.038 \text{ GeV}, \quad (11.33)$$
$$M_{\mathrm{z}} = 91.1876 \pm 0.0021\text{GeV}. \quad (11.34)$$

From (11.31) and Section 4.9, we identify

$$\phi_0 g_2/\sqrt{2} = M_{\mathrm{w}}, \quad (11.35)$$
$$\phi_0\left(g_1^2 + g_2^2\right)^{1/2}/\sqrt{2} = M_{\mathrm{z}}. \quad (11.36)$$

Then, from (11.27), and neglecting quantum corrections to the mass ratio,

$$\cos\theta_{\mathrm{w}} = M_{\mathrm{w}}/M_{\mathrm{z}} = 0.8810 \pm 0.0016. \quad (11.37a)$$

It is usual to quote the value of $\sin^2\theta_{\mathrm{w}}$, which will appear in later calculations. The estimate above would suggest

$$\sin^2\theta_{\mathrm{w}} = 0.23120 \pm 0.00015.$$

The uncertainty arises mainly from uncertainty in M_{w}. Other ways of estimating $\sin^2\theta_{\mathrm{w}}$ exist and the accepted value (in 1996) was

$$\sin^2\theta_{\mathrm{w}} = 0.2315 \pm 0.0004. \quad (11.37b)$$

We shall adopt this value in subsequent calculations.

The $W^\pm$ bosons are found experimentally to carry charge $\pm e$. In (11.31) the gauge derivative is

$$D_\mu W_\nu{}^+ = (\partial_\mu + ig_2\sin\theta_{\mathrm{w}} A_\mu)W_\nu{}^+,$$

so that from the coupling to the electromagnetic field A_μ and (11.27) we can identify

$$e = g_2\sin\theta_{\mathrm{w}} = g_1\cos\theta_{\mathrm{w}}. \quad (11.38)$$

The fields W_μ^1, W_μ^2, and Z_μ have free field expansions similar to (4.15) but with three polarisation states (see Section 4.9). As a quantum field W_μ^+ destroys W^+ bosons and creates W^- bosons; W_μ^- destroys W^- bosons and creates W^+ bosons.

There remains the scalar Higgs field $h(x)$. The vacuum state expectation value ϕ_0 of the Higgs field is, from (11.35),

$$\phi_0 = \frac{\sqrt{2}M_{\mathrm{W}}}{g_2} = \frac{\sqrt{2}M_{\mathrm{W}}\sin\theta_{\mathrm{W}}}{e} = 180\,\mathrm{GeV}. \tag{11.39}$$

The only parameter not fixed from experiment is the mass $M_{\mathrm{H}} = \sqrt{2}m$ of the Higgs boson. No Higgs boson has yet been identified experimentally, though its existence is, apparently, an essential part of the Standard Model. The failure so far of experimental searches to find the Higgs boson suggests $M_{\mathrm{H}} >$ 64 GeV. Recent experimental and theoretical studies suggest an M_{H} close to this limit.

The requirements of $U(1)$ and $SU(2)$ symmetry, followed by $SU(2)$ symmetry breaking, have generated the electromagnetic field, the massive vector $W^\pm$ and Z boson fields, and the scalar Higgs field, in a remarkably economical way. In the next chapter, we add lepton fermion fields to these boson fields, to obtain the richness of the Weinberg–Salam electroweak theory.

Problems

11.1 Show that the W'_μ defined by (11.6) are Hermitian and have zero trace. (Use the expression (B.9) of Appendix B: $\mathbf{U} = \cos\alpha\mathbf{I} + i\sin\alpha(\hat{\alpha}\cdot\tau)$.)

11.2 Verify that the expressions (11.13) and (11.16) for $\mathcal{L}_{\mathrm{dyn}}$ are equivalent.

11.3 Verify that the last two terms on the right-hand side of (11.31) correspond to a pair of massive charged vector boson fields.

11.4 Show that the Higgs boson can decay to two photons, in the third order of perturbation theory. Draw the appropriate Feynman graph.

11.5 Under an $SU(2)$ transformation, $\Phi \to \Phi'$ where

$$\begin{pmatrix} \Phi'_A \\ \Phi'_B \end{pmatrix} = \mathbf{U}\begin{pmatrix} \Phi_A \\ \Phi_B \end{pmatrix}.$$

Using (B.9), show that $\tau^2 U^* = U\tau^2$. Hence show that

$$\begin{pmatrix} \Phi'^*_{\mathrm{B}} \\ -\Phi'^*_{\mathrm{A}} \end{pmatrix} = \mathbf{U}\begin{pmatrix} \Phi^*_{\mathrm{B}} \\ -\Phi^*_{\mathrm{A}} \end{pmatrix}.$$

11.6 Show that the $SU(2)$ matrix $\mathbf{U} = e^{i\tau\alpha}$ with $\alpha = \alpha(\sin\phi, \cos\phi, 0)$ is

$$\mathbf{U} = \begin{pmatrix} \cos\alpha & e^{i\phi}\sin\alpha \\ -e^{-i\phi}\sin\alpha & \cos\alpha \end{pmatrix}.$$

Show that under the $SU(2)$ transformation $\Phi' = \mathbf{U}\Phi$, the two-component complex field

$$\Phi = \begin{pmatrix} \Phi_A \\ \Phi_B \end{pmatrix} = \begin{pmatrix} a e^{i\delta} \\ b e^{i\gamma} \end{pmatrix}$$

can be put in the form

$$\Phi' = \begin{pmatrix} \Phi'_A \\ \Phi'_B \end{pmatrix} = \begin{pmatrix} 0 \\ e^{i\gamma}\sqrt{a^2+b^2} \end{pmatrix},$$

taking $\phi = (\delta - \gamma)$ and $\alpha = -\tan^{-1}(a/b)$. Show that Φ' can then be put in the standard form (11.23) by a further $SU(2)$ transformation with $\alpha = \gamma(0, 0, 1)$.

12

The Weinberg–Salam electroweak theory for leptons

We shall now couple the lepton fields to all the gauge boson fields: the electromagnetic field, the W^+ and W^- fields, and the Z field. We know that at low energies the theory must reproduce the phenomenology of Chapter 9. This consideration and the principles of $U(1) \times SU(2)$ local gauge symmetry determine the couplings uniquely.

We have seen how the Higgs mechanism gives mass to the $\mathrm{W}^{\pm}$ and Z bosons. To give mass to the charged leptons: the electron, the muon, the tau, they too must be coupled to the Higgs field. We shall finally arrive at the Weinberg–Salam unified theory of the electroweak interaction.

12.1 Lepton doublets and the Weinberg–Salam theory

We shall first construct a Lagrangian density for lepton fields that is invariant under $U(1)$ and $SU(2)$ transformations. The left-handed electron spinor e_{L} and the electron neutrino spinor ν_{eL} are put together in an $SU(2)$ doublet, like the Higgs fields in equation (11.1),

$$\mathbf{L} = \begin{pmatrix} \nu_{\mathrm{eL}} \\ e_{\mathrm{L}} \end{pmatrix} = \begin{pmatrix} \mathrm{L_A} \\ \mathrm{L_B} \end{pmatrix}. \tag{12.1}$$

We are now again specialising our notation; two-component left-handed and right-handed spinors were denoted by ψ_{L} and ψ_{R}, respectively, in Chapter 6. Under an $SU(2)$ transformation, this doublet transforms in exactly the same way as the Higgs doublet:

$$\mathbf{L} \to \mathbf{L}' = \mathbf{UL}. \tag{12.2}$$

Since $SU(2)$ transformations mix the two spinor fields making up the doublet, to maintain Lorentz invariance only fields with the same Lorentz transformation properties can be combined together into a doublet.

From the phenomenology of Chapter 9 the right-handed lepton fields do not couple to the W boson field so that e_R and ν_{eR} are invariant under $SU(2)$ transformations:

$$e_R \to e'_R = e_R. \quad \nu_{eR} \to \nu'_{eR} = \nu_{eR}. \tag{12.3}$$

To be consistent with the transformation rule (12.2), all $SU(2)$ gauge derivatives must be of the same form, $\partial_\mu + i(g_2/2)\mathbf{W}_\mu$, where $g_2 \sin\theta_w = e$, as in (11.8) and (11.38). This is a consequence of the non-Abelian nature of the group $SU(2)$. However, there is no similar constraint on the coupling constant to the $U(1)$ gauge field B_μ. (See Problem 12.1.) We may take

$$D_\mu \mathbf{L} = [\partial_\mu + i(g_2/2)\mathbf{W}_\mu + i(g'/2)B_\mu)]\mathbf{L}, \tag{12.4}$$

where g' remains at our disposal. We must choose g' so that the neutrino is neutral and the electron has charge $-e$. The terms in $D_\mu \mathbf{L}$ which couple to the electromagnetic field A_μ are linear combinations of W_μ^3 and B_μ. Using (11.7) and (11.29) the terms in A_μ are

$$\begin{pmatrix} \partial_\mu + \{i(g_2/2)\sin\theta_w + i(g'/2)\cos\theta_w\}A_\mu, & 0 \\ 0, & \partial_\mu + \{-i(g_2/2)\sin\theta_w + i(g'/2)\cos\theta_w\}A_\mu \end{pmatrix} \begin{pmatrix} \nu_{eL} \\ e_L \end{pmatrix}.$$

The gauge derivatives $\partial_\mu \nu_{eL}$ and $(\partial_\mu - ieA_\mu)e_L$ which leave the neutrino electrically neutral but impart electric charge $-e$ to e_L, are obtained with the choice

$$g' \cos\theta_w = -g_2 \sin\theta_w = -e.$$

The complete gauge derivative of the left-handed fields is then

$$D_\mu \mathbf{L} = \begin{pmatrix} \partial_\mu + i(e/\sin 2\theta_w)Z_\mu, & i\{e/(\sqrt{2}\sin\theta_w)\}W_\mu^+ \\ i\{e/(\sqrt{2}\sin\theta_w)\}W_\mu^-, & \partial_\mu - ieA_\mu - ie\cot(2\theta_w)Z_\mu \end{pmatrix} \begin{pmatrix} \nu_{eL} \\ e_L \end{pmatrix} \tag{12.5}$$

where we have used (11.7), (11.17) and (11.29).

The gauge derivative of e_R must be of the form

$$D_\mu e_R = [\partial_\mu + i(g''/2)B_\mu]e_R. \tag{12.6a}$$

Since the electron has charge $-e$ we take $g'' = -2e/\cos\theta_w = -2g_1$, (see (11.38)) so that, using (11.29) again,

$$D_\mu e_R = [(\partial_\mu - ieA_\mu) + ie\tan\theta_w Z_\mu]e_R. \tag{12.6b}$$

With $g'' = -2g_1$ and $g' = -g_1$, it can easily be checked that, under a local $U(1) \times SU(2)$ transformation

$$\mathbf{L} \to \mathbf{L}' = e^{i\theta(x)}\mathbf{U}(x)\mathbf{L},$$
$$e_R \to e'_R = e^{2i\theta(x)}e_R,$$

the gauge derivatives satisfy

$$D_\mu{}'\mathrm{L}' = (\partial_\mu + \mathrm{i}(g_2/2)\mathbf{W}_\mu{}' + \mathrm{i}(g'/2)B_\mu{}')\mathbf{L}' = \mathrm{e}^{\mathrm{i}\theta}\mathbf{U}D_\mu\mathbf{L}$$
$$D_\mu{}'e_\mathrm{R}{}' = (\partial_\mu + \mathrm{i}(g''/2)B_\mu{}')e_R{}' = \mathrm{e}^{2\mathrm{i}\theta}\mathrm{D}_\mu e_\mathrm{R},$$

where the fields B_μ and $\mathbf{W}_\mu$ transform as in (11.4b) and (11.6).

We can now construct a gauge invariant and Lorentz invariant expression for the dynamical part of the Lagrangian density for the electron and the electron neutrino:

$$\mathcal{L}^e_{\mathrm{dyn}} = \mathbf{L}^\dagger\tilde{\sigma}^\mu \mathrm{i}D_\mu\mathbf{L} + e^\dagger_\mathrm{R}\sigma^\mu \mathrm{i}D_\mu e_\mathrm{R} + \nu^+_{\mathrm{eR}}\sigma^\mu \mathrm{i}\partial_\mu\nu_{e\mathrm{R}}. \tag{12.7}$$

The gauge invariance follows from our construction of the gauge derivatives, and the Lorentz invariance from the spinor properties set out in Section 5.4. (Remember that the $\tilde{\sigma}_\mu$ matrices act on the spinor indices, whereas the $SU(2)$ transformation acts independently on the components of the doublet of spinor fields.) Note that besides the interaction with the electromagnetic field we have fully determined, from the factor $D_\mu\mathbf{L}$, all the interactions with the heavy vector bosons.

Finally, we must give mass to the charged leptons. A gauge and Lorentz invariant contribution to the Lagrangian density that will impart mass to the electron but leave the neutrino massless is (neutrino mass will be introduced in Chapter 19)

$$\begin{aligned}\mathcal{L}^e_{\mathrm{mass}} &= -c_\mathrm{e}[(\mathbf{L}^\dagger\Phi)e_\mathrm{R} + e^\dagger_\mathrm{R}(\Phi^\dagger\mathbf{L})] \\ &= -c_\mathrm{e}[(\nu^\dagger_\mathrm{L}\Phi_\mathrm{A} + e^\dagger_\mathrm{L}\Phi_\mathrm{B})e_\mathrm{R} + e^\dagger_\mathrm{R}(\Phi^\dagger_\mathrm{A}\nu_\mathrm{L} + \Phi^\dagger_\mathrm{B}e_\mathrm{L})],\end{aligned} \tag{12.8}$$

where Φ is the Higgs doublet field and c_e is a dimensionless coupling constant. After symmetry breaking (see (11.23)), $\mathcal{L}^e_{\mathrm{mass}}$ becomes

$$\mathcal{L}^e_{\mathrm{mass}} = -c_\mathrm{e}\phi_0(e^\dagger_\mathrm{L}e_R + e^\dagger_\mathrm{R}e_\mathrm{L}) - \frac{c_\mathrm{e}h}{\sqrt{2}}\left(e^\dagger_\mathrm{L}e_\mathrm{R} + e^\dagger_\mathrm{R}e_\mathrm{L}\right). \tag{12.9}$$

Comparing this with the Dirac Lagrangian density (5.12), we identify $c_\mathrm{e}\phi_0$ with the electron mass m_e. Introducing mass by following the principles of symmetry has left us no option but to introduce an interaction between the electron field and the Higgs field $h(x)$. Hence the coupling constant to the Higgs field is

$$\frac{c_\mathrm{e}}{\sqrt{2}} = \frac{m_\mathrm{e}}{\sqrt{2}\phi_0} = 2.01 \times 10^{-6} \tag{12.10}$$

(using (11.39)). It is just as well that c_e is small: we do not want this term to upset the calculations of QED!

The total Lagrangian density $\mathcal{L}^e$ for the electron and its neutrino is given by (12.7) and (12.8):

$$\mathcal{L}^e = \mathcal{L}^e_{\mathrm{dyn}} + \mathcal{L}^e_{\mathrm{mass}}. \tag{12.11}$$

From $\mathcal{L}^e$ we can pick out the terms

$$\begin{aligned}\mathcal{L}^e_{\text{Dirac}} &= \nu^\dagger_{\text{eL}}\tilde{\sigma}^\mu \text{i}(\partial_\mu \nu_{\text{eL}}) + e^\dagger_{\text{L}}\tilde{\sigma}^\mu \text{i}(\partial_\mu - \text{i}eA_\mu)e_{\text{L}} + \nu^\dagger_{\text{eR}}\sigma^\mu \text{i}\partial_\mu \nu_{\text{eR}} \\ &\quad + e^\dagger_{\text{R}}\sigma^\mu \text{i}(\partial_\mu - \text{i}eA_\mu)e_{\text{R}} - m_{\text{e}}\left(e^\dagger_{\text{L}}e_{\text{R}} + e^\dagger_{\text{R}}e_{\text{L}}\right),\end{aligned} \tag{12.12}$$

which correspond to the expressions we found in Chapter 6 and Chapter 7 for a Dirac massless neutrino, and a Dirac electron of mass m_{e} and charge $-e$ in an electromagnetic field.

The Lagrangian densities $\mathcal{L}^\mu$ and $\mathcal{L}^\tau$ for the muon and tau leptons and their neutrinos differ from (12.11) only in their mass parameters and, hence, their couplings to the Higgs field:

$$\frac{c_\mu}{\sqrt{2}} = \frac{m_\mu}{\sqrt{2}\phi_0} = 4.15 \times 10^{-4}, \quad \frac{c_\tau}{\sqrt{2}} = \frac{m_\tau}{\sqrt{2}\phi_0} = 6.98 \times 10^{-3}. \tag{12.13}$$

The coupling constant g_2 of the $SU(2)$ gauge theory, or, equivalently, the Weinberg angle θ_{w} (see (11.38)), which determines the coupling to the $W^\pm$ and Z fields, must be the same for all leptons, a feature of the theory that is forced on us by the $SU(2)$ group, and that is known as *lepton universality.*

The complete Lagrangian density $\mathcal{L}^{\text{ws}}$ of the Weinberg–Salam theory (Weinberg, 1967; Salam, 1968) is the sum of the lepton contributions, and the boson contributions given by (11.31) and (11.32):

$$\mathcal{L}^{\text{ws}} = \mathcal{L}^{\text{e}} + \mathcal{L}^{\mu} + \mathcal{L}^{\tau} + \mathcal{L}^{\text{bosons}}, \tag{12.14}$$

The form of $\mathcal{L}^{\text{ws}}$ has been determined by considerations of symmetry: invariance under Lorentz transformations, and under $U(1)$ and $SU(2)$ transformations. Massive bosons and leptons appear through the Higgs mechanism of local symmetry breaking. It has been proved by t'Hooft (1976), who introduced radically new methods of analysis, that the theory is renormalisable. We shall see in Chapter 13 that there is a great body of data that supports it.

12.2 Lepton coupling to the $W^\pm$

The coupling of the electron and the electron neutrino to the W^+ and W^- gauge fields is given by the appropriate terms in (12.5) and (12.7), which are

$$\begin{aligned}\mathcal{L}_{\text{ew}} &= -\left(g_2/\sqrt{2}\right)\nu^\dagger_{\text{eL}}\tilde{\sigma}^\mu e_{\text{L}} W^+_\mu - \left(g_2/\sqrt{2}\right)e^\dagger_{\text{L}}\tilde{\sigma}^\mu \nu_{\text{eL}} W^-_\mu \\ &= -\left(g_2/\sqrt{2}\right)[j^{\mu\dagger}_{\text{e}} W^+_\mu + j^\mu_{\text{e}} W^-_\mu].\end{aligned} \tag{12.15}$$

The right-handed fields do not contribute to this interaction. As in Chapter 9 the currents are defined as

$$j^\mu_{\text{e}} = e^\dagger_{\text{L}}\tilde{\sigma}^\mu \nu_{\text{eL}}, \quad j^{\mu\dagger}_{\text{e}} = \nu^\dagger_{\text{eL}}\tilde{\sigma}^\mu e_{\text{L}}. \tag{12.16}$$

There are similar muon and tau currents, giving a total lepton current

$$j^{\mu} = \left(e_{\rm L}^{\dagger}\tilde{\sigma}^{\mu}\nu_{e\rm L} + \mu_{\rm L}^{\dagger}\tilde{\sigma}^{\mu}\nu_{\mu\rm L} + \tau_{\rm L}^{\dagger}\tilde{\sigma}^{\mu}\nu_{\tau\rm L}\right), \tag{12.17}$$

and total interaction Lagrangian density

$$\mathcal{L}_{lW} = -(g_2/\sqrt{2})\left[j^{\mu\dagger}W_{\mu}^{+} + j^{\mu}W_{\mu}^{-}\right]. \tag{12.18}$$

The effective $\mathcal{L}_{\rm lepton}$ used in the discussion of muon decay in Section 9.4 can be obtained as the low energy limit of the Weinberg–Salam theory. Since the mass $M_{\rm w}$ is so large, at low energies the term $M_{\rm w}^2 W_{\mu}^{-}W^{+\mu}$ in (11.31) dominates in the W contribution to the Lagrangian density, and

$$\mathcal{L}_{\rm w} \approx M_{\rm w}^2 W_{\mu}^{-}W^{+\mu} - \left(g_2/\sqrt{2}\right)[j^{\mu\dagger}W_{\mu}^{\dagger} + j^{\mu}W_{\mu}^{-}]. \tag{12.19}$$

Physical field configurations correspond to stationary values of the action. Varying W_{μ}^{+} and W_{μ}^{-} independently gives the field equations

$$M_{\rm w}^2 W_{\mu}^{-} = \left(g_2/\sqrt{2}\right)j_{\mu}^{\dagger}, \quad M_{\rm w}^2 W_{\mu}^{+} = \left(g_2/\sqrt{2}\right)j_{\mu}, \tag{12.20}$$

and using these in (12.19) gives

$$\mathcal{L}_{w} \approx -\frac{1}{2}g_2^2 M_{\rm w}^{-2} j_{\mu}^{\dagger}j^{\mu}. \tag{12.21}$$

$\mathcal{L}_{\rm w}$ is equivalent to the effective $\mathcal{L}_{\rm lepton}$ of (9.8) if we make the identification

$$G_{\rm F} = \frac{g_2^2}{4\sqrt{2}M_{\rm w}^2} = \frac{e^2}{4\sqrt{2}M_{\rm w}^2\sin^2\theta_{\rm w}}. \tag{12.22}$$

Taking $M_{\rm w} = 80.33\,$Gev, $M_{\rm z} = 91.187\,$GeV, $\sin^2\theta_{\rm w} = 1 - M_{\rm w}^2/M_z^2$, gives $G_{\rm F} = 1.12\times 10^{-5}\,\text{GeV}^{-2}$, which is in good agreement with the accepted experimental value, $1.166\times 10^{-5}\,\text{GeV}^{-2}$. Historically, the knowledge of $G_{\rm F}$, together with an estimate of $\theta_{\rm w}$ (see Section 13.1) was used to predict the masses of the $\rm W^{\pm}$ and Z bosons, and the CERN proton–antiproton collider was then built to find them.

12.3 Lepton coupling to the Z

The coupling of the leptons to the Z field can be extracted from the terms involving Z_{μ} in (12.7):

$$\begin{aligned}\mathcal{L}_{e\rm Z} = &-\nu_{e\rm L}^{\dagger}\tilde{\sigma}^{\mu}\nu_{e\rm L}\left(\frac{e}{\sin(2\theta_{\rm w})}\right)Z_{\mu} + e_{\rm L}^{\dagger}\tilde{\sigma}^{\mu}e_{\rm L}\left(\frac{e\cos(2\theta_{\rm w})}{\sin(2\theta_{\rm w})}\right)Z_{\mu}\\ &-e_{\rm R}^{\dagger}\sigma^{\mu}e_{\rm R}(e\tan\theta_{\rm w})Z_{\mu} \quad \text{(using (12.5) and (12.6b))}\\ = &\frac{-e}{\sin(2\theta_{\rm w})}(j_{\rm neutral})_{\mu}Z^{\mu},\end{aligned}$$

where

$$(j_{\text{neutral}})^{\mu} = \nu_{\text{eL}}^{\dagger}\tilde{\sigma}^{\mu}\nu_{\text{eL}} - \cos(2\theta_{\text{w}})e_{\text{L}}^{\dagger}\tilde{\sigma}^{\mu}e_{\text{L}} + 2\sin^2\theta_{\text{w}}e_{\text{R}}^{\dagger}\sigma^{\mu}e_{\text{R}}. \tag{12.23}$$

There are similar expressions for $\mathcal{L}_{\mu\text{z}}$ and $\mathcal{L}_{\tau\text{z}}$. Note that the right-handed charged lepton fields also couple to the Z field but not the right-handed neutrino.

The low energy limit of $\mathcal{L}_{\text{z}}$ may be obtained in the same way as we obtained the low energy limit $\mathcal{L}_{\text{w}}$ in Section 12.2, with the same identification of coupling constants, and is identical with the effective Lagrangian density (9.15) if, comparing (12.23) with (9.17),

$$c_{\text{A}} = -\frac{1}{2}, \quad c_{\text{V}} = -\frac{1}{2} + 2\sin^2\theta_{\text{w}}. \tag{12.24}$$

The low energy muon neutrino–electron elastic scattering cross-sections calculated from the effective Lagrangian density are

$$\sigma(\nu_{\mu} + \text{e}^- \to \nu_{\mu} + \text{e}^-) = \frac{G_{\text{F}}^2 s}{\pi}\left[\frac{4}{3}\sin^4\theta_{\text{w}} - \sin^2\theta_{\text{w}} + \frac{1}{4}\right], \tag{12.25}$$

$$\sigma(\bar{\nu}_{\mu} + \text{e}^- \to \bar{\nu}_{\mu} + \text{e}^-) = \frac{G_{\text{F}}^2 s}{\pi}\left[\frac{4}{3}\sin^4\theta_{\text{w}} - \frac{1}{3}\sin^2\theta_{\text{w}} + \frac{1}{12}\right], \tag{12.26}$$

where s is the square of the centre of mass energy and $E_{\nu} \gg m_{\text{e}}$ (see Perkins, 1987, p. 327).

These low energy ($\ll M_{\text{Z}}, M_{\text{W}}$) cross-sections have been measured at CERN (CHARM II Collaboration, 1994), and their ratio yields an estimate for $\sin^2\theta_{\text{w}} = 0.2324 \pm 0.0083$.

The Fermi constant G_{F} is also known experimentally from low energy phenomena, and e is of course well known. Hence within the framework of the Weinberg–Salam theory the masses of the Z and $\text{W}^{\pm}$ gauge bosons can be estimated from low energy data alone, using (12.22) and (11.37). (Earlier estimates of $\sin^2\theta_{\text{w}}$ came from neutrino–nuclear scattering.)

12.4 Conservation of lepton number and conservation of charge

The Weinberg–Salam Lagrangian density $\mathcal{L}^{\text{WS}}$ has also further independent global $U(1)$ symmetries. It is invariant under the $U(1)$ transformation $\mathbf{L}_{\text{e}} \to \text{e}^{\text{i}\alpha}\mathbf{L}_{\text{e}}$, $e_{\text{R}} \to \text{e}^{\text{i}\alpha}e_{\text{R}}$, where α is a constant phase (see (12.7) and (12.9)). Using the device (by now familiar) of varying α so that $\alpha \to \alpha + \delta\alpha(x)$, where $\delta\alpha$ is space and time dependent, the first-order variation in the action comes from the dynamical part of

$\mathcal{L}^e_{\rm dyn}$ (equation (12.7)), and is

$$\begin{aligned}\delta S &= -\int \mathbf{L}^\dagger \tilde{\sigma}^\mu \mathbf{L} \partial_\mu(\delta\alpha)\, \mathrm{d}^4x - \int e_{\rm R}^\dagger \sigma^\mu e_{\rm R} \partial_\mu(\delta\alpha)\, \mathrm{d}^4x \\ &= \int \left[\partial_\mu(\mathbf{L}^\dagger \tilde{\sigma}^\mu \mathbf{L}) + \partial_\mu\left(e_{\rm R}^\dagger \sigma^\mu e_{\rm R}\right)\right](\delta\alpha)\, \mathrm{d}^4x,\end{aligned}$$

on integrating by parts. Setting $\delta S = 0$ for arbitrary $\delta\alpha$ yields

$$\partial_\mu\left(\nu_L^\dagger \tilde{\sigma}^\mu \nu_{\rm L} + e_{\rm L}^\dagger \tilde{\sigma}^\mu e_{\rm L}\right) + \partial_\mu\left(e_{\rm R}^+ \sigma^\mu e_{\rm R}\right) = 0,$$

or

$$\partial_\mu\left(J_{\rm e}^\mu\right) = 0, \tag{12.27}$$

where

$$\begin{aligned} J_{\rm e}^0 &= \nu_{\rm L}^\dagger \nu_{\rm L} + e_{\rm L}^\dagger e_{\rm L} + e_{\rm R}^\dagger e_{\rm R}, \\ J_{\rm e}^i &= \nu_{\rm L}^\dagger \tilde{\sigma}^i \nu_{\rm L} + e_{\rm L}^\dagger \tilde{\sigma}^i e_{\rm L} + e_{\rm R}^\dagger \sigma^i e_{\rm R}. \end{aligned} \tag{12.28}$$

Equation (12.28), which we may write as

$$\frac{\partial J_e^0}{\partial t} + \nabla \cdot \mathbf{J}_{\rm e} = 0, \tag{12.29}$$

expresses the conservation of electron lepton number. Similar $U(1)$ transformations applied to the muon and tau parts of $\mathcal{L}_{\rm ws}$ give the conservation of muon lepton number, and tau lepton number. We will see in Chapter 19 that the inclusion of Dirac neutrino mass into the Standard Model reduces these three conservation laws to one.

As in Chapters 4 and 5, the inhomogeneous Maxwell equations can be obtained by varying A_μ. There are contributions to the electric current from the charged $W^\pm$ fields, as well as from the charged leptons. Conservation of charge follows from Maxwell's equations, but can be obtained more directly from the $U(1)$ symmetry apparent in each term of the Weinberg–Salam Lagrangian density (12.14):

$$\begin{aligned} &e_{\rm L} \to \mathrm{e}^{\mathrm{i}\alpha} e_{\rm L},\ e_{\rm R} \to \mathrm{e}^{\mathrm{i}\alpha} e_{\rm R};\ \mu_{\rm L} \to e^{\mathrm{i}\alpha} \mu_{\rm L},\ \mu_{\rm R} \to \mathrm{e}^{\mathrm{i}\alpha} \mu_R;\ \tau_{\rm L} \\ &\to \mathrm{e}^{\mathrm{i}\alpha} \tau_{\rm L},\ \tau_{\rm R} \to \mathrm{e}^{\mathrm{i}\alpha} \tau_{\rm R};\ W_\mu^+ \to \mathrm{e}^{-\mathrm{i}\alpha} W_\mu^+,\ W_\mu^- \to \mathrm{e}^{\mathrm{i}\alpha} W_\mu^-. \end{aligned} \tag{12.30}$$

12.5 *CP* symmetry

We saw in Chapter 5 (equation (5.27)) that under space inversion a left-handed spinor $\psi_{\rm L}$ transforms into a right-handed spinor $\psi_{\rm R}$, and *vice versa.* The Weinberg–Salam Lagrangian does not have space inversion symmetry, since only the left-hand components of the lepton wave functions are coupled to the $SU(2)$ gauge field $\mathbf{W}_\mu$.

We also discussed in Chapter 7 the operation of charge conjugation,

$$\psi_{\mathrm{L}}^{C} = -\mathrm{i}\sigma^2\psi_{\mathrm{R}}^*, \quad \psi_{\mathrm{R}}^{C} = \mathrm{i}\sigma^2\psi_{\mathrm{L}}^*,$$

which relates solutions of the Dirac equation for particles to solutions for antiparticles. In the Weinberg–Salam theory there is no charge symmetry.

The Weinberg–Salam Lagrangian does exhibit a symmetry under the combined *CP* (charge conjugation, parity) operation. This symmetry implies that the physics of particles described in a right-handed coordinate system is the same as the physics of antiparticles described in a left-handed coordinate system.

Under the combined *CP* operation, lepton fields transform according to

$$\psi_{\mathrm{L}}^{CP} = -\mathrm{i}\sigma^2\psi_{\mathrm{L}}^*, \quad \psi_{\mathrm{R}}^{CP} = \mathrm{i}\sigma^2\psi_{\mathrm{R}}^*. \tag{12.31}$$

The other fields in the electroweak theory transform as set out below:

Higgs field: $\begin{pmatrix}\Phi_{\mathrm{A}}^{CP}\\ \Phi_{\mathrm{B}}^{CP}\end{pmatrix} = \begin{pmatrix}\Phi_{\mathrm{A}}^{*}\\ \Phi_{\mathrm{B}}^{*}\end{pmatrix}.$

$U(1)$ gauge fields: $B_0^{CP} = -B_0$, $B_i^{CP} = B_i$.

$SU(2)$ gauge fields:

$$\begin{pmatrix} W_0^3 & W_0^1 - \mathrm{i}W_0^2 \\ W_0^1 + \mathrm{i}W_0^2 & -W_0^3 \end{pmatrix}^{CP} = -\begin{pmatrix} W_0^3 & W_0^1 + \mathrm{i}W_0^2 \\ W_0^1 - \mathrm{i}W_0^2 & -W_0^3 \end{pmatrix},$$

$$\begin{pmatrix} W_i^3 & W_i^1 - \mathrm{i}W_i^2 \\ W_i^1 + \mathrm{i}W_i^2 & -W_i^3 \end{pmatrix}^{CP} = \begin{pmatrix} W_i^3 & W_i^1 + \mathrm{i}W_i^2 \\ W_i^1 - \mathrm{i}W_i^2 & -W_i^3 \end{pmatrix}.$$

It follows that

$$\begin{aligned} W_0^{+CP} &= -W_0^-, & W_i^{+CP} &= W_i^-, \\ Z_0^{CP} &= -Z_0, & Z_i^{CP} &= Z_i, \\ A_0^{CP} &= -A_0, & A_i^{CP} &= A_i. \end{aligned} \tag{12.32}$$

Space derivatives of fields are replaced by their negatives.

To show that the Lagrangian density is invariant under these transformations requires some care. We demonstrate it here for just one term, but one which involves all the necessary steps in the complete argument, and we leave the remaining terms to the reader. Consider then the term from the expression (12.7)

$$e_{\mathrm{R}}^{\dagger}\sigma^{\mu}\mathrm{i}[\partial_\mu + \mathrm{i}(g''/2)B_\mu]e_{\mathrm{R}} = l, \text{ say.}$$

Replacing the fields by their CP transforms, and ∂_i by $-\partial_i$, gives

$$l^{CP} = e_{\mathrm{R}}^{\mathrm{T}}(\sigma^{\mu})^{\mathrm{T}}\mathrm{i}[\partial_{\mu} - \mathrm{i}(g''/2)B_{\mu}]e_{\mathrm{R}}^{*},$$

where we have used the results

$$(\sigma^2)^2 = 1, \quad \sigma^2\sigma^i\sigma^2 = -(\sigma^i)^{\mathrm{T}}.$$

The operators ∂_{μ} now act on the conjugate fields. In fact l^{CP} is not identical to l, but differs from it only by a sum of total derivatives and, as explained in Section 3.1, a total derivative is of no consequence. If we add to l^{CP} the terms $-\mathrm{i}\partial_{\mu}[e_{\mathrm{R}}^{\mathrm{T}}(\sigma^{\mu})^{\mathrm{T}}e_{\mathrm{R}}^{*}]$ we obtain

$$-\mathrm{i}\left(\partial_{\mu}e_{\mathrm{R}}^{\mathrm{T}}\right)(\sigma^{\mu})^{\mathrm{T}}e_{\mathrm{R}}^{*} + \left(g''/2\right)B_{\mu}e_{\mathrm{R}}^{\mathrm{T}}(\sigma^{\mu})^{\mathrm{T}}e_{\mathrm{R}}^{*}.$$

Transposing this expression introduces another minus sign, since e_{R} and $e_{\mathrm{R}}{}^{\dagger}$ are fermion fields and hence anticommute. We then recover l.

12.6 Mass terms in $\mathcal{L}$: an attempted generalisation

For later use, when the theory is extended to quarks, we finish this chapter by contemplating a possible generalisation of our Lagrangian density. The coupling of the three lepton families to the Higgs field was taken to be

$$\mathcal{L}_{\mathrm{mass}} = -\sum_{i=1}^{3} c_i\left[\left(\mathbf{L}_i^{\dagger}\Phi\right)r_i + r_i^{\dagger}\left(\Phi^{\dagger}\mathbf{L}_i\right)\right],$$

where the sum is over the three lepton families, and we have modified the notation of (12.8) in an obvious way. We might have taken a more general coupling,

$$\mathcal{L}_{\mathrm{mass}}^{\mathrm{gen}} = -\sum\left[G_{ij}\left(\mathbf{L}_i^{\dagger}\Phi\right)r_j + G_{ij}^{*}r_j^{\dagger}\left(\Phi^{\dagger}\mathbf{L}_i\right)\right].$$

This preserves the $U(1) \times SU(2)$ symmetry with G_{ij} any 3×3 complex matrix.

We wish to show that this form has no essential difference from that already introduced. This is because an arbitrary complex matrix can always be put into real diagonal form with the help of two unitary matrices, $\mathbf{U}_{\mathrm{L}}$ and $\mathbf{U}_{\mathrm{R}}$ (Appendix A):

$$\mathbf{G} = \mathbf{U}_{\mathrm{L}}{}^{\dagger}\mathbf{C}\mathbf{U}_{\mathrm{R}},$$

with $C_{ij} = 0$ for $i \neq j$.

$\mathbf{U}_{\mathrm{L}}$ and $\mathbf{U}_{\mathrm{R}}$ are in general unique, except that both may be multiplied on the left by the same 'phase factor' matrix

$$\begin{pmatrix} e^{i\alpha_1} & 0 & 0 \\ 0 & e^{i\alpha_2} & 0 \\ 0 & 0 & e^{i\alpha_3} \end{pmatrix}.$$

If we define $r_i{}' = U_{\mathrm{R}ij} r_j$, $L_i{}' = U_{\mathrm{L}ij} L_j$ we recover the original form for the coupling to the Higgs field. Since the dynamical terms in the Lagrangian density are of the same form after these unitary transformations (Problem 12.5), $\mathcal{L}^{\mathrm{gen}}_{\mathrm{mass}}$ is just a more complicated expression of the same physics. The three phase factors $\exp(\mathrm{i}\alpha_k)$ correspond to the three $U(1)$ symmetries which lead to electron, muon, and tau number conservation.

Problems

12.1 Set the fields W_μ to be zero, and consider the dynamical Lagrangian density

$$\mathcal{L}_1 = \mathbf{L}^\dagger \tilde{\sigma}^\mu \mathrm{i} \left(\partial_\mu + \mathrm{i}(g'/2)\, B_\mu\right) \mathbf{L}.$$

With the gauge transformation (11.4b),

$$B_\mu \to B_\mu{}' = B_\mu + (2/g_1)\, \partial_\mu \theta,$$

show that $\mathcal{L}_1$ is invariant if $\mathbf{L}$ transforms as

$$\mathbf{L} \to \mathbf{L}' = \exp[-\mathrm{i}(g'/g_1)\theta]\mathbf{L}.$$

Now set the fields B_μ to be zero, and consider

$$\mathcal{L}_2 = \mathbf{L}^\dagger \tilde{\sigma}^\mu \mathrm{i}(\partial_\mu + \mathrm{i}(g'/2)\mathbf{W}_\mu)\mathbf{L}.$$

With the gauge transformation (11.6),

$$\mathbf{W}_\mu \to \mathbf{W}_\mu{}' = \mathbf{U}\mathbf{W}_\mu\mathbf{U}^\dagger + (2\mathrm{i}/g_2)(\partial_\mu \mathbf{U})\mathbf{U}^\dagger,$$

show that $\mathcal{L}_2$, can be made invariant only if

$$\mathbf{L} \to \mathbf{L}' = \mathbf{U}\mathbf{L} \quad \text{and} \quad g' = g_2.$$

12.2 Show that, to conform with the mathematical structure of Chapter 11, if two fields are to be put together in an $SU(2)$ doublet then they must differ by e in electric charge.

12.3 Inspection of (12.9) shows that the Higgs boson can decay into an e^+e^- pair. Show that, in the rest frame of the Higgs particle, the electron and positron must have equal and opposite momenta and the same helicity (i.e. both positive or both negative).

Show that the final density of momentum states for the decay is

$$\rho(E_f) = \frac{V}{(2\pi)^2} p_e E_e,$$

where p_e and E_e are the momentum and energy of the electron.

Calculate the matrix elements for the transition, and hence show that to lowest order in perturbation theory,

$$\text{total decay rate} = \frac{c_e^2}{16\pi} m_H \left(\frac{v_e}{c}\right)^3,$$

where v_e is the electron velocity.

12.4 Show that the ratio of the leptonic partial width of the Higgs particle to its mass is approximately

$$\frac{1}{16\pi}\left(\frac{m_\tau}{\phi_0}\right)^2 \approx 2 \times 10^{-6}.$$

12.5 Verify that the unitary transformations of Section 12.6 preserve the form of the dynamical terms in the Lagrangian density.

13

Experimental tests of the Weinberg–Salam theory

13.1 The search for the gauge bosons

We saw in the preceding chapter that the low energy limit of the electroweak Weinberg–Salam theory reduces to the successful phenomenology of Chapter 9. There is no reason to doubt that the Weinberg–Salam theory describes all low energy β decays, but it also describes very much more. The pathological cross-section of equation (9.14) is modified to

$$\sigma(\nu_\mu e^- \rightarrow \mu^- \nu_e) = \frac{G_F^2}{\pi} \left(\frac{\left(s - m_\mu^2 \right)^2}{s[1 + \left(s - m_\mu^2 \right) / M_w^2]} \right). \tag{13.1}$$

At high energies $\gg M_w$, this expression tends to $G_F^2 M_w^2 / \pi = 1.08 \times 10^{-10}$ b. It is a renormalisable theory, so that quantum corrections can be calculated. At high energies these corrections become increasingly important (at the few per cent level).

   The clearest test of the theory is the observation of the conjectured gauge bosons, the W$^\pm$ and Z. These were discovered at CERN in 1983, using a specially constructed proton–antiproton collider, with a centre of mass energy of 540 GeV. It was very important for the successful identification of the new particles that their masses and decay characteristics had already been well estimated within the theory. The masses depend on G_F, e and the Weinberg angle θ_w (equations (11.37) and (12.22)). The values of G_F and e were well established, and estimates of θ_w were available from careful observations of neutral current events. We saw in Section 12.3 that the $e\nu_\mu \rightarrow e\nu_\mu$ and $e\bar{\nu}_\mu \rightarrow e\bar{\nu}_\mu$ cross-sections are sensitive to θ_w. Similarly, the cross-sections for ν and $\bar{\nu}$ scattering from nuclei depend on θ_w, as we shall see in more detail in Chapter 14. Since the centre of mass energy available in neutrino–nuclear scattering is much greater than in neutrino–electron scattering (equation (9.13)) and the cross-sections increase with energy, it was the neutral

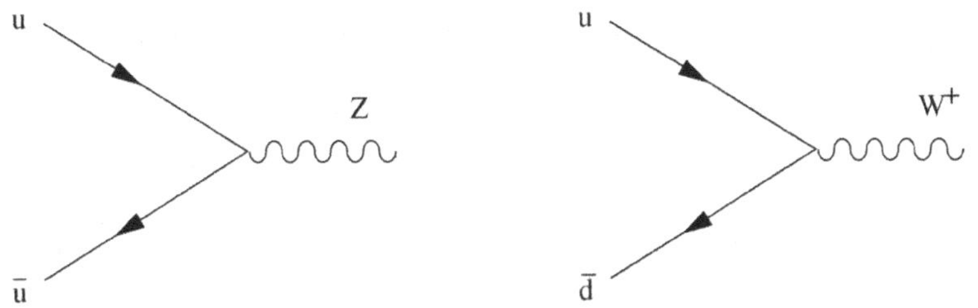

Figure 13.1 Quark–antiquark annihilation is the principal process contributing to W and Z production in proton–antiproton collisions at present day collider energies.

current experiments on nuclei which gave an estimate of $\theta_{\rm w}$, and this estimate was in fact close to the presently accepted value. The experimental physicists knew what to look for!

The successful identification of the new particles also relied on estimates of the likely production cross-sections of the particles. We have not yet discussed how quarks interact with the $W^\pm$ and Z bosons, but we shall see in Chapter 14 that the interactions are similar to the interactions of leptons with the gauge bosons. Two of the processes that contribute to Z and W^+ production are sketched in Fig. 13.1. The outgoing proton and antiproton remnants materialise as complicated jets of particles moving in directions closely correlated with the original proton and antiproton directions. It is a fortunate circumstance for identification that the decay products of the gauge bosons are frequently well separated from the particles in the remnants (Problem 13.1).

The quark–antiquark pair responsible for gauge boson production carry only a fraction of the original 540 GeV of energy, and the 540 GeV design parameter allowed for this effect. The important analysis of the partition of the energy of a beam particle between its constituents is discussed in Appendix D.

13.2 The W± bosons

The results of these experiments at CERN and subsequent experiments dramatically confirmed the theoretical expectations. The charged $W^\pm$ bosons have a mass

$$M_{\rm w} = 80.425 \pm 0.038\,\text{GeV},$$

and their decay rates to lepton pairs are measured to be

$$\begin{aligned}\Gamma(W^+ \to e^+\nu_e) &= 228 \pm 6\,\text{MeV},\\ \Gamma(W^+ \to \mu^+\nu_\mu) &= 225 \pm 9\,\text{MeV},\\ \Gamma(W^+ \to \tau^+\nu_\tau) &= 228 \pm 11\,\text{MeV},\end{aligned}$$

and $\Gamma(W^+ \to e^+\nu_e) = \Gamma(W^- \to e^-\bar{\nu}_e)$, etc.

To lowest order in perturbation theory, and neglecting terms in $(m_{lepton}/M_w)^2$, these partial widths are all equal in the Standard Model and

$$\Gamma(W^+ \rightarrow e^+\nu) = \frac{G_F M_W^3}{6\pi\sqrt{2}} = 226 \pm 1 \, \text{MeV}, \tag{13.2}$$

(Problem 13.3) in good agreement with the experimental data.

13.3 The Z boson

The experiments that revealed the charged $W^\pm$ bosons also revealed the neutral Z boson, but the mass of the Z boson and its decay rates are now known far more accurately than those of the $W^\pm$ bosons. In 1989, two e^+e^- colliders were opened: LEP at CERN and SLC at Stanford. In these machines, the electrons and positrons have equal energies and opposite momenta, and the centre of mass energy can be tuned to lie at and around the mass of the Z. Typical resonant cross-sections for particle production are shown in Fig. 13.2, and corresponding Feynman diagrams in Fig. 13.3. At the peak energy, Z bosons at rest are copiously produced by e^+e^- annihilation. These very clean events have given precise data on the properties of the Z. The mass of the Z is

$$M_z = 91.1876 \pm 0.0021 \, \text{GeV},$$

and partial decay widths to charged lepton–antilepton pairs are

$$\Gamma(Z \rightarrow e^+e^-) = 83.91 \pm 0.20 \, \text{MeV},$$
$$\Gamma(Z \rightarrow \mu^+\mu^-) = 83.99 \pm 0.35 \, \text{MeV},$$
$$\Gamma(Z \rightarrow \tau^+\tau^-) = 84.09 \pm 0.40 \, \text{MeV}.$$

The total decay width, which includes decays to hadrons and the $\nu\bar{\nu}$ pairs, is $\Gamma \, (\text{total}) = 2495 \pm 2 \, \text{MeV}$.

The theoretical partial widths for decay to charged lepton pairs depend on the Weinberg angle θ_w. To lowest order and neglecting terms in $(m_{lepton}/M_z)^2$, the partial widths are all equal and

$$\Gamma(Z \rightarrow e^+e^-) = \frac{G_F M_z^{\,3}}{12\sqrt{2}\pi}\left[\left(1 - 2\sin^2\theta_w\right)^2 + 4\sin^4\theta_w\right]. \tag{13.3}$$

Taking the accepted value of $\sin^2\theta_w = 0.2312$, this gives, to lowest order,

$$\Gamma(Z \rightarrow e^+e^-) = 83.4 \, \text{MeV}.$$

Again, there is remarkable agreement between theory and experiment.

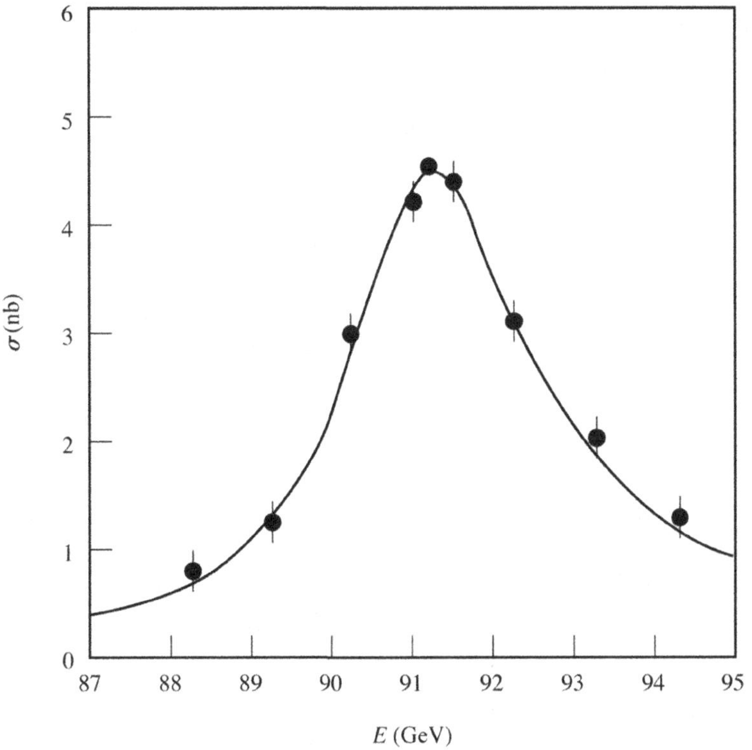

Figure 13.2 The cross-section $\sigma(e^+e^- \rightarrow e^+e^- + \mu^+\mu^- + \tau^+\tau^-)$ as a function of E the initiating e^+e^- centre of mass energy. The experimental data were presented at the 25th International Conference on High Energy Physics in Singapore in 1990 by the ALEPH collaboration of CERN. The curve is the prediction of the Standard Model but with parameters such as the Z mass as variables determined by the data (see Hansen (1991)).

13.4 The number of lepton families

For the decay rates to neutrino–antineutrino pairs, the Standard Model gives

$$\Gamma(Z \rightarrow \nu_e\bar{\nu}_e) = \Gamma(Z \rightarrow \nu_\mu\bar{\nu}_\mu) = \Gamma(Z \rightarrow \nu_\tau\bar{\nu}_\tau) = \frac{G_F M_Z^3}{12\sqrt{2}\pi} = 165.9\,\text{MeV}. \tag{13.4}$$

Hence the partial width for decay to any neutrino–antineutrino pair is

$$3\Gamma(Z \rightarrow \nu_e\bar{\nu}_e) = 497.6\,\text{MeV}.$$

This can be compared with the partial width Γ(invisible) associated with e^+e^- pairs annihilating without trace, since neutrinos and antineutrinos are the only particles that will escape unseen by the particle detectors.

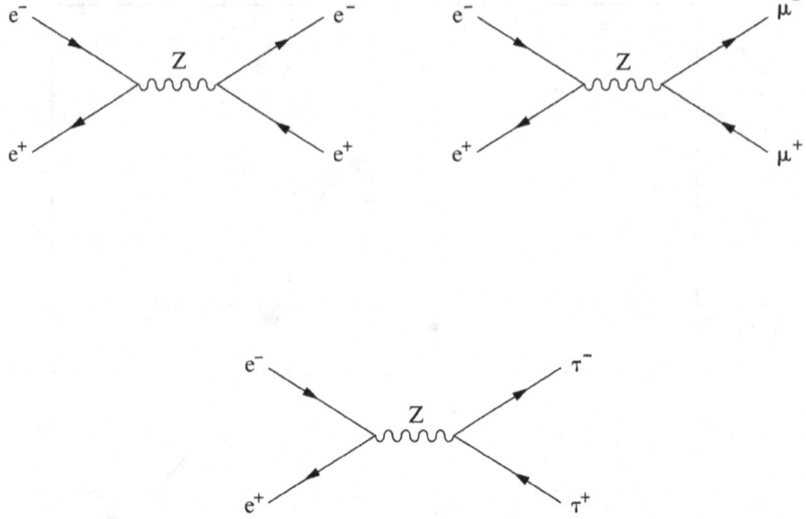

Figure 13.3 The basic Feynman graphs that describe the processes of Fig. 13.2. The fitting curve indudes additional graphs that give the Z resonance its width and graphs that describe accompanying electromagnetic processes.

Experimentally, it is found that

$$\Gamma(\text{invisible}) = 498.3 \pm 4.2\,\text{MeV}.$$

The agreement with the Standard Model value is a striking confirmation of the theory. It implies that there are no more light neutrino types and rules out there being any more 'standard' lepton doublets in Nature than the three already known. This is a result of fundamental significance.

13.5 The measurement of partial widths

In view of the importance of the partial widths for Z decay, we shall sketch how they are obtained from the experimental results. The cross-section for e^+e^- elastic scattering at small angles is dominated by photon exchange, even around the Z resonance, and is well known from QED. This small angle elastic scattering of the beam particles is constantly monitored during data taking, and the cross-section for any other process, for example $e^+e^- \to \mu^+\mu^-$, is then obtained from the measured rate of $\mu^+\mu^-$ production relative to the rate of e^+e^- small angle scattering. This, essentially, is how the graphs of Fig. 13.2 are arrived at. We give now a much simplified analysis that indicates how the partial widths are extracted.

Assume that the cross-sections are described by a simple Breit–Wigner formula. For example,

$$\sigma\left(e^+e^- \to \mu^+\mu^-\right) = \frac{3\pi}{M_z^2}\frac{\Gamma_{ee}\Gamma_{\mu\mu}}{(E - M_z)^2 + \Gamma^2/4}, \qquad (13.5)$$

$$\sigma\left(e^+e^- \to \text{hadrons}\right) = \frac{3\pi}{M_z^2}\frac{\Gamma_{ee}\Gamma_{had}}{(E - M_z)^2 + \Gamma^2/4}. \qquad (13.6)$$

(The factor 3 is a spin factor.)

M_z and the total decay width Γ can be found from the position and width of the experimental peak. Then, taking $\Gamma_{ee} = \Gamma_{\mu\mu}$, the ratio Γ_{ee}/Γ can be found from the peak of the cross-section $\sigma\left(e^+e^- \to \mu^+\mu^-\right)$ at $E = M_z$, using (13.5):

$$\frac{\Gamma_{ee}}{\Gamma} = \left(\frac{M_z^2\sigma\left(e^+e^- \to \mu^+\mu^- \text{ at } E = M_Z\right)}{12\pi}\right)^{1/2}.$$

Using this result, the ratio Γ_{had}/Γ follows from the peak of the cross-section $\sigma(e^+e^- \to \text{hadrons})$. From (13.6),

$$\frac{\Gamma_{had}}{\Gamma} = \frac{M_z^2}{12\pi}\frac{\Gamma}{\Gamma_{ee}}\sigma\left(e^+e^- \to \text{hadrons at } E = M_z\right).$$

To obtain Γ(invisible), we take

$$\Gamma\,(\text{invisible}) = \Gamma - 3\Gamma_{ee} - \Gamma_{had}.$$

In reality the data have to be treated very much more carefully than is implied above. In particular electromagnetic effects during the collision process distort the simple Breit–Wigner shape, and appropriate corrections are applied in the actual analysis.

Figure 13.4 shows the result of such a more sophisticated fit, compared with Standard Model predictions assuming two, three and four types of massless neutrinos. The data unequivocally require three.

13.6 Left–right production cross-section asymmetry and lepton decay asymmetry of the Z boson

Other details of the Weinberg–Salam theory can be tested with e^+e^- colliders. Much work has been done at Stanford with the SLC beam energies tuned to the Z boson mass. The beam intensities at SLC were lower than those at the CERN collider, but the SLC had an advantage in that the electron beam can be polarised along the beam direction so that the relative proportions of positive and negative helicity electrons can be changed. We have seen in Chapter 7 that, at high energies, negative

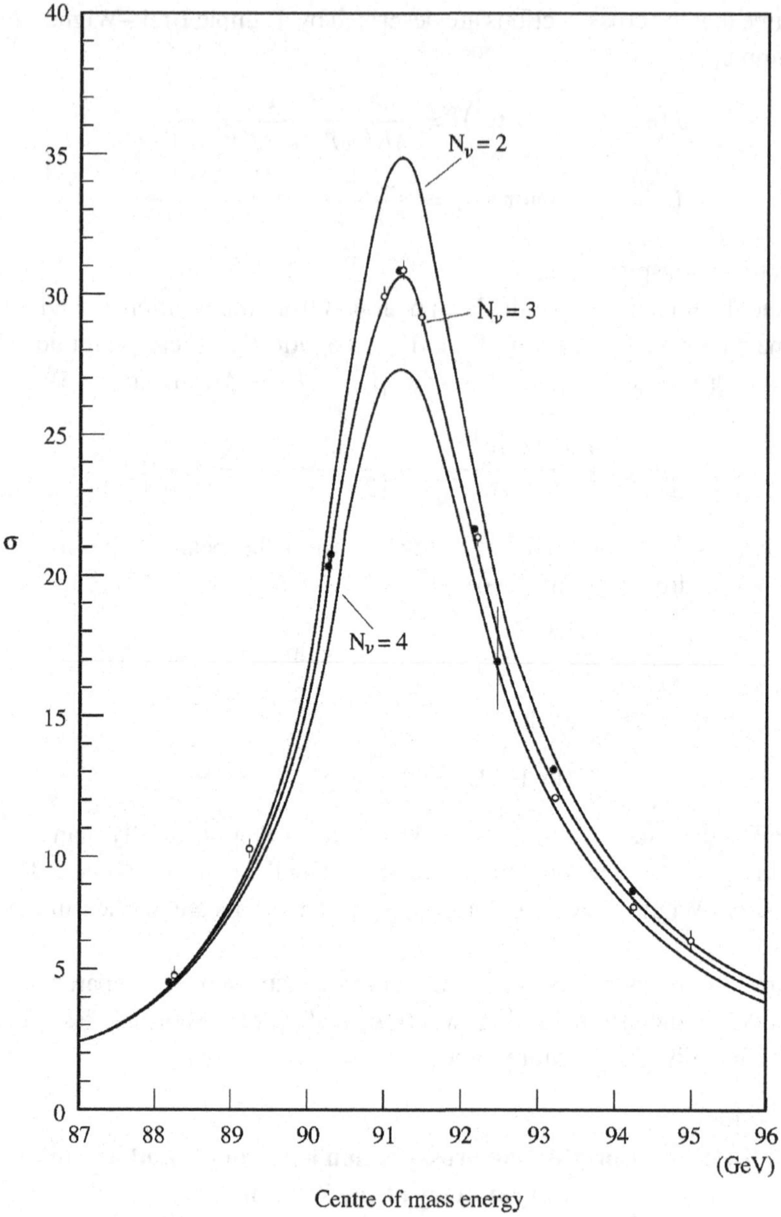

Figure 13.4 The cross-section $\sigma\left(e^+e^- \to \text{hadrons}\right)$ as a function of E the initiating e^+e^- centre of mass energy. The experimental data were presented at the 25th International Conference on High Energy Physics in Singapore in 1990 by the OPAL collaboration of CERN. The data are compared with the predictions of the Standard Model but with two, three and four neutrino types. Three light neutrino types are clearly favoured (see Mori (1991)).

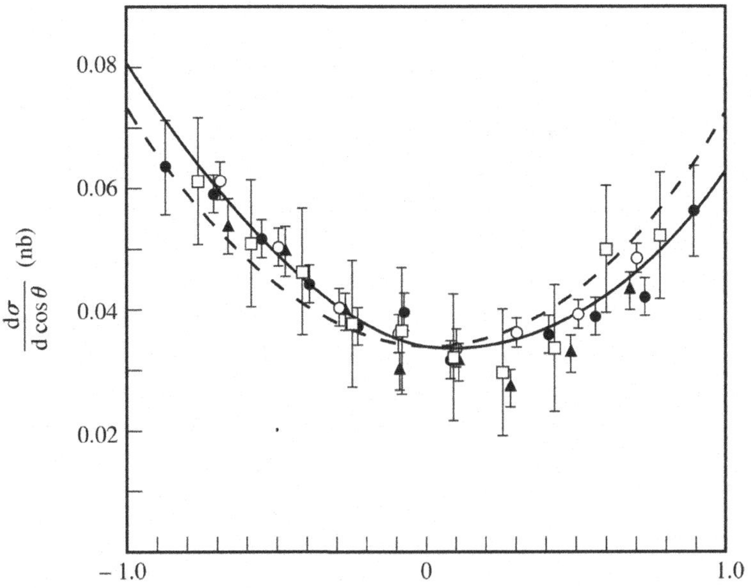

Figure 13.5 The differential cross-section $d\sigma\left(e^+e^- \rightarrow \mu^+\mu^-\right)/d\cos\theta$. The data were taken at DESY at an e^+e^- centre of mass energy of 30 GeV. The dashed line is the prediction of quantum electrodynamics alone, the full line fits the data and shows the modification due to the presence of the Z boson which gives this interference effect (R. Marshall, Rutherford Appleton Laboratory Report RAL 89–021).

helicity electrons and positive helicity positrons are associated with left-handed fields, positive helicity electrons and negative helicity positrons are associated with right-handed fields. It follows from the form of the interaction term (12.33) in the Weinberg–Salam Lagrangian that in interacting with an unpolarised positron beam (equal numbers of positive helicity and negative helicity positrons) the cross-section σ_L for Z production by a negative helicity electron is proportional to $(\cos 2\theta_w)^2$ and the cross-section σ_R for Z production by a positive helicity electron is proportional to $\left(2\sin^2\theta_w\right)^2$. The constants of proportionality are the same so that the left–right cross-section asymmetry is, to lowest order,

$$A_{LR} = \frac{\sigma_L - \sigma_R}{\sigma_L + \sigma_R} = \frac{(\cos 2\theta_w)^2 - \left(2\sin^2\theta_w\right)^2}{(\cos 2\theta_w)^2 + \left(2\sin^2\theta_w\right)^2} = \frac{2\left(1 - 4\sin^2\theta_w\right)}{1 + \left(1 - 4\sin^2\theta_w\right)^2}.$$

From the measurements at SLC (Fero, 1994) it is calculated that $A_{LR} = 0.1628 \pm 0.0099$, which gives an estimate

$$\sin^2\theta_w = 0.2292 \pm 0.0013.$$

This estimate does not depend on the ratio $M_{\mathrm{w}}/M_{\mathrm{z}}$, since the $\mathrm{W}^{\pm}$ bosons are not involved.

At CERN and at a previous $\mathrm{e}^{+}\mathrm{e}^{-}$ collider at DESY in Hamburg the electron beams had no longitudinal polarisation. Nevertheless if a Z boson is formed its spin is aligned with the direction of the electron beam with probability proportional to $[2\sin^2\theta_{\mathrm{w}}]^2$, and anti-aligned with probability proportional to $[\cos 2\theta_{\mathrm{w}}]^2$, giving it a mean polarisation in the direction of the beam of $-A_{\mathrm{LR}}$.

When the Z decays to a lepton–antilepton pair, the direction of the lepton is correlated with the direction of the Z spin. The polarisation of the Z therefore gives a forward–backward asymmetry in the angular distribution of the leptons.

The competing process of lepton production through the electromagnetic interaction does give a symmetrical angular distribution. The observed asymmetry depends on the interference between Z and γ processes, and is energy dependent. Figure 13.5 shows the angular distribution of leptons with respect to the electron beam distribution at a centre of mass energy $E = 30$ GeV (which is below M_{z}). This data was taken at DESY and gave an estimate of $\sin^2\theta_{\mathrm{w}} = 0.212 \pm 0.014$. This is another impressive confirmation of the overall consistency of the Weinberg–Salam theory.

Problems

13.1 $\mathrm{W}^{\pm}$ bosons are produced when a beam of high energy protons is in head-on collision with a beam of antiprotons. The W boson momenta are strongly aligned with the beams. The transverse component of momentum given to the W is small. Neglecting this component, and assuming that in the W rest frame there is an isotropic distribution of decay products, show that in a decay to a charged lepton and a neutrino, the root mean square transverse lepton momentum is approximately $M_{\mathrm{w}}/\sqrt{6} = 33$ GeV.

Events with large transverse momenta are rare, and their observation allows W production to be identified. (Note that the transverse momenta are unchanged by a Lorentz boost of the W in the beam direction.)

13.2 From the interaction term in (12.23) of the Z boson with an electron–positron pair, show that in head-on unpolarised $\mathrm{e}^{+}\mathrm{e}^{-}$ collisions, the probability of the Z boson spin being aligned with the electron beam is proportional to $\left(2\sin^2\theta_{\mathrm{w}}\right)^2$, and of being antialigned is proportional to $(\cos 2\theta_{\mathrm{w}})^2$.

13.3 Neglecting lepton mass terms, obtain the partial widths (13.2), (13.3) and (13.4).

13.4 Recalculate (13.3), taking $\cos\theta_{\mathrm{w}} = M_{\mathrm{w}}/M_{\mathrm{z}}$.

14

The electromagnetic and weak interactions of quarks

In the Standard Model it is the quarks' colour that is the source of their strong interaction. In this chapter we shall consider only the electromagnetic and weak interactions of quarks, and colour will not enter. The theory will be constructed in close analogy with the electroweak theory for leptons set out in Chapter 12. The theory for quarks is not as well founded in experiment as the theory for leptons. This is because quarks cannot be isolated from hadrons. Experiments can only be performed on composite quark systems, and the basic Lagrangian density is obscured at low energies by the strong interactions. At higher energies, and especially through the hadronic decays of the Z bosons, the electroweak physics of the isolated quarks can to some extent be discerned. In Chapter 15 some of the relevant experimental data on these decays will be described.

14.1 Construction of the Lagrangian density

At low energies, the model has to describe decays like

$$\mathrm{n} \to \mathrm{p} + \mathrm{e}^- + \bar{\nu}_\mathrm{e}$$

or, at quark level,

$$\mathrm{d} \to \mathrm{u} + \mathrm{e}^- + \bar{\nu}_\mathrm{e}.$$

This decay is mediated by the W boson. Comparing it with muon decay,

$$\mu^- \to \nu_\mu + \mathrm{e}^- + \bar{\nu}_\mathrm{e},$$

which is also mediated by the W boson, suggests that the left-handed components u_L and d_L of the quark fields should be put together in an $SU(2)$ doublet,

$$L = \begin{pmatrix} u_\mathrm{L} \\ d_\mathrm{L} \end{pmatrix}, \tag{14.1}$$

while u_R and d_R are, like ν_R and e_R, unchanged by $SU(2)$ transformations. We shall see that this simple assignment would be correct if Nature had provided us with only one type of up quark, and only one type of down quark.

With such an assignment there is no freedom in the construction of the weak interaction. There is only one way to make the dynamical part of the quark Lagrangian density gauge invariant. The coupling to the field $\mathbf{W}_\mu$ is uniquely determined by $SU(2)$ symmetry and the coupling to the field B_μ is fixed by the quark electric charges: $2e/3$ on the u quark, $-e/3$ on the d quark. Hence

$$\begin{aligned} \mathcal{L}_{\text{dyn}} = {} & \mathrm{L}^\dagger \tilde{\sigma}^\mu \mathrm{i}[\partial_\mu + (\mathrm{i}g_2/2)\mathbf{W}_\mu + (\mathrm{i}g_1/6)B_\mu]\mathbf{L} \\ & + \mu_R^\dagger \sigma^\mu \mathrm{i}[d_\mu + (2\mathrm{i}g_1/3)B_\mu]u_R \\ & + d_R^\dagger \sigma^\mu \mathrm{i}[\partial_\mu - (ig_1/3)B_\mu]d_R, \end{aligned} \tag{14.2}$$

where $g_2 \sin\theta_w = g_1 \cos\theta_w = e$.

To conform with the transformation laws (11.4b) and (11.6) on the gauge fields, the $U(1) \times SU(2)$ transformation of the quark fields must be

$$\begin{aligned} \mathbf{L} \to \mathbf{L}' &= \mathrm{e}^{-\mathrm{i}\theta(x)/3}\mathbf{UL}, \\ u_R \to u_R{}' &= \mathrm{e}^{-4\mathrm{i}\theta(x)/3}u_R, \\ d_R \to d_R{}' &= \mathrm{e}^{2\mathrm{i}\theta(x)/3}d_R. \end{aligned} \tag{14.3}$$

Using (11.17) and (11.29), $\mathcal{L}_{\text{dyn}}$ can be written in terms of the fields $W_\mu^\pm$, Z_μ and A_μ and becomes

$$\begin{aligned} \mathcal{L}_{\text{dyn}} = {} & \mathrm{L}^\dagger \tilde{\sigma}^\mu i \begin{pmatrix} \partial_\mu + \dfrac{2\mathrm{i}e}{3}A_\mu + \dfrac{\mathrm{i}e}{3\sin 2\theta_w}(1 + 2\cos 2\theta_w)Z_\mu, & \dfrac{\mathrm{i}e}{\sqrt{2}\sin\theta_w}W_\mu^+ \\ \dfrac{\mathrm{i}e}{\sqrt{2}\sin\theta_w}W_\mu^-, & \partial_\mu - \dfrac{\mathrm{i}e}{3}A_\mu - \dfrac{\mathrm{i}e}{3\sin 2\theta_w}(2 + \cos 2\theta_w)Z_\mu \end{pmatrix} \mathrm{L} \\ & + u_R^\dagger \sigma^\mu \mathrm{i}\left[\partial_\mu + \frac{2\mathrm{i}e}{3}A_\mu - \frac{2\mathrm{i}e}{3}\tan\theta_w Z_\mu\right]u_R \\ & + d_R^\dagger \sigma^\mu i\left[\partial_\mu - \frac{\mathrm{i}e}{3}A_\mu + \frac{\mathrm{i}e}{3}\tan\theta_w Z_\mu\right]d_R. \end{aligned} \tag{14.4}$$

However, the Standard Model postulates three families, or generations, of quarks. We therefore introduce *three* left-handed $SU(2)$ doublets:

$$\begin{pmatrix} u_{L1} \\ d_{L1} \end{pmatrix}, \begin{pmatrix} u_{L2} \\ d_{L2} \end{pmatrix}, \begin{pmatrix} u_{L3} \\ d_{L3} \end{pmatrix},$$

and six right-handed singlets: $u_{R1}, d_{R1}; u_{R2}, d_{R2}; u_{R3}, d_{R3}$. For a more compact notation we shall denote these by

$$\mathrm{L}_k = \begin{pmatrix} u_{Lk} \\ d_{Lk} \end{pmatrix}, \quad u_{Rk}, \; d_{Rk} \text{ with } k = 1, 2, 3.$$

As in the lepton case, we take the dynamical part of the total quark Lagrangian as a sum:

$$\mathcal{L}_{\text{dyn}}(\text{quark}) = \sum_{k=1}^{3} \mathcal{L}_{\text{dyn}}(\text{u}_\text{k}, \text{d}_\text{k}). \tag{14.5}$$

14.2 Quark masses and the Kobayashi–Maskawa mixing matrix

To retain renormalisability we must retain gauge symmetry, and give mass to the quarks by coupling to the Higgs field as in Chapter 12 where we gave mass to the leptons. For the d_k quarks this is straightforward. The most general form we might consider that preserves the gauge symmetries is

$$\mathcal{L}_{\text{Higgs}}(\text{d}) = -\sum \left[G^{\text{d}}_{ij}(\mathbf{L}^{\dagger}_{i}\mathbf{\Phi})d_{\text{R}j} + G^{\text{d}*}_{ij} d^{\dagger}_{\text{R}j}(\mathbf{\Phi}^{\dagger}\mathbf{L}_i)\right], \tag{14.6}$$

as we discussed in the lepton case in Section 12.6. After the symmetry breaking of the Higgs field Φ, this gives the mass term for the d-type quarks:

$$\mathcal{L}_{\text{mass}}(\text{d}) = -\phi_0 \sum \left[G^{\text{d}}_{ij} d^{\dagger}_{\text{L}i} d_{\text{R}j} + G^{\text{d}*}_{ij} d^{\dagger}_{\text{R}j} d_{\text{L}i}\right]. \tag{14.7}$$

A priori, G^{d}_{ij} is an arbitrary 3×3 complex matrix. As we remarked in Section 12.6, such a matrix can always be put into real diagonal form with the help of two unitary matrices, so that we can write

$$\phi_0 \mathbf{G}^{\text{d}} = \mathbf{D}^{\dagger}_{\text{L}} \mathbf{m}^{d} \mathbf{D}_{\mathbf{R}},$$

where $\mathbf{m}^d$ is a real diagonal matrix, and $\mathbf{D}_\text{L}$, $\mathbf{D}_\mathbf{R}$ are unitary matrices. If the diagonal elements are distinct, as appears experimentally to be the case, $\mathbf{D}_\text{L}$, $\mathbf{D}_\mathbf{R}$ are unique, except that both may be multiplied on the left by the same phase-factor matrix

$$\begin{pmatrix} e^{i\alpha_1} & 0 & 0 \\ 0 & e^{i\alpha_2} & 0 \\ 0 & 0 & e^{i\alpha_3} \end{pmatrix}. \tag{14.8}$$

In the Standard Model as set out in Chapter 12, the neutrinos were taken to have zero mass. However, for the u-type quarks, which are here making up a left-handed doublet, we need a mass term. For this purpose we introduce the 2×2 matrix in $SU(2)$ space

$$\varepsilon = \begin{pmatrix} \varepsilon_{AA} & \varepsilon_{AB} \\ \varepsilon_{BA} & \varepsilon_{BB} \end{pmatrix} = \begin{pmatrix} 0 & 1 \\ -1 & 0 \end{pmatrix}.$$

A suitable $SU(2)$ invariant expression which we can construct from the doublets $\mathbf{\Phi}$ and $\mathbf{L}_\text{i}$ is $\left(\mathbf{\Phi}^\text{T}\, \varepsilon\, \mathbf{L}_\text{i}\right)$, where $\mathbf{\Phi}^\text{T} = (\mathbf{\Phi}_\text{A}, \mathbf{\Phi}_\text{B})$ is the transpose of $\mathbf{\Phi}$ (Problem 14.3).

We then take

$$\mathcal{L}_{\text{Higgs}}(u) = -\sum_{ij}\left[G^{\text{u}}_{ij}(\mathbf{L}^{\dagger}_{i}\,\varepsilon\,\mathbf{\Phi}^{*})u_{\text{R}j} - G^{\text{u}*}_{ij}\,u^{\dagger}_{\text{R}j}(\mathbf{\Phi}^{\text{T}}\,\varepsilon\,\mathbf{L}_{i})\right] \tag{14.9}$$

where G^{u}_{ij} is another complex 3×3 matrix. On symmetry breaking, this gives the u-quarks mass term

$$\mathcal{L}_{\text{mass}}(u) = -\phi_0 \sum\left[G^{\text{u}}_{ij}u^{\dagger}_{\text{L}i}u_{\text{R}j} + G^{\text{u}*}_{ij}u^{\dagger}_{\text{R}j}u_{\text{L}i}\right], \tag{14.10}$$

which is, as we might expect, similar to (14.7), and likewise preserves the gauge symmetries. It can be brought into real diagonal form in a similar way:

$$\phi_0\mathbf{G}^{\text{u}} = \mathbf{U}_{\text{L}}{}^{\dagger}\mathbf{m}^{\text{u}}\mathbf{U}_{\text{R}},$$

where $\mathbf{U}_{\text{L}}$ and $\mathbf{U}_{\text{R}}$ are unitary matrices, and $\mathbf{m}^{u}$ is diagonal.

$\mathbf{U}_{\text{L}}$ and $\mathbf{U}_{\text{R}}$ may be both multiplied on the left by a phase factor matrix, say

$$\begin{pmatrix} e^{i\beta_1} & 0 & 0 \\ 0 & e^{i\beta_2} & 0 \\ 0 & 0 & e^{i\beta_3} \end{pmatrix}.$$

The theory is most directly described in terms of the 'true' quark fields, for which the mass matrices are diagonal, so that we define the six quark fields:

$$\begin{aligned} d'_{\text{L}i} &= D_{\text{L}ij}d_{\text{L}j}, \quad d'_{\text{R}i} = D_{\text{R}ij}d_{\text{R}j}, \\ u'_{\text{L}i} &= U_{\text{L}ij}u_{\text{L}j}, \quad u'_{\text{R}i} = U_{\text{R}ij}u_{\text{R}j}. \end{aligned} \tag{14.11}$$

The quark mass contribution to $\mathcal{L}$ becomes:

$$\mathcal{L}_{\text{mass}}(\text{quarks}) = -\sum_{i=1}^{3}\left[m^{\text{d}}_{i}\left(d'^{\dagger}_{\text{L}i}d'_{\text{R}i} + d'^{\dagger}_{\text{R}i}d'_{\text{L}i}\right) + m^{\text{u}}_{i}\left(u'^{\dagger}_{\text{L}i}u'_{\text{R}i} + u'^{\dagger}_{\text{R}i}u'_{\text{L}i}\right)\right]. \tag{14.12a}$$

We identify the Dirac spinors

$$\begin{pmatrix} u'_{\text{L}1} \\ u'_{\text{R}1} \end{pmatrix}, \begin{pmatrix} u'_{\text{L}2} \\ u'_{\text{R}2} \end{pmatrix}, \begin{pmatrix} u'_{\text{L}3} \\ u'_{\text{R}3} \end{pmatrix}$$

with the u, c and t quarks, respectively, and the Dirac spinors

$$\begin{pmatrix} d'_{\text{L}1} \\ d'_{\text{R}1} \end{pmatrix}, \begin{pmatrix} d'_{\text{L}2} \\ d'_{\text{R}2} \end{pmatrix}, \begin{pmatrix} d'_{\text{L}3} \\ d'_{\text{R}3} \end{pmatrix}$$

with the d, s and b quarks, so that we might rewrite (14.12a) as

$$\begin{aligned} \mathcal{L}_{\text{mass}}(\text{quarks}) = &-\left[m^{d}\left(d^{\dagger}_{\text{L}}d_{\text{R}} + d^{\dagger}_{\text{R}}d_{\text{L}}\right) + m^{u}\left(u^{\dagger}_{\text{L}}u_{\text{R}} + u^{\dagger}_{\text{R}}u_{\text{L}}\right)\right] \\ &-\left[m^{s}\left(s^{\dagger}_{\text{L}}s_{\text{R}} + s^{\dagger}_{\text{R}}s_{\text{L}}\right) + m^{c}\left(c^{\dagger}_{\text{L}}c_{\text{R}} + c^{\dagger}_{\text{R}}c_{\text{L}}\right)\right] \\ &-\left[m^{b}\left(b^{\dagger}_{\text{L}}b_{\text{R}} + b^{\dagger}_{\text{R}}b_{\text{L}}\right) + m^{t}\left(t^{\dagger}_{\text{L}}t_{\text{R}} + t^{\dagger}_{\text{R}}t_{\text{L}}\right)\right]. \end{aligned} \tag{14.12b}$$

The terms in (14.12b) correspond to six Dirac fermions.

We have dropped the primes, and for the remainder of the book u_k and d_k, for $k = 1, 2, 3$, will denote true quark fields.

In the $\mathcal{L}_{\text{dyn}}$ given by (14.2) and (14.5), the 'diagonal' terms do not mix u-type and d-type quarks and are invariant under the unitary transformations (14.11). However, the terms that arise from the off-diagonal elements of the matrix $\mathbf{W}_\mu$, mix u and d quarks through their coupling to the $W^\pm$ boson fields, and these terms are profoundly changed.

The diagonal terms give $\mathcal{L}_{q\text{Dirac}}$ and $\mathcal{L}_{qz}$ that parallel the expressions (12.12) and (12.23) of the lepton theory of Chapter 12. The complete electroweak Lagrangian density for the quarks is

$$\mathcal{L}_q = \mathcal{L}_{q\text{Dirac}} + \mathcal{L}_{qz} + \mathcal{L}_{q\text{w}} + \mathcal{L}_{q\text{H}}$$

where

$$\begin{aligned}\mathcal{L}_{q\text{Dirac}} = \sum_i & \left[u^\dagger_{\text{L}i}\tilde{\sigma}^\mu \text{i}\{\partial_\mu + \text{i}(2e/3)A_\mu\}u_{\text{L}i} + u^\dagger_{\text{R}i}\sigma^\mu \text{i}\{\partial_\mu + \text{i}(2e/3)A_\mu\}u_{\text{R}i}\right] \\ & + \left[d^\dagger_{\text{L}i}\tilde{\sigma}^\mu \text{i}\{\partial_\mu - \text{i}(e/3)A_\mu\}d_{\text{L}i} + d^\dagger_{\text{R}i}\sigma^\mu \text{i}\{\partial_\mu - \text{i}(e/3)A_\mu\}d_{\text{R}i}\right] + \mathcal{L}_{q\text{mass}}\end{aligned} \tag{14.13}$$

$$\begin{aligned}\mathcal{L}_{qz} = \sum_i \Bigg[& -u^\dagger_{\text{L}i}\tilde{\sigma}^\mu u_{\text{L}i}\left(\frac{e}{\sin(2\theta_\text{w})}\right) Z_\mu(1 - (4/3)\sin^2\theta_\text{w}) \\ & + u^\dagger_{\text{R}i}\sigma^\mu u_{\text{R}i}\left(\frac{e}{\sin(2\theta_\text{w})}\right) Z_\mu \frac{4}{3}\sin^2\theta_\text{w} \\ & + d^\dagger_{\text{L}i}\tilde{\sigma}^\mu d_{\text{L}i}\left(\frac{e}{\sin(2\theta_\text{w})}\right) Z_\mu(1 - (2/3)\sin^2\theta_\text{w}) \\ & - d^\dagger_{\text{R}i}\sigma^\mu d_{\text{R}i}\left(\frac{e}{\sin(2\theta_\text{w})}\right) Z_\mu \frac{2}{3}\sin^2\theta_\text{w}\Bigg].\end{aligned} \tag{14.14}$$

In the $\mathcal{L}_{q\text{w}}$, part of the Lagrangian density, the terms

$$-\frac{e}{\sqrt{2}\sin\theta_\text{w}}\sum_i\left[u^\dagger_{\text{L}i}\tilde{\sigma}^\mu d_{\text{L}i}W^+_\mu + d^\dagger_{\text{L}i}\tilde{\sigma}^\mu u_{\text{L}i}W^-_\mu\right],$$

when written in terms of the 'true' quark fields given by (14.11), become

$$\begin{aligned}\mathcal{L}_{q\text{w}} = & -\frac{e}{\sqrt{2}\sin\theta_\text{w}}\left(u^\dagger_\text{L}, c^\dagger_\text{L}, t^\dagger_\text{L}\right)\begin{pmatrix} V_{\text{ud}} & V_{\text{us}} & V_{\text{ub}} \\ V_{\text{cd}} & V_{\text{cs}} & V_{\text{cb}} \\ V_{\text{td}} & V_{\text{ts}} & V_{\text{tb}}\end{pmatrix}\begin{pmatrix}\tilde{\sigma}^\mu d_\text{L} \\ \tilde{\sigma}^\mu s_\text{L} \\ \tilde{\sigma}^\mu b_\text{L}\end{pmatrix} W^+_\mu \\ & + \text{Hermitian conjugate},\end{aligned} \tag{14.15}$$

where $\mathbf{V} = \mathbf{U}_\text{L}\mathbf{D}^\dagger_\text{L}$.

Since the product of two unitary matrices is unitary, $\mathbf{V}$ is a 3×3 unitary matrix. The elements of $\mathbf{V}$ are not determined within the theory. It is in this matrix that another four of the parameters of the Standard Model reside. An

$n \times n$ unitary matrix is specified by n^2 parameters (Appendix A), so we apparently have nine parameters to be measured experimentally. However, five of these can be absorbed into the non-physical phases of the quark fields, through the phase-factor matrices associated with $\mathbf{D}_\mathrm{L}$ (see (14.8)) and $\mathbf{U}_\mathrm{L}$. (There are five, rather than six, non-physical phases since only phase differences appear in $\mathbf{V}$. For example $V_{\mathrm{ud}} = \exp[\mathrm{i}(\beta_\mathrm{u} - \alpha_\mathrm{d})]\, V^0_{\mathrm{ud}}$.)

When the quark phase factors have been extracted, the resulting matrix $\mathbf{V}^0$ is dependent on four physical parameters. It is called the *Kobayashi–Maskawa* (*KM*) *matrix* (Kobayashi and Maskawa, 1973).

14.3 The parameterisation of the KM matrix

A 3×3 rotation matrix is also a unitary matrix. A more general unitary matrix can be constructed as a product of rotation matrices and unitary matrices made up of phase factors. There is no unique parameterisation of the KM matrix by this method. That advocated by the Particle Data Group is

$$\begin{aligned}\mathbf{V} &= \begin{pmatrix} 1 & 0 & 0 \\ 0 & c_{23} & s_{23} \\ 0 & -s_{23} & c_{23} \end{pmatrix} \begin{pmatrix} \mathrm{e}^{-\mathrm{i}\delta/2} & 0 & 0 \\ 0 & 1 & 0 \\ 0 & 0 & e^{\mathrm{i}\delta/2} \end{pmatrix} \begin{pmatrix} c_{13} & 0 & s_{13} \\ 0 & 1 & 0 \\ -s_{13} & 0 & c_{13} \end{pmatrix} \\ &\quad \times \begin{pmatrix} \mathrm{e}^{\mathrm{i}\delta/2} & 0 & 0 \\ 0 & 1 & 0 \\ 0 & 0 & \mathrm{e}^{-\mathrm{i}\delta/2} \end{pmatrix} \begin{pmatrix} c_{12} & s_{12} & 0 \\ -s_{12} & c_{12} & 0 \\ 0 & 0 & 1 \end{pmatrix} \\ &= \begin{pmatrix} c_{12}c_{13} & s_{12}c_{13} & s_{13}\mathrm{e}^{-\mathrm{i}\delta} \\ -s_{12}c_{23} - c_{12}s_{23}s_{13}\mathrm{e}^{\mathrm{i}\delta} & c_{12}c_{23} - s_{12}s_{23}s_{13}\mathrm{e}^{\mathrm{i}\delta} & s_{23}c_{13} \\ s_{12}s_{23} - c_{12}c_{23}s_{13}\mathrm{e}^{\mathrm{i}\delta} & -c_{12}s_{23} - s_{12}c_{23}s_{13}\mathrm{e}^{\mathrm{i}\delta} & c_{23}c_{13} \end{pmatrix}\end{aligned} \tag{14.16}$$

where $c_{ij} = \cos\theta_{ij}$, $s_{ij} = \sin\theta_{ij}$. The four parameters are the three rotation angles $\theta_{12}, \theta_{23}, \theta_{13}$, and the phase δ.

Evidently, if $s_{13} = 0$ or $\sin\delta = 0$ then $\mathbf{V}$ is real. Less evidently, if $s_{12} = 0$ then $\mathbf{V}$ is made real by redefining the quark fields

$$\mathrm{e}^{\mathrm{i}\delta}u_1 \to u_1, \quad \mathrm{e}^{\mathrm{i}\delta}d_1 \to d_1,$$

and if $s_{23} = 0$ then $\mathbf{V}$ is made real by redefining

$$\mathrm{e}^{-\mathrm{i}\delta}u_3 \to u_3, \quad \mathrm{e}^{-\mathrm{i}\delta}d_3 \to d_3,$$

as the reader may verify.

A general redefinition of the quark phases,

$$d_i \to e^{i\alpha_i} d_i, \; u_i \to e^{i\beta_i} u_i,$$

will change the matrix elements of **V** by

$$V_{ij} \to e^{i(\alpha_i - \beta_j)} V_{ij}. \tag{14.17}$$

Using this freedom, the three rotation angles can be chosen all to lie in the first quadrant.

Jarlskog (1985) gives an important necessary and sufficient condition for determining whether, given a unitary matrix **V**, it is possible to make it real by such changes. She considers the imaginary part of any one of the nine products, $V_{ij} V_{kl} V^*_{kj} V^*_{il}$ with $i \neq k$ and $j \neq l$, for example

$$\text{Im}\left(V_{11} V_{22} V^*_{21} V^*_{12}\right) = J \text{ say}. \tag{14.18}$$

J is invariant under a general phase change (14.17), so that if J is not zero then it cannot be made so, and hence **V** cannot be made real. All nine quantities are equal to $\pm\, J$. In the parameterisation of equation (14.16),

$$J = c_{12} c^2_{13} c_{23} s_{12} s_{13} s_{23} \sin\delta. \tag{14.19}$$

(The conditions already obtained for the reality of the KM matrix are contained in the condition $J = 0$.)

Having fixed the KM matrix there remains only one global $U(1)$ symmetry which leaves it unchanged. All six quark fields, left and right, can be multiplied by the same phase factor. As a consequence, only the total quark number current and hence the total quark number is conserved. At the macroscopic level this is observed as baryon number conservation.

14.4 *CP* symmetry and the KM matrix

We shall now show that, if the KM matrix cannot be made real by a redefinition of the quark phases, the Standard Model does not have *CP* (change conjugation, parity) symmetry.

We saw in Section 12.5 that the Weinberg–Salam electroweak theory is invariant under the *CP* operation. Similarly, *CP* is a symmetry of every term in the Standard Model of the weak and electromagnetic interactions of quarks, except for those terms that give the interaction between the quarks and the W bosons. These are the terms that involve the KM matrix.

The *CP* transforms of the *W* fields are defined in equation (12.32):

$$W_0^{+CP} = -W_0^-, \quad W_i^{+CP} = W_i^-,$$

and the quark fields transform like all fermion fields:

$$q_{\rm L}^{CP} = -i\sigma^2 q_{\rm L}^*, \; q_{\rm R}^{CP} = i\sigma^2 q_{\rm R}^*.$$

To show how *CP* symmetry is violated, we consider the terms (14.15), which we write as

$$(-e/\sqrt{2}\sin\theta_{\rm w})\sum_{i,j}\left[u_{{\rm L}i}^\dagger\tilde{\sigma}^\mu V_{ij}d_{{\rm L}j}W_\mu^+ + d_{{\rm L}j}^\dagger\tilde{\sigma}^\mu V_{ij}^* u_{{\rm L}i}W_\mu^-\right]$$
$$(i = \text{u, c, t};\ j = \text{d, s, b})$$

Replacing the fields by their CP transforms gives

$$(-e/\sqrt{2}\sin\theta_{\rm w})\sum_{ij}\left[-u_{{\rm L}i}^{\rm T}(\tilde{\sigma}^\mu)^{\rm T}V_{ij}d_{{\rm L}j}^* W_\mu^- - d_{{\rm L}j}^{\rm T}(\tilde{\sigma}^\mu)^{\rm T}V_{ij}^* u_{{\rm L}i}^* W_\mu^+\right]$$

where, as in Section 12.5, we have used the results

$$(\sigma^2)^2 = 1, \quad \sigma^2\sigma^i\sigma^2 = -(\sigma^i)^{\rm T}.$$

On transposing this expression with respect to the spinor indices we introduce a minus sign from the anticommuting fermion fields, and obtain the CP transformed expression

$$(-e/\sqrt{2}\sin\theta_{\rm w})\sum_{i,j}\left[d_{{\rm L}j}^\dagger\tilde{\sigma}^\mu V_{ij}u_{{\rm L}i}W_\mu^- + d_{{\rm L}i}^\dagger\tilde{\sigma}^\mu V_{ij}^* d_{{\rm L}j}W_\mu^+\right].$$

This is the same as the original term if and only if V_{ij} is real for all i,j.

Experimental evidence for the breakdown of *CP* symmetry first became apparent in 1964, in the decay of the $\rm K^0$ ($\rm d\bar{s}$) meson. We shall discuss this decay and its implications in Chapter 18, where we consider what is known experimentally about the parameters of the KM matrix. It is an interesting fact that *CP*-violating effects in the Standard Model are proportional to J.

14.5 The weak interaction in the low energy limit

Combining the results of Chapter 12 (equation (12.18)) with those of the present chapter (equation (14.15)), we have the complete interaction of the W bosons with all the fermions, both leptons and quarks, of the Standard Model:

$$\mathcal{L}_{\rm Wint} = (-e/\sqrt{2}\sin\theta_{\rm w})\left[j^{\mu\dagger}W_\mu^+ + j^\mu W_\mu^-\right]$$

where

$$j^\mu = \sum_{\rm leptons} e_{{\rm L}l}^\dagger\tilde{\sigma}^\mu\nu_{{\rm L}l} + \sum_{ij} d_{{\rm L}j}^\dagger\tilde{\sigma}^\mu u_{{\rm L}i}V_{ij}^* \; (i = \text{u, c, t};\ j = \text{d, s, b}). \tag{14.20}$$

Note that we have suppressed colour indices in this chapter. The labels *i,j* on the quark spinors in (14.15) carry with them implied colour indices which are also summed over.

By eliminating the *W* field as in Section 12.2, we obtain the low energy effective interaction

$$\mathcal{L}_{\text{Weff}} = -2\sqrt{2}G_F j^\dagger_\mu j^\mu. \tag{14.21}$$

For example, the part of this effective interaction which is basically responsible for all nuclear β decays involves the electron field and the u and d quarks ($i = j = 1$):

$$\mathcal{L}_{\text{eff}} = -2\sqrt{2}G_{\text{F}}\left[g_{\mu\nu}v^\dagger_{\text{eL}}\tilde{\sigma}^\mu e_{\text{L}} d^\dagger_{\text{L}}\tilde{\sigma}^\nu u_{\text{L}} V^*_{\text{ud}}\right] + \text{Hermitian conjugate} \tag{14.22}$$

That part of the effective interaction responsible for the decay $K^0 \to \pi^+\pi^-$ $\left(\bar{s} \to u + \bar{u} + \bar{d}\right)$ is

$$\mathcal{L}_{\text{eff}} = -2\sqrt{2}G_{\text{F}}\left[g_{\mu\nu}s^\dagger_{\text{L}}\tilde{\sigma}^\mu u_{\text{L}} u^\dagger_{\text{L}}\tilde{\sigma}^\nu d_{\text{L}}\, V^*_{\text{us}} V_{\text{ud}}\right]. \tag{14.23}$$

We have also the complete interaction of the Z boson with all the fermions. Combining (12.23) with (14.14) gives

$$\mathcal{L}_{\text{Zint}} = \frac{-e}{\sin(2\theta_{\text{w}})}(j_{\text{neutral}})_\mu Z^\mu \tag{14.24}$$

where

$$\begin{aligned}(j_{\text{neutral}})^\mu = & \sum_{\text{leptons}}\left[v^\dagger_{\text{L}l}\tilde{\sigma}^\mu v_{\text{L}l} - \cos(2\theta_{\text{w}})e^\dagger_{\text{L}l}\tilde{\sigma}^\mu e_{\text{L}l}\right.\\ & \left. + 2\sin^2\theta_{\text{w}} e^\dagger_{\text{R}}\sigma^\mu e_R\right]\\ & + \sum_i\left[u^\dagger_{\text{L}i}\tilde{\sigma}^\mu u_{\text{L}i}\left(1 - \frac{4}{3}\sin^2\theta_{\text{w}}\right) - u^\dagger_{\text{R}i}\sigma^\mu u_{\text{R}i}\left(\frac{4}{3}\sin^2\theta_{\text{w}}\right)\right.\\ & \left. - d^\dagger_{\text{L}i}\tilde{\sigma}^\mu d_{\text{L}i}\left(1 - \frac{2}{3}\sin^2\theta_{\text{w}}\right) + d^\dagger_{\text{R}i}\sigma^\mu d_{\text{R}i}\left(\frac{2}{3}\sin^2\theta_{\text{w}}\right)\right].\end{aligned}$$

By eliminating the Z field, we obtain the low energy effective interaction

$$\mathcal{L}_{\text{Zeff}} = -\left(G_{\text{F}}/\sqrt{2}\right)(j_{\text{neutral}})_\mu(j_{\text{neutral}})^\mu. \tag{14.25}$$

Problems

14.1 Verify that the transformations (14.3) along with (11.4b) and (11.6) leave $\mathcal{L}_{\text{dyn}}$ invariant.

14.2 Obtain $\mathcal{L}_{\text{qZ}}$, (equation (14.14)) from (14.4).

14.3 Show that $(\Phi\, \varepsilon\, L)$ is an $SU(2)$ invariant. (Show that $U^{\text{T}}\, \varepsilon\, U = \varepsilon \det(U)$)

14.4 Write down the interaction Lagrangian density between the quark fields and the Higgs field, which appears in (14.6) and (14.9).

Estimate the coupling constant c_{t} between the Higgs field and the top quark.

14.5 Which terms in (14.20) and (14.21) are responsible for the meson decays

$$\text{K}^+\,(\text{u}\bar{\text{s}}) \to \mu^+ + \nu_\mu,$$
$$\text{D}^+\,(\text{c}\bar{\text{d}}) \to \overline{\text{K}^0}\,(\bar{\text{d}}\text{s}) + \text{e}^+ + \nu_\text{e},$$
$$\text{B}^+\,(\text{u}\bar{\text{b}}) \to \overline{\text{D}^0}\,(\bar{\text{c}}\text{u}) + \pi^+\,(\text{u}\bar{\text{d}})?$$

Sketch appropriate quark diagrams.

14.6 There are no ‘flavour changing neutral currents’, i.e. there are no terms in the neutral current of (14.24) that involve a change of quark flavour. Draw Feynman diagrams from higher orders of perturbation theory that simulate the flavour changing neutral current decays

$$\text{b} \to \text{s} + \gamma, \quad \text{b} \to \text{s} + \text{e}^+ + \text{e}^-.$$

15

The hadronic decays of the Z and W bosons

In Chapter 13 we described the results on the leptonic decays of the Z boson, obtained from experiments using e^+e^- colliders. These results are in striking agreement with the predictions of the Weinberg–Salam electroweak model. In this chapter, we shall consider some of the wealth of data that has been accumulated at CERN and SLAC on the hadronic decays of the Z, and we shall find equally striking agreement between experiment and theory.

15.1 Hadronic decays of the Z

In the Standard Model, a hadronic decay of the Z is most likely to be triggered by an initial decay to a quark–antiquark pair. The subsequent hadrons produced are mostly confined to two *jets*, back-to-back in the Z rest frame and made up of stable, or long lived, particles (see Fig. 15.1). The precise details of the processes involved in the creation of a jet are not fully understood.

The momentum of a jet may be defined as the total momentum of the particles associated with it, and may be presumed to be equal to the momentum of the initiating quark or antiquark. The Z has sufficient rest energy to decay to any quark–antiquark pair other than a $t\bar{t}$ pair, but it has so far not been possible to identify jets as arising specifically from u, d or s quarks, or their antiquarks. However, many b quark jets can be identified with some confidence from the recognition of B mesons ($b\bar{u}$, $b\bar{d}$), which have a high probability of being produced in b quark jets, and a low probability of being produced in other jets. Similarly, $\bar{B}$ mesons are used to identify $\bar{b}$ jets. The observation of charmed hadrons in jets has likewise been used to identify jets arising from c quarks and $\bar{c}$ antiquarks.

Associating the observed jets with the initiating quarks, comparisons can be made with the Standard Model predictions of Z decay rates to quark–antiquark pairs. We shall first consider the decay of a Z that is in a definite spin state. The interaction Lagrangian (14.4) has the same form for the d, s and b quarks, and in

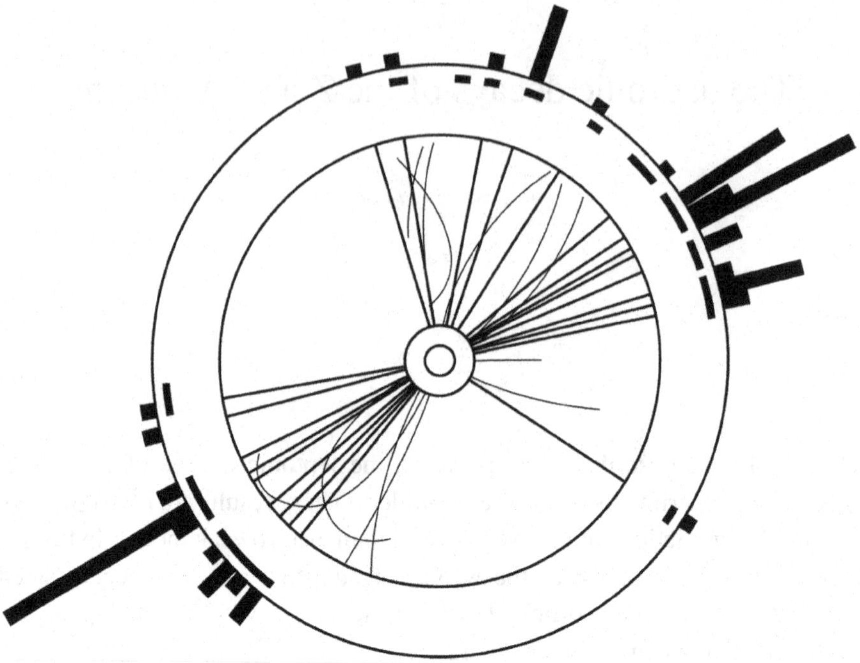

Figure 15.1 A Z hadronic decay recorded by the OPAL detector at CERN. The charged particle tracks can be seen in the inner region. The dark bands around the outer circle indicate the angular distribution of energy deposited in the outer calorimeter. The figure gives a projection of the event onto a plane perpendicular to the beam axis (see Dydak (1990)).

the lowest order of perturbation theory gives a differential decay rate into a $\mathrm{d}_k\bar{\mathrm{d}}_k$ pair ($\mathrm{d}_1 = \mathrm{d}$, $\mathrm{d}_2 = \mathrm{s}$, $\mathrm{d}_3 = \mathrm{b}$)

$$\frac{\mathrm{d}\Gamma(\mathrm{d}_k\bar{\mathrm{d}}_k)}{\mathrm{d}\cos\theta} = \frac{3G_\mathrm{F}{M_Z}^3}{32\sqrt{2}\pi}\left[\left(1-\frac{2}{3}\sin^2\theta_w\right)^2(1-\cos\theta)^2 + \left(\frac{2}{3}\sin^2\theta_\mathrm{w}\right)^2(1+\cos\theta)^2\right], \tag{15.1}$$

where θ is the angle between the direction of the d_k quark momentum and the direction of the Z spin. Similarly, the decay rate to a $\mathrm{u}\bar{\mathrm{u}}$ or $\mathrm{c}\bar{\mathrm{c}}$ pair is

$$\frac{\mathrm{d}\Gamma(\mathrm{u}_k\bar{\mathrm{u}}_k)}{\mathrm{d}\cos\theta} = \frac{3G_\mathrm{F}{M_Z}^3}{32\sqrt{2}\pi}\left[\left(1-\frac{4}{3}\sin^2\theta_w\right)^2(1-\cos\theta)^2 + \left(\frac{4}{3}\sin^2\theta_w\right)^2(1+\cos\theta)^2\right]. \tag{15.2}$$

The colour factor of 3 is included in these rates. Terms in m_q/M_Z are neglected. Integrating over θ gives the total decay rates

$$\Gamma(\mathrm{d}_k\bar{\mathrm{d}}_k) = \frac{G_F M_Z{}^3}{4\sqrt{2}\pi}\left[1 - \frac{4}{3}\sin^2\theta_\mathrm{w} + \frac{8}{9}\sin^4\theta_\mathrm{w}\right] = 0.3677\,\mathrm{GeV}, \quad (15.3)$$

$$\Gamma(\mathrm{u}_k\bar{\mathrm{u}}_k) = \frac{G_F M_Z{}^3}{4\sqrt{2}\pi}\left[1 - \frac{8}{3}\sin^2\theta_\mathrm{w} + \frac{32}{9}\sin^4\theta_\mathrm{w}\right] = 0.2853\,\mathrm{GeV}. \quad (15.4)$$

These numbers are obtained taking $\sin^2\theta_\mathrm{w} = 0.2315$ (see Section 11.4). Adding the decay rates to all pairs gives a total decay rate

$$\Gamma_{\mathrm{q}\bar{\mathrm{q}}} = 1.6737\,\mathrm{GeV}.$$

This lowest order calculation is in quite good agreement with the experimental total hadronic decay rate, which is

$$\Gamma_{\mathrm{experiment}} = 1.741 \pm 0.006\,\mathrm{GeV}.$$

At the high energy of the Z boson, the effects of the strong interaction can be estimated with some confidence (Chapter 17). When additional gluon radiation is taken into account, the theoretical $\Gamma_{\mathrm{q}\bar{\mathrm{q}}}$ is modified by a factor $f = 1.038$, and gives

$$\Gamma_{\mathrm{theoretical}} = f\Gamma_{\mathrm{q}\bar{\mathrm{q}}} = 1.737\,\mathrm{GeV},$$

in very close agreement with experiment.

The identification of $\mathrm{b}\bar{\mathrm{b}}$ jets and (less precisely) $\mathrm{c}\bar{\mathrm{c}}$ jets enables these partial decay modes also to be compared with the Standard Model. The estimates from experiment are

$$\Gamma(\mathrm{b}\bar{\mathrm{b}}) = 0.385 \pm 0.006\,\mathrm{GeV},$$
$$\Gamma(\mathrm{c}\bar{\mathrm{c}}) = 0.275 \pm 0.025\,\mathrm{GeV}.$$

The Standard Model values, (15.3) and (15.4) corrected by the factor *f*, are

$$\Gamma(\mathrm{b}\bar{\mathrm{b}})\,(\mathrm{theoretical}) = 0.3817\,\mathrm{GeV},$$
$$\Gamma(\mathrm{c}\bar{\mathrm{c}})\,(\mathrm{theoretical}) = 0.2961\,\mathrm{GeV}.$$

The agreement between theory and experiment is satisfactory.

15.2 Asymmetry in quark production

We noted in Section 13.6 that the SLC electron beam can be polarised to produce Z bosons with a much higher degree of polarisation than those produced at CERN by unpolarised beams. From (15.1) there is a forward–backward asymmetry, with respect to the Z spin direction, in the angular distribution of b quarks in a $b\bar{b}$ pair produced by Z decay, given by

$$\begin{aligned}\frac{\Delta\Gamma}{\Gamma} &= \frac{\Gamma(0<\theta<\pi/2)-\Gamma(\pi/2<\theta<\pi)}{\Gamma(0<\theta<\pi/2)+\Gamma(\pi/2<\theta<\pi)} \\ &= -\frac{3}{4}\left(\frac{1-(4/3)\sin^2\theta_{\rm w}}{1-(4/3)\sin^2\theta_{\rm w}+(8/9)\sin^4\theta_{\rm w}}\right).\end{aligned}$$

Taking $\sin^2\theta_{\rm w} = 0.2315$ gives $\Delta\Gamma/\Gamma = -0.7016$. At the peak of the Z mass distribution electromagnetic interference effects are very small, and one can expect a forward–backward asymmetry in the b quark jets relative to the electron beam direction. Measurements of b quark jets at SLC give a value of $\Delta\Gamma/\Gamma = -0.630 \pm 0.075$ (Prescott, 1996).

At LEP the Zs produced in e^+e^- collisions are polarised along the direction of the electron beam with polarisation P, to give a forward–backward asymmetry of b quark jets with respect to the electron beam direction of

$$A^b_{\rm FB} = P\frac{\Delta\Gamma}{\Gamma}.$$

From Section 13.6, taking $\sin^2\theta_{\rm w} = 0.2315$ gives $P = -A_{\rm LR} = -0.148$, so that

$$A^b_{\rm FB}(\text{theory}) = 0.104.$$

The experimental value (Renton, 1996) is

$$A^b_{\rm FB}(\text{experimental}) = 0.0997 \pm 0.0031.$$

The corresponding numbers for the c quark jets are

$$\begin{aligned}A^c_{\rm FB}(\text{theory}) &= 0.0719,\\ A^c_{\rm FB}(\text{experimental}) &= 0.0729 \pm 0.0058.\end{aligned}$$

Again the Standard Model and experiment are in accord.

A significant aspect of these asymmetry measurements is that an assignment of the right-handed rather than the left-handed quark fields to the $SU(2)$ doublet would lead to an asymmetry of opposite sign. (The total widths would be unaffected.) The results vindicate the left-handed assignment.

15.3 Hadronic decays of the $W^\pm$

The e^+e^- colliders give a clean source of Z bosons, but there is as yet no clean source of $W^\pm$ bosons. Consequently the experimental data on $W^\pm$ decays is less precise than that for Z decay. The hadronic decays of a $W^\pm$ are, in its rest frame, like those of the Z: principally into two back-to-back jets, which are interpreted as the signatures of the initiating quark–antiquark pairs.

Consider for example the decay of the W^+ to a quark u_i ($u_1 = u$, $u_2 = c$) and an antiquark $\bar{d}_j$ ($\bar{d}_1 = \bar{d}$, $\bar{d}_2 = \bar{s}$, $\bar{d}_3 = \bar{b}$). The coupling of the W^+ to the quark fields is given by $\mathcal{L}_{qw}$ (equation (14.15)), and depends on the elements V_{ij} of the Kobayashi–Maskawa matrix. In the lowest order of perturbation theory, and neglecting quark masses, the differential decay rate to a pair $u_i\bar{d}_j$ is

$$\frac{d\Gamma_{ij}}{d\cos\theta} = \frac{3G_F M_W{}^3}{16\sqrt{2}\pi}|V_{ij}|^2(1-\cos\theta)^2, \tag{15.5}$$

where θ is the angle between the direction of the u_i momentum and the direction of the W^+ spin. Integrating over θ gives the total decay rate

$$\Gamma(W^+ \to u_i\bar{d}_j) = \frac{G_F M_w{}^3}{2\sqrt{2}\pi}|V_{ij}|^2 = (0.677 \pm 0.006)|V_{ij}|^2\ \text{GeV}. \tag{15.6}$$

There is no data that resolves both initiating quark jets, so that we have no information from W decay on individual components of the KM matrix. However, we can sum over j, and since the KM matrix is unitary

$$\sum_{j=1}^{3}|V_{ij}|^2 = \sum_{j=1}^{3} V_{ij}V_{ij}^* = \sum_{j=1}^{3} V_{ij}V_{ji}{}^\dagger = 1 \quad \text{for } i = 1, 2, 3.$$

Then summing over the possible u_i, the u and c quarks, and including the factor f, we have

$$\Gamma(\text{all possible } q\bar{q}' \text{ pairs}) = \frac{G_F M_W{}^3 f}{\sqrt{2}\pi} = 1.41 \pm 0.008\ \text{GeV}.$$

This value is in close agreement with the observed hadronic decay rate of the W^+:

$$\Gamma(\text{hadronic}) = 1.44 \pm 0.04\ \text{GeV}.$$

Also, c quark jets can be identified with some confidence. From the above we would expect

$$\frac{\Gamma(\text{all possible } c\bar{q}' \text{ pairs})}{\Gamma(\text{all possible } q\bar{q}' \text{ pairs})} = 0.5$$

close to the measured value 0.51 ± 0.08.

In conclusion, it would seem that we have no reason to doubt the efficacy of the Standard Model in describing the interactions of the Z and $W^\pm$ bosons with both leptons and quarks. The details of the KM matrix V_{ij} remain undetermined by these experiments, but it does pass two tests of unitarity. We have to rely on lower energy hadron physics to investigate the KM matrix more thoroughly, as will be discussed in Chapter 18.

Problems

15.1 Obtain the decay rates (15.3), (15.4) and (15.6). Note that quark masses have been neglected in these expressions (cf. Problem 13.3).

16

The theory of strong interactions: quantum chromodynamics

The basic features of the quark model of hadrons were set out in Chapter 1. Quarks carry a colour index, and interact with the gluon fields which mediate the strong interaction.

We have seen that in the Standard Model the electromagnetic interaction and the weak interaction are well described by gauge theories. In the Standard Model the strong interaction also is described by a gauge theory. In this chapter we show how this is done. The theory is known as *quantum chromodynamics* (QCD) and has the remarkable property that in the theory quarks are confined, as appears to be the case experimentally (Section 1.4). In this chapter we concentrate exclusively on the strong interaction. The electromagnetic and weak interactions of quarks are neglected.

16.1 A local $SU(3)$ gauge theory

In QCD, we have three fields for each flavour of quark. These are put into so-called *colour triplets*. For example the u quark is associated with the triplet

$$\mathbf{u} = \begin{pmatrix} \mathrm{u_r} \\ \mathrm{u_g} \\ \mathrm{u_b} \end{pmatrix},$$

where $\mathrm{u_r}$, $\mathrm{u_g}$, $\mathrm{u_b}$ are four-component Dirac spinors, and the subscripts r, g, b label the colour states (red, green, blue, say).

We then postulate that the theory is invariant under a local $SU(3)$ transformation

$$\mathbf{q} \to \mathbf{q}' = \mathbf{U}\mathbf{q} \tag{16.1a}$$

where $\mathbf{q}$ is any quark triplet, and $\mathbf{U}$ is any space- and time-dependent element of the group $SU(3)$. The mathematical steps follow those of the $SU(2)$ theory of the weak interaction of leptons. We introduce a 3×3 matrix gauge field $\mathbf{G}_\mu$, which is the analogue of the matrix field $\mathbf{W}_\mu$ of the electroweak theory. Under an $SU(3)$ transformation,

$$\mathbf{G}_\mu \to \mathbf{G}_\mu{}' = \mathbf{U}\mathbf{G}_\mu\mathbf{U}^\dagger + (\mathrm{i}/g)(\partial_\mu \mathbf{U})\mathbf{U}^\dagger. \tag{16.1b}$$

We define

$$D_\mu \mathbf{q} = (\partial_\mu + \mathrm{i}g\mathbf{G}_\mu)\mathbf{q}. \tag{16.2}$$

It follows that under a local $SU(3)$ transformation

$$D_\mu{}'\mathbf{q}' = \mathbf{U}D_\mu \mathbf{q} \tag{16.3}$$

where $D_\mu{}'\mathbf{q}' = (\partial_\mu + igG_\mu{}')\mathbf{q}'$. The parameter g that appears in these equations is the *strong coupling constant.*

$\mathbf{G}_\mu$ is taken to be Hermitian and traceless, like $\mathbf{W}_\mu$ in the electroweak theory, and hence it can be expressed in terms of the eight matrices λ_a set out in Appendix B, Section B.7:

$$\mathbf{G}_\mu = \frac{1}{2}\sum_{a=1}^{8} G^a_\mu \lambda_a \tag{16.4}$$

where the coefficients $G^a_\mu(x)$ are eight real independent gluon gauge fields. (The factor $\frac{1}{2}$ is conventional.)

The Yang–Mills construction (cf. Section 11.2),

$$\mathbf{G}_{\mu\nu} = \partial_\mu \mathbf{G}_\nu - \partial_\nu \mathbf{G}_\mu + \mathrm{i}g(\mathbf{G}_\mu\mathbf{G}_\nu - \mathbf{G}_\nu\mathbf{G}_\mu), \tag{16.5}$$

leads to the result that, under $SU(3)$ transformations of the form (16.1b),

$$\mathbf{G}'_{\mu\nu} = \mathbf{U}\mathbf{G}_{\mu\nu}\mathbf{U}^+. \tag{16.6}$$

The gluon Lagrangian density is taken to be

$$\mathcal{L}_{\text{gluon}} = -\frac{1}{2}\mathrm{Tr}[\mathbf{G}_{\mu\nu}\mathbf{G}^{\mu\nu}]. \tag{16.7}$$

It follows from (16.16) and the cyclic invariance of the trace that $\mathcal{L}_{\text{gluon}}$ is gauge invariant.

We can expand $\mathbf{G}_{\mu\nu}$ in terms of its 'components',

$$\mathbf{G}_{\mu\nu} = \frac{1}{2}\sum_{a=1}^{8} G^a_{\mu\nu}\lambda_a, \tag{16.8}$$

using equation (B.27) of Appendix B. Hence, using also the property (B.28), that

$$\mathrm{Tr}(\lambda_a\lambda_b) = 2\delta_{ab},$$

the gluon Lagrangian density becomes

$$\mathcal{L}_{\text{gluon}} = -\frac{1}{4}\sum_{a=1}^{8} G^a_{\mu\nu}G^{a\mu\nu}. \tag{16.9}$$

The quark Lagrangian density is taken to be of the standard Dirac form (equation (7.7)):

$$\mathcal{L}_{\text{quark}} = \sum_{f=1}^{6} [\bar{\mathbf{q}}_f \mathrm{i}\gamma^\mu(\partial_\mu + \mathrm{i}g\mathbf{G}_\mu)\mathbf{q}_f - m_f\bar{\mathbf{q}}_f\mathbf{q}_f], \tag{16.10}$$

where the sum is over all flavours of quark and m_f are the 'true' quark masses defined in Section 14.2. $\mathcal{L}_{\text{quark}}$ is evidently invariant under an $SU(3)$ transformation (using (16.3)). The reader should note here the very compact notation that has been developed: as well as the explicit sum over flavours, there are sums over colour indices and sums over the indices of the four-component Dirac spinor and γ matrices. It is perhaps instructive for the reader to write out the expression in full.

The total strong interaction Lagrangian density is

$$\mathcal{L}_{\text{strong}} = \mathcal{L}_{\text{gluon}} + \mathcal{L}_{\text{quark}}. \tag{16.11}$$

The eight gluon gauge fields have no mass terms. There is no direct coupling of the gluon fields to the Higgs field. The Higgs field is relevant in that it gives mass to the quarks. The field equations follow from Hamilton's principle of stationary action. For the six quark triplets we easily obtain (cf. Section 5.5)

$$(\mathrm{i}\gamma^\mu D_\mu - m_f)\mathbf{q}_f = 0. \tag{16.12}$$

For the eight gluon fields, variation of the Lagrangian density with respect to the field G^a_ν gives (cf. Section 4.2)

$$\partial_\mu G^{a\mu\nu} = j^{a\nu} \tag{16.13}$$

where

$$j^{a\nu} = g[f_{abc}G^b_\mu G^{c\mu\nu} + \sum_f \bar{\mathbf{q}}_f\gamma^\nu(\lambda_a/2)\mathbf{q}_f]. \tag{16.14}$$

Here f_{abc} are the $SU(3)$ structure constants, defined by

$$[\lambda_a, \lambda_b] = \lambda_a\lambda_b - \lambda_b\lambda_a = 2i\sum_{c=1}^{8} f_{abc}\lambda_c. \tag{16.15}$$

(See Appendix B, Section B.7.) Their appearance here stems from the definition (16.5) of $G_{\mu\nu}$.

Since $\mathbf{G}^{\mu\nu} = -\mathbf{G}^{\nu\mu}$ it follows that

$$\partial_\nu j^{a\nu} = 0, \tag{16.16}$$

and we have eight conserved currents. These are the Noether currents, which are a consequence of the $SU(3)$ symmetry taken as a global symmetry. We therefore have eight constants of the motion, associated with the time-independent operators

$$Q^a = \int j^{a0} \mathrm{d}^3\mathbf{x}. \tag{16.17}$$

The field equations, and in particular the gluon field equations, are non-linear, like the equations of the electroweak theory. It is clear from (16.14) that both the quarks and the gluon fields themselves contribute to the currents $j^{a\nu}$ which are the sources of the gluon fields. The quarks interact through the mediation of the gluon fields; the gluon fields are also self-interacting.

Since the gluon fields are massless we might anticipate colour forces to be long range, which appears inconsistent with the short range of the strong interaction. However, the fields are known to be *confining* on a length scale greater than about 10^{-15} m $= 1$ fm: neither free quarks nor free 'gluons' have ever been observed.

In the electroweak theory, the 'free field' approximation in which all coupling constants are set to zero is the basis for the successful perturbation calculations we have seen in the preceding chapters. The free field approximation for quarks and gluons is not a good starting point for calculations in QCD, except on the scale of very small distances (≤ 0.1 fm) or very high energies (> 10 GeV). For low energy physics, the equations of the theory are analytically highly intractable. Even the vacuum state is characterised by complicated field configurations that have so far defied analysis. There is no analytical proof of confinement. Confinement is not displayed in perturbation theory, but numerical simulations demonstrate convincingly that QCD has this necessary property for an acceptable theory.

16.2 Colour gauge transformations on baryons and mesons

Since colour symmetry plays such an important part in the theory of strong interactions, it is natural to ask why it is not readily apparent in the particles, baryons and mesons, formed from quarks by the strong interaction. Here we attempt to answer that question.

In Section 1.4 we asserted that baryons are essentially made up of three quarks, and mesons are essentially quark–antiquark pairs. We shall denote a three-quark state in which quark 1 is in colour state i, quark 2 is in colour state j, and quark 3 is in colour state k by $|i, j, k\rangle$, and take the colour indices to be the numbers 1, 2, 3. We have suppressed all other aspects (position, spin, flavour) of the quarks. In

Section 1.7 we saw that the Pauli principle required baryon states to be antisymmetric in the interchange of colour indices. The only antisymmetric combination of colour states we can construct is

$$|\text{state}\rangle = (1/\sqrt{6})\varepsilon_{ijk}|i, j, k\rangle, \tag{16.18}$$

where ε_{ijk} is defined by:

$$\varepsilon_{123} = \varepsilon_{231} = \varepsilon_{312} = -\varepsilon_{132} = -\varepsilon_{321} = -\varepsilon_{213} = 1,$$

and $\varepsilon_{ijk} = 0$ if any two of *i, j, k* are the same. $(1/\sqrt{6})$ is a normalisation factor.

How does this state transform under a colour $SU(3)$ transformation? We restrict the discussion to a global (space- and time-independent) transformation, since a baryon is an object extended in space. We consider the quark fields to be transformed by $\mathbf{q} \to \mathbf{q}' = \mathbf{U}\mathbf{q}$. In quantum field theory, these fields destroy quarks and create antiquarks. It follows that under the transformation the baryon state (16.18) will transform as $|\,\text{state}\rangle \to |\,\text{state}\rangle' = (1/\sqrt{6})|a,b,c\rangle U^*_{ai}U^*_{bj}U^*_{ck}\varepsilon_{ijk}$. But $\varepsilon_{ijk}U^*_{ai}U^*_{bj}U^*_{ck} = \varepsilon_{abc} \det \mathbf{U}^* = \varepsilon_{abc}$, since the determinant of an $SU(3)$ matrix is 1. Thus we have the important result that under an $SU(3)$ transformation, $|\text{state}\rangle' = |\text{state}\rangle$. The transformation of the state is a trivial multiplication by unity. The state is said to be a *colour singlet.*

Turning now to the mesons, we denote a state of a quark, colour *i*, and an antiquark of colour *j* by $|i, \bar{j}\rangle$. Again, we have suppressed all other aspects of the quarks. Meson states are linear combinations

$$|\text{mesons}\rangle = (1/\sqrt{3})(|1, \bar{1}\rangle + |2, \bar{2}\rangle + |3, \bar{3}\rangle). \tag{16.19}$$

Under an $SU(3)$ transformation,

$$|\text{meson}\rangle \to |\text{meson}\rangle' = (1/\sqrt{3})|a, \bar{b}\rangle U^*_{ai}U_{bi}.$$

But $U^*_{ai}U_{bi} = U_{bi}U^\dagger_{ia} = \delta_{ab}$, so that

$$|\text{meson}\rangle' = |\text{meson}\rangle.$$

The meson states, like the baryon states, are colour singlets.

In the quark model, we see that colour transformations have no effect on the observed particles. It can also be shown that the eight gluon colour operators Q^a, defined by (16.17), give zero when they act on these states. Thus the $SU(3)$ symmetry is well hidden by Nature: the particles are blind to the transformation of colour symmetry. These observations can be related to lattice QCD, in which calculations indicate that all the allowed states of the theory have this property.

16.3 Lattice QCD and asymptotic freedom

Numerical simulations of QCD replace continuous space-time by a finite but large four-dimensional space and time lattice of points. The quark and gluon fields are only defined at these points. Sophisticated computer programs have been written that are capable of handling the lattice. Gluon fields are commuting boson fields. The quark fields are anticommuting fermion fields and pose a technically much more difficult numerical problem. In fact the first lattice calculations were done neglecting all quark fields, even those of the light u and d quarks, and thus excluding all effects of virtual quark pair creation and annihilation. In this so-called *quenched approximation* the Lagrangian density it taken to be the $\mathcal{L}_{\text{gluon}}$ of (16.9). $\mathcal{L}_{\text{gluon}}$ displays confinement at distances greater than about a fermi.

At shorter distances, less than about 0.2 fermi, both $\mathcal{L}_{\text{gluon}}$ and the full QCD Lagrangian density display another important property, known as *asymptotic freedom.* The *effective* strong interaction coupling constant becomes so small at short distances that quarks and gluons can be considered as approximately free, and their interactions can be treated in perturbation theory.

To set the scene for the discussion of the effective 'running' strong interaction coupling constant, we first discuss the case of electromagnetism.

At atomic distances $\sim 10^{-10}$ m, the electrostatic interaction between an electron and a positron is given by the Coulomb energy $V(r) = -e^2/4\pi r$. In the lowest order of perturbation theory, the amplitude for electron–positron Coulomb scattering is proportional to the Fourier transform $V(Q^2)$ of $V(r)$,

$$V(Q^2) = \int V(r)\mathrm{e}^{\mathrm{i}\mathbf{Q}\cdot\mathbf{r}}\mathrm{d}^3r = -e^2/Q^2, \tag{16.20}$$

where Q is the momentum transfer in the centre of mass system.

In QED, this result is modified by quantum corrections: virtual $\mathrm{e}^+\mathrm{e}^-$ pairs created from the vacuum are polarised by the electric field of a charge, so that its measured charge at atomic distances is a 'bare' charge screened by virtual $\mathrm{e}^+\mathrm{e}^-$ pairs. At short distances the screening is reduced, so that the effective charge is greater. Perturbation calculations in QED that include vacuum polarisation effects (Fig. 16.1) show that at large Q^2, (16.20) is modified to

$$V(Q^2) = -\frac{e^2}{Q^2}\frac{1}{1-(e^2/12\pi^2)\ln(Q^2/4m^2)} \tag{16.21}$$

where m is the electron mass. This result holds for large $Q^2 \gg 4m^2$ (but not so large Q^2 that the denominator vanishes!). Thus at large Q^2 we have an effective coupling constant

$$\alpha(Q^2) = \frac{e^2(Q^2)}{4\pi} = \frac{(e^2/4\pi)}{1-(e^2/12\pi^2)\ln(Q^2/4m^2)}, \tag{16.22}$$

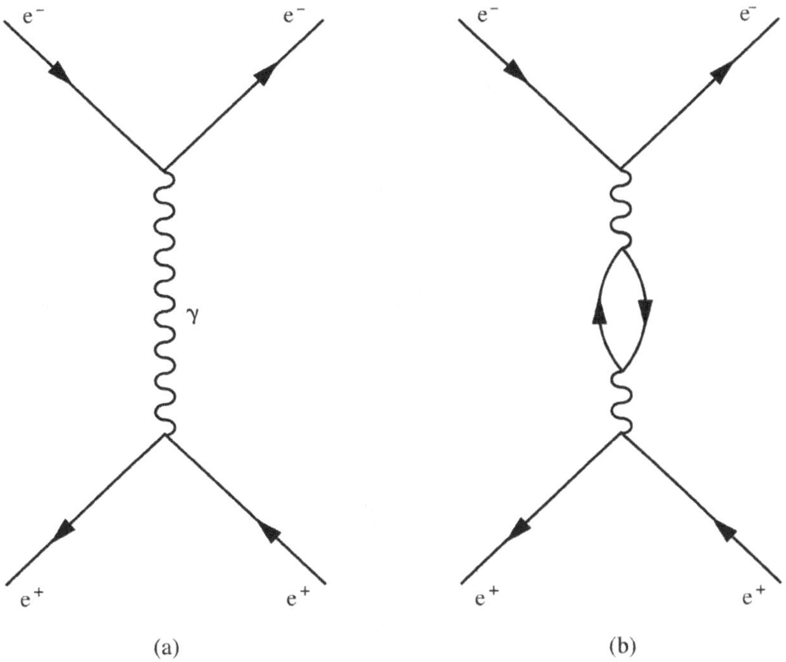

Figure 16.1 (a) The lowest order Feynman diagram representing single photon exchange. The corresponding perturbation calculation reproduces the result of (16.20). (b) The lowest order modification due to vacuum polarisation. Including this effect gives, at large Q^2/m^2, the result of (16.21).

which increases as Q^2 increases (or, equivalently, as we probe shorter distances). Because $e^2/12\pi^2 \approx 10^{-3}$ the effects of vacuum polarisation are small, but in atomic physics they have been calculated and measured with high precision.

Similar vacuum polarisation effects occur in QCD, but the coupling is much larger and the consequences are more dramatic. If the scattering of a quark and an antiquark is calculated to the same order of perturbation theory as that used to obtain (16.22), then at large Q^2 the effective strong coupling constant $\alpha_s(Q^2)$ is (see Close, 1979, p. 217)

$$\alpha_s(Q^2) = \frac{g^2(Q^2)}{4\pi} = \frac{g^2/4\pi}{1 + (g^2/16\pi^2)[11 - (2/3)n_{\mathrm{f}}]\ln(Q^2/\lambda^2)}. \tag{16.23}$$

In this expression λ is a parameter with the dimensions of energy that replaces the electron mass appearing in QED. It is a necessary parameter associated with the renormalisation scheme. n_{f} is the effective number of quark flavours. For very large $Q^2 >$ (mass of the top quark)^2, $n_{\mathrm{f}} = 6$, but n_{f} is smaller at smaller Q^2. The important point to note is that $(11 - (2/3)n_{\mathrm{f}})$ is a positive number. Thus, in contrast to what happens in QED, $g(Q^2)$ decreases as Q^2 increases, and this is the basis of

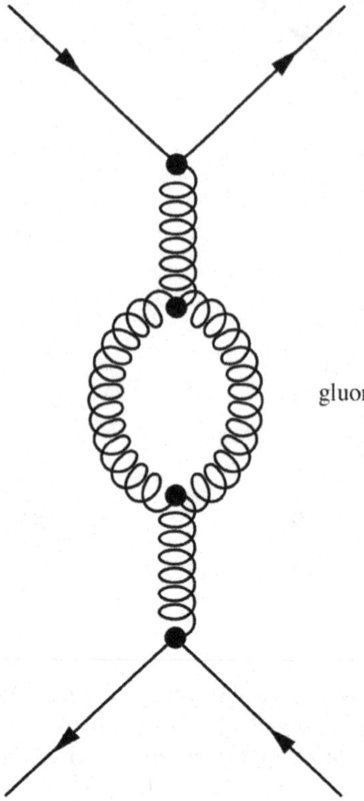

Figure 16.2 There are Feynman graphs similar to those of Fig. 16.1 but for gluon exchange between quarks and antiquarks. An additional lowest order contribution to vacuum polarisation is associated with this Feynman graph coming from the gluon self-coupling.

asymptotic freedom. As with QED the fermions contribute with a negative sign, but their contribution is outweighed by the virtual gluons that contribute the number 11. The difference is due to the presence of gluon loops in QCD (Fig. 16.2). This property of QCD was discovered by Gross and Wilczek (1973) and Politzer (1973).

Although renormalisation seems to necessitate the introduction of a second, dimensioned, parameter λ, the effective coupling constant is in fact dependent on only one parameter. We can set

$$\frac{1}{g^2} - \frac{1}{16\pi^2}[11 - (2/3)n_f] \ln \lambda^2 = -\frac{1}{16\pi^2}[11 - (2/3)n_\mathrm{f}] \ln \Lambda^2, \tag{16.24}$$

thus defining Λ, and then

$$\alpha_s(Q^2) = \frac{g^2(Q^2)}{4\pi} = \frac{4\pi}{[11 - (2/3)n_f]\ln(Q^2/\Lambda^2)}. \qquad (16.25)$$

This remarkable feature survives in all orders of perturbation theory. Higher terms in the expansion of $\alpha_s(Q^2)$ are given in, for example, Particle Data Group (2005).

Λ is well defined in the limit of large Q^2, and it is standard practice to regard the one parameter Λ, rather than the two parameters g and λ, as the fundamental constant of QCD, which must be determined from experiment. It is also interesting to note that we have replaced a dimensionless parameter g by a dimensioned one, Λ. Asymptotic freedom is displayed since $\alpha_s(Q^2) \to 0$ as $Q^2 \to \infty$. It is clear from (16.25) that perturbation theory breaks down at $Q^2 = \Lambda^2$, when the effective coupling constant becomes infinite. Small values of Q^2 are associated with large distances, and the length scale Λ^{-1} is called the *confinement length.*

16.4 The quark–antiquark interaction at short distances

In QED, single photon exchange between an electron and a positron gives the Coulomb potential

$$V(r) = \frac{1}{(2\pi)^3}\int V(Q^2)\mathrm{e}^{-\mathrm{i}Q\cdot r}\mathrm{d}^3Q = \frac{e^2}{4\pi r} = -\frac{\alpha}{r},$$

where $V(Q^2) = -e^2/Q^2$ and α is the fine-structure constant. In QCD perturbation theory, single photon exchange is replaced by the sum of eight single gluon exchanges. To lowest order, the Coulomb-like potential between a quark and an antiquark in a colour singlet state and at a distance r apart may be shown to be (see Leader and Predazzi, 1982, p. 175)

$$V_{\mathrm{QCD}}(r) = -\sum_a \frac{g^2}{4\pi r}\frac{1}{3}\frac{\lambda_{aij}}{2}\frac{\lambda_{aji}}{2} = -\sum_a \frac{g^2}{4\pi r}\frac{1}{12}\mathrm{Tr}(\lambda_a\lambda_a) = -\frac{4}{3}\frac{g^2}{4\pi r}. \qquad (16.26)$$

The factor (1/3) is from the normalisation of the colour singlet state (see (16.19)). With quantum corrections, the effective potential at short distances becomes

$$V_{\mathrm{QCD}} = -\frac{4}{3}\frac{\alpha_s(r)}{r},$$

where

$$\frac{\alpha_s(r)}{r} = \frac{4\pi}{(2\pi)^3}\int \frac{\alpha_s(Q^2)}{Q^2}\mathrm{e}^{-\mathrm{i}Q\cdot r}\mathrm{d}^3Q. \qquad (16.27)$$

This is a significant result for the charmonium $c\bar{c}$ and bottomonium $b\bar{b}$ systems, in which the heavy quark and antiquark are slowly moving. In these systems the colour Coulomb energy is the main contribution to the potential energy: colour magnetic effects are of relative order v/c. The behaviour of $\alpha_s(Q^2)$ at large Q^2 gives the dominant contribution to $V_{\mathrm{QCD}}(r)$ at small r (Problem 16.5). We shall return to charmonium and bottomonium in Chapter 17.

16.5 The conservation of quarks

In addition to the $SU(3)$ local colour symmetry, the Lagrangian density (16.11) has six global $U(1)$ symmetries:

$$\mathbf{q}_{\mathrm{f}} \to \mathbf{q}_{\mathrm{f}}{}' = \exp(\mathrm{i}\alpha_{\mathrm{f}})\mathbf{q}_{\mathrm{f}}. \tag{16.28}$$

In the Standard Model these remain global and are not elevated into local gauge symmetries. They imply conservation of quark number for each flavour of quark. Thus the strong interaction does not change quark flavour. Regarding mesons and baryons, the K^+, for example, which can be denoted $\mathrm{K}(\mathrm{u}\bar{\mathrm{s}})$ has u quark number 1 and s quark number -1, the proton P (uud) has u quark number 2 and d quark number 1. Only the weak interaction, as exemplified in weak decays, can change quark flavour. Including the weak interaction, and in particular that part involving the Kobayashi–Maskawa mixing matrix, the six $U(1)$ symmetries reduce to one. Individual quark flavour numbers are not conserved, and only the overall quark number remains constant.

16.6 Isospin symmetry

The estimated masses of the u quark ($1.5\,\mathrm{MeV} < m_{\mathrm{u}} < 4\,\mathrm{MeV}$) and d quark ($4\,\mathrm{MeV} < m_{\mathrm{d}} < 8\,\mathrm{MeV}$) are small compared with those of the s quark ($100\,\mathrm{MeV} < m_{\mathrm{s}} < 300\,\mathrm{MeV}$) and the heavy c, b and t quarks. The masses of the u and d quarks are also small compared with those of the lightest hadrons: the π^0 has a mass $\sim 135\,\mathrm{MeV}$ and the proton has a mass $\sim 938\,\mathrm{MeV}$. At low energies we may therefore neglect all but the u and d quarks, and consider the Lagrangian density to be, as a first approximation,

$$\mathcal{L}_{\mathrm{ud}} = \bar{\mathbf{u}}\mathrm{i}\gamma^{\mu}(\partial_{\mu} + \mathrm{i}g\mathbf{G}_{\mu})\mathbf{u} + \bar{\mathbf{d}}\mathrm{i}\gamma^{\mu}(\partial_{\mu} + \mathrm{i}g\mathbf{G}_{\mu})\mathbf{d} - m_{\mathrm{u}}\bar{\mathbf{u}}\mathbf{u} - m_{\mathrm{d}}\bar{\mathbf{d}}\mathbf{d} \tag{16.29}$$

where here $\mathbf{G}_{\mu}$ is the gluon field matrix, evaluated from the field equations (16.13) with all but the **u** and **d** quark fields neglected. The fields **u** and **d** in (16.29) are triplets of Dirac fermion fields; colour indices and Dirac indices have been suppressed.

We now combine the **u** and **d** fields into an *isospin doublet*,

$$\mathbf{D}(x) = \begin{pmatrix} \mathbf{u}(x) \\ \mathbf{d}(x) \end{pmatrix} \tag{16.30}$$

and we can write

$$\mathcal{L}_{\mathrm{ud}} = \bar{\mathbf{D}}\mathrm{i}\gamma^{\mu}(\partial_{\mu} + \mathrm{i}g\mathbf{G}_{\mu})\mathbf{D} - (1/2)(m_{\mathrm{u}} + m_{\mathrm{d}})\bar{\mathbf{D}}\mathbf{D} - (1/2)(m_{\mathrm{u}} - m_{\mathrm{d}})\bar{\mathbf{D}}\tau_3\mathbf{D} \tag{16.31}$$

where

$$\tau_3 = \begin{pmatrix} 1 & 0 \\ 0 & -1 \end{pmatrix} \quad \text{and} \quad \bar{\mathbf{D}} = (\mathrm{u}^{+}\gamma^{0}, \mathrm{d}^{\dagger}\gamma^{0}).$$

$\mathcal{L}_{\mathrm{ud}}$ is invariant under a global $U(1)$ transformation

$$\mathbf{D} \to \mathbf{D}' = \exp(-\mathrm{i}\alpha^{0})\mathbf{D}, \tag{16.32}$$

which leads (cf. Section 4.1) to the conserved quark current

$$J^{\mu} = \bar{\mathbf{D}}\gamma^{\mu}\mathbf{D} = \bar{\mathbf{u}}\gamma^{\mu}\mathbf{u} + \bar{\mathbf{d}}\gamma^{\mu}\mathbf{d}. \tag{16.33}$$

It is also invariant under a global $U(1)$ transformation

$$\mathbf{D} \to \mathbf{D}' = \exp(-\mathrm{i}\alpha^{3}\tau^{3})\mathbf{D} \tag{16.34}$$

which leads to the conserved current

$$J_3{}^{\mu} = \bar{\mathbf{D}}\gamma^{\mu}\tau^{3}\mathbf{D} = \bar{\mathbf{u}}\gamma^{\mu}\mathbf{u} - \mathbf{d}\gamma^{\mu}\mathbf{d}. \tag{16.35}$$

(16.33) and (16.35) show that this Lagrangian density (16.31) conserves both u and d quark numbers separately.

So-called *isospin symmetry* appears if we neglect the mass difference $(m_{\mathrm{u}} - m_{\mathrm{d}})$. The resulting, simplified, Lagrangian density is invariant under the global $SU(2)$ transformation

$$\mathbf{D} \to \mathbf{D}' = \exp(-\mathrm{i}\alpha^{k}\tau^{k})\mathbf{D} \tag{16.36}$$

where the τ^k are the generators of the group $SU(2)$ (Appendix B, Section B.3). In addition to the conserved current (16.35) we now have also the conserved currents

$$J_1^{\mu} = \bar{\mathbf{D}}\gamma^{\mu}\tau^{1}\mathbf{D}, \quad J_2^{\mu} = \bar{\mathbf{D}}\gamma^{\mu}\tau^{2}\mathbf{D} \tag{16.37}$$

and the corresponding time-independent quantities

$$\int \mathbf{D}^{\dagger}\tau^{k}\mathbf{D}\,\mathrm{d}^{3}x, \quad k = 1,2,3. \tag{16.38}$$

$SU(2)$ transformations are equivalent to rotations in a three-dimensional 'isospin space'. In analogy with the intrinsic angular momentum operator $\mathbf{S} = (1/2)\boldsymbol{\sigma}$, we define the isospin operator $\mathbf{I} = (1/2)\boldsymbol{\tau}$; then

$$\mathbf{I}^2 = I_1{}^2 + I_2{}^2 + I_3{}^2 = (3/4)\begin{pmatrix} 1 & 0 \\ 0 & 1 \end{pmatrix} = \frac{1}{2}\left(\frac{1}{2}+1\right)\begin{pmatrix} 1 & 0 \\ 0 & 1 \end{pmatrix}.$$

A u quark state is an eigenstate of $\mathbf{I}^2$ and I_3 with $I = 1/2$, $I_3 = 1/2$, and a d quark state is an eigenstate with $I = 1/2$, $I_3 = -1/2$. The mathematics of isospin is identical to the mathematics of angular momentum, and the formalism of isospin is very useful in understanding and classifying hadron states, as indicated in Chapter 1. We see here its origin in QCD, with the neglect of the u − d mass difference and the electromagnetic and weak interactions.

16.7 Chiral symmetry

If we neglect entirely the quark masses, further approximate symmetries arise. These are of interest in particle physics. The Lagrangian density (16.31) may be written in terms of the left-handed and right-handed isospin doublets $\mathbf{L} = (1/2)(1-\gamma^5)\mathbf{D}$ and $\mathbf{R} = (1/2)(1+\gamma^5)\mathbf{D}$. Neglecting the mass terms it becomes

$$\mathcal{L} = \mathbf{L}^\dagger \mathrm{i}\tilde{\sigma}^\mu(\partial_\mu + \mathrm{i}g\mathbf{G}_\mu)\mathbf{L} + \mathbf{R}^\dagger \mathrm{i}\sigma^\mu(\partial_\mu + \mathrm{i}g\mathbf{G}_\mu)\mathbf{R}. \tag{16.39}$$

$\mathbf{L}$ and $\mathbf{R}$ are now doublets of two-component spinors, and there are eight conserved currents:

$$\mathbf{L}^\dagger\tilde{\sigma}^\mu\mathbf{L}, \quad \mathbf{L}^\dagger\tilde{\sigma}^\mu\tau^k\mathbf{L}, \quad \mathbf{R}^\dagger\sigma^\mu\mathbf{R}, \quad \mathbf{R}^\dagger\sigma^\mu\tau^k\mathbf{R}, \quad k = 1, 2, 3.$$

An important observation is that the currents $\mathbf{L}^\dagger\tilde{\sigma}^\mu\tau^1\mathbf{L}$ and $\mathbf{L}^\dagger\tilde{\sigma}^\mu\tau^2\mathbf{L}$ couple to the $W^\pm$ boson fields in the Lagrangian density (14.15), and appear in the effective Lagrangian density (14.22). The relevant quark factor in (14.15) is $\mathbf{u}_\mathrm{L}^\dagger\tilde{\sigma}^\mu\mathbf{d}_\mathrm{L}V_{\mathrm{ud}}$, and we may write

$$\begin{aligned} \mathbf{u}_\mathrm{L}^\dagger\tilde{\sigma}^\mu\mathbf{d}_\mathrm{L} &= \mathbf{L}^\dagger\tilde{\sigma}^\mu(1/2)(\tau^1 + \mathrm{i}\tau^2)\mathbf{L}, \\ \mathbf{d}_\mathrm{L}^\dagger\tilde{\sigma}^\mu\mathbf{u}_\mathrm{L} &= \mathbf{L}^\dagger\tilde{\sigma}^\mu(1/2)(\tau^1 - \mathrm{i}\tau^2)\mathbf{L}. \end{aligned} \tag{16.40}$$

This observation gives insight into the nature of the effective Lagrangian for β decay, as we shall see in Chapter 18.

The independent symmetry transformations

$$\mathbf{L} \to \mathbf{L}' = \exp[-\mathrm{i}(\alpha^0 + \alpha^k\tau^k)]\mathbf{L}, \quad \mathbf{R} \to \mathbf{R}$$

and

$$\mathbf{R} \to \mathbf{R}' = \exp[-\mathrm{i}(\beta^0 + \beta^k\tau^k)]\mathbf{R}, \quad \mathbf{L} \to \mathbf{L}$$

may be written in terms of Dirac spinors as

$$\mathbf{D} \to \mathbf{D}' = \exp[-\mathrm{i}(\alpha^0 + \alpha^k \tau^k)(1/2)(1-\gamma^5)]\mathbf{D}, \qquad (16.41)$$
$$\mathbf{D} \to \mathbf{D}' = \exp[-\mathrm{i}(\beta^0 + \beta^k \tau^k)(1/2)(1+\gamma^5)]\mathbf{D}, \qquad (16.42)$$

respectively.

The eight independent symmetry operations can also be taken as

$$\mathbf{D} \to \mathbf{D}' = \exp[-\mathrm{i}(\alpha'^0 + \alpha'^k \tau^k)]\mathbf{D} \qquad (16.43)$$

which give conservation of quark number and isospin, and

$$\mathbf{D} \to \mathbf{D}' = \mathbf{exp}[-\mathrm{i}(\beta'^0 + \beta'^k \tau^k)\gamma^5]\mathbf{D} \qquad (16.44)$$

The last four are known as the *chiral symmetries.*

Problems

16.1 Show that

$$G^a_{\mu\nu} = (\partial_\mu G^a_\nu - \partial_\nu G^a_\mu) - g \sum_{b,c} f_{abc} G^b_\mu G^c_\nu .$$

16.2 Using Problem 16.1, show that the gluon self-coupling terms in the Lagrangian density (16.9) are

$$\mathcal{L}_{\text{int}} = g(\partial_\mu G^a_\nu f_{abc} G^{b\mu} G^{c\nu}) - (g^2/4) f_{abc} f_{ade} G^b_\mu G^c_\nu G^{d\mu} G^{e\nu}.$$

16.3 Verify the expression (16.14) for the current $j^{a\nu}$.

16.4 Estimate the value of Q for which $V(Q^2)$ of equation (16.21) becomes infinite.

16.5 From (16.27) show that

$$\alpha_s(r) = \frac{2}{\pi} \int_0^\infty \alpha_s(x^2/r^2) \frac{\sin x}{x}\, \mathrm{d}x.$$

(Note that the expression (16.25) for $\alpha_s(x^2/r^2)$ is only valid for $x > \Lambda r$, but for small r this range may be anticipated to give the main contribution to the integral.)

17

Quantum chromodynamics: calculations

Calculations in QCD have been made in two ways: lattice simulations at low energies, and perturbative calculations at high energies. In this chapter we outline some of the results obtained.

17.1 Lattice QCD and confinement

It was pointed out in Section 16.1 that, at low energies, a non-perturbative approach to QCD is needed. 'Lattice QCD' is such an approach. The gluon fields are defined on a four-dimensional lattice of points $(n^\mu, \mathbf{n})a$, where a is the lattice spacing and the n^μ are integers. Field derivatives are replaced by discrete differences. This gives a 'lattice regularised' QCD. The lattice spacing corresponds to an ultraviolet cut-off, since wavelengths $< 2a$ cannot be described on the lattice. A lattice does not have full rotational symmetry in space, but it is believed that nevertheless continuum QCD corresponds to the limit $a \to 0$. Current computing power allows lattices of $\sim(36)^4$ points. The range of the strong nuclear force is $\sim$ 1 fm. To fit such a distance comfortably on the lattice, we can anticipate that we shall not want a to be much less than $(2\text{fm})/36 = 0.056\text{fm}$ (and $\hbar c/a > 3.5\,\text{GeV}$).

In the high energy perturbation theory described in Section 16.3, the renormalisation parameter λ and the dimensionless coupling parameter g are combined to give a single physical parameter, Λ, having the dimensions of energy. The relationship between the effective coupling constant $\alpha_s(Q^2)$ and Λ in the lowest order of perturbation theory is given by (16.25). In lattice QCD, the unphysical lattice parameter a and the dimensionless coupling parameter $g(a)$ combine to give a single physical parameter Λ_{latt}, having the dimensions of energy. In the lowest order of 'lattice' perturbation theory, as $a \to 0$ then $g(a) \to 0$,

$$g^2(a) = \frac{-16\pi^2}{11\,\ln\left(a^2\Lambda_{\text{latt}}^2\right)} \tag{17.1}$$

(see Hasenfratz and Hasenfratz, 1985).

Λ_{latt} is independent of a in the limit $a \to 0$. This remarkable feature of the theory is called *dimensional transmutation.*

Equation (17.1) may be compared with (16.25) with n_{f} set equal to zero. It can be shown theoretically (Dashen and Gross, 1981) that

$$\frac{\Lambda_{\text{latt}}}{\Lambda} = \text{constant} \approx \frac{1}{30}. \tag{17.2}$$

The precise value of the constant depends on the renormalisation scheme in which Λ is defined, and the number of quark flavours included. Λ_{latt}, or equivalently Λ, is to be determined from experiment. We shall see in Section 17.3 that Λ is known to be $\sim 300\,\text{MeV}$, so that $\Lambda_{\text{latt}} \sim 10\,\text{MeV}$. We can then infer from equation (17.1) that for $a \sim 0.056\,\text{fm}$, the coupling constant g should be of order 1.

Lattice QCD calculations have been made to compute the potential energy of a fixed quark and an antiquark in a colour singlet state, as a function of their separation distance. The form of this potential at short distances was discussed in Section 16.4. Non-perturbative lattice calculations have been made in the quenched approximation, excluding effects of virtual quark pair creation.

In the lattice calculations, distances are measured in units of a, and energies in units of $(1/a)$. A coupling constant g is chosen, and the quark and antiquark are localised on lattice sites that are spatially fixed at a distance apart of $r = |\mathbf{n}|a$, where $\mathbf{n}$ is a set of three integers. The field energy $E(r)$ generated by the quark–antiquark pair is computed for a sequence of separation distances, and is found to be of the form

$$E(r) = 2A + Kr - \frac{4}{3}\frac{\alpha_{\text{latt}}(r)}{r}, \tag{17.3}$$

where A and K are constants, and the factor (4/3) has been inserted to facilitate comparison with the perturbation results of Section 16.4. The constant $2A$ can be interpreted as a contribution to the rest energies of the quark and antiquark, and is absorbed into their notional masses to leave an effective potential energy

$$V(r) = Kr - \frac{4}{3}\frac{\alpha_{\text{latt}}(r)}{r}. \tag{17.4}$$

The results of such a calculation by Bali and Schilling (1993) using a $(32)^4$ lattice are shown in Fig. 17.1. In this calculation $g = 0.97$. The term Kr dominates at large distances. The constant K is called the *string tension.* In quenched QCD on a lattice, with g fixed, there is only one energy parameter a^{-1} (or Λ_{latt}). Hence

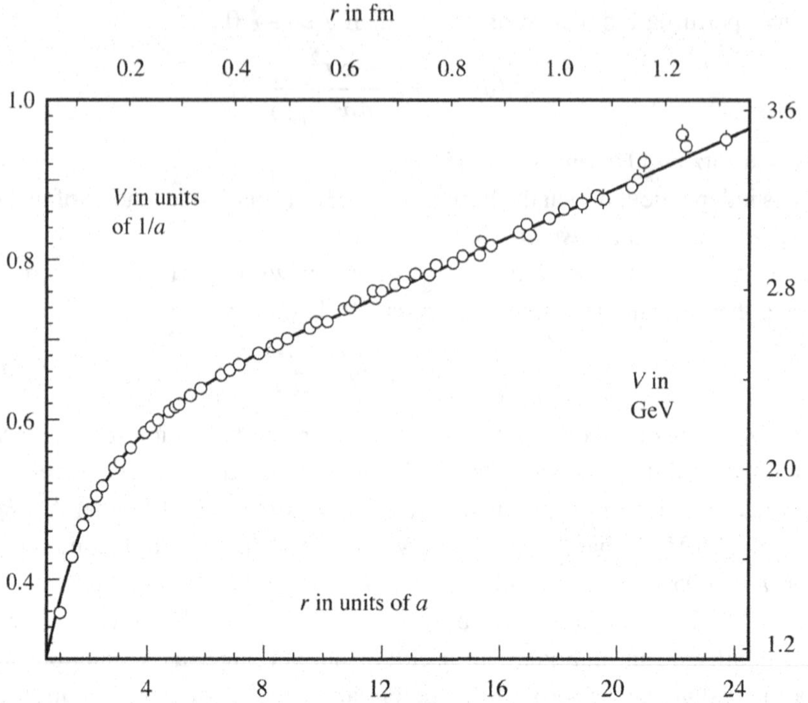

Figure 17.1 The colour singlet quark–antiquark potential as computed on a lattice. For a fixed value of the coupling constant g (of order 1) $V(r)$ is computed in lattice units (r in units of a, V in units of $1/a$). The computed points are fitted with a curve of the form

$$V(r) = 2A + Kr - (c/r) + (f/r^2).$$

In this example g was fixed at 0.97. The calculation determined $K = 0.0148$; K is the string tension in units of $1/a^2$. The phenomenology of $c\bar{c}$ and $b\,\bar{b}$ quark systems suggests $K \approx (440\,\text{MeV})^2$. Taking this value determines $a = 0.055$ fm and $1/a = 3.58$ GeV. It also determines one point on the curve $g(a)$ as a function of a. The calculations must be repeated to compute a for several values of g to check the extent to which the asymptotic form, like equation (17.1), is obeyed (Λ_{latt} is independent of a) in order to be confident of the continuum limit (Bali and Schilling, 1993).

K has the dimensions of a^{-2}. Bali and Schilling (1993) find $K = 0.01475(29)a^{-2}$. In Chapter 1, Fig. (1.5) shows the experimental spectra of the heavy quark systems charmonium (c, $\bar{c}$) and bottomonium (b, $\bar{b}$). Many fits to these spectra have been made using a Schrödinger equation with an interaction potential of the form (17.3). In the lowest energy states of heavy quark systems, the quark and antiquark are slowly moving, so that a non-relativistic approximation is reasonable. The spectra are well fitted with $K = (440\,\text{MeV})^2 = 1\,\text{GeV fm}^{-1}$, $\alpha(r) = \text{constant} = 0.39$.

Taking $K = (440\,\text{MeV})^2$ fixes the lattice spacing $a = 0.0544\,\text{fm}$, and $a^{-1} = 3.62\,\text{GeV}$.

Equation (17.1) could now be used to estimate Λ_{latt}. However, this equation (and more sophisticated extensions to higher orders of lattice perturbation theory) hold only in the limit $a \to 0$. To extract Λ_{latt} reliably, the calculations must be repeated for different values of g. The corresponding values of a follow from the string tension. The limit Λ_{latt} as $a \to 0$ may then be estimated. Bali and Schilling (1993) found $\sqrt{K}/\Lambda_{\text{latt}} = 51.9^{+1.6}_{-1.8}$, which is consistent with the value $\sqrt{K}/\Lambda_{\text{latt}} = 49.6$ (3.8) estimated by Booth *et al.* (1992) from results on a $(36)^4$ lattice. Taking $\sqrt{K} = 440\,\text{MeV}$ gives $\Lambda_{\text{latt}} \approx 8.5\,\text{MeV}$, and from (17.2) $\Lambda \approx 255\,\text{MeV}$.

At small r the attractive Coulomb-like term dominates. It is found that $\alpha_{\text{latt}}(r)$ is a slowly varying function of r that decreases with decreasing r, as expected from perturbation theory (Section 16.3). The potential of Fig. 17.1 is well fitted with

$$\alpha_{\text{latt}}(r) = 0.236 - (0.0031\,\text{fm})/r.$$

This is to be compared with the value of $\alpha = e^2/4\pi \approx 1/137$ of QED.

It is interesting to note that the linearly rising term in the potential is computed in the quenched approximation. If quantum fluctuating quark fields were to be included, the large potential energy available at large separation distances of the fixed quark and antiquark pair would produce pairs of quarks and antiquarks. A quark would migrate to the neighbourhood of the fixed antiquark to form a colour singlet, and an antiquark would similarly form another singlet with the fixed quark, resulting in two well separated mesons.

17.2 Lattice QCD and hadrons

Systems of quarks and antiquarks held together by the associated gluon field are called *hadrons* (see Section 1.4). For example, the proton, the only stable hadron, has up quark number two and down quark number one. Other systems, for example mesons, are held together only transiently by their gluon field. As well as these so-called *valence quarks* that define a system, a hadron contains quark–antiquark pairs excited by the gluon field, and known as *sea quarks.*

So far, in our discussion of hadrons and confinement, sea quarks have been neglected. Convincing calculations of hadron properties require their inclusion especially $u\bar{u}$. $d\bar{d}$ and $s\bar{s}$ pairs which because of their small masses with respect to Λ_{QCD} are readily excited by the gluon field Since the first edition of this book, much progress in lattice QCD has been made to include these pairs.

Quarks on the lattice require the introduction of quark masses. In the work of Davies *et al.* (2004) calculations are made with $m_u = m_d$ (the isospin symmetry limit: see Section 16.6). A mean mass $(m_u + m_d)/2$ is introduced along with the masses m_s, m_c, m_b, and the strong coupling constant g: five parameters in all. With a fixed value of g the lattice spacing a and the four quark masses are determined by fitting the five experimentally determined masses $m(b\bar{b}1s) = 9.460\,\text{GeV}$, $m(b\bar{b}2s) = 10.023\,\text{GeV}$ (see Figure 1.5), $m_\pi = 0.139\,\text{GeV}$, $m_K = 0.496\,\text{GeV}$ and $m_D = 1.867\,\text{GeV}$. The D^+ meson $D(c\bar{s})$ is the ground state of the $c\bar{s}$ valence quark system.

As in Section 17.1 the lattice spacing a is a function of g and so also are the quark masses. The calculations have to be repeated for different values of g to extract Λ_{latt} and $g(a)$ and the four quark masses which are also taken to be functions of a. They can also be regarded as function of energy, $\hbar c/a$. The fact that the strong coupling constant and quark masses are functions of the energy at which they are measured is a natural feature of QCD. The calculations give, at an energy of 2 GeV for the light quarks

$$\begin{aligned}\left(\frac{m_u + m_d}{2}\right)(2\,\text{GeV}) &= 3.2 \pm 0.4\,\text{MeV}\\ m_s\,(2\,\text{GeV}) &= 87 \pm 8\,\text{MeV}\\ m_c &= 1.1 \pm 0.1\,\text{GeV}\\ m_b &= 4.25 \pm 0.15\,\text{GeV}\end{aligned}$$

and

$$\alpha_s\,(M_z) = 0.121 \pm 0.003.$$

m_c and m_b are quoted at their own mass scale and it is conventional to quote α_s at the scale of the Z boson. To find the parameters at different scales their energy dependence is given by equations like (16.25).

Having values for the parameters of QCD its validity can be tested by confronting independent experimental data with calculations. At present one is confined to single hadrons that are stable to the strong interaction. Unstable particles or those that are close to instability tend to fluctuate outside the lattice boundaries. Also the baryons, and in particular the proton and neutron that carry u and d valence quarks can not yet be reliably handled on the lattice. Nevertheless many particle properties lend themselves to lattice calculations and the success in fitting data is impressive. Figure 17.2 shows results taken from Davies *et al.* (2004). Ten calculations are compared with experiment. The results are expressed as the calculated divided by the experimental value. The experimental values are accurately known and the errors that bracket the mean values indicate the estimated accuracy of the calculation. It seems that with present computing power, theory and experiment agree to better

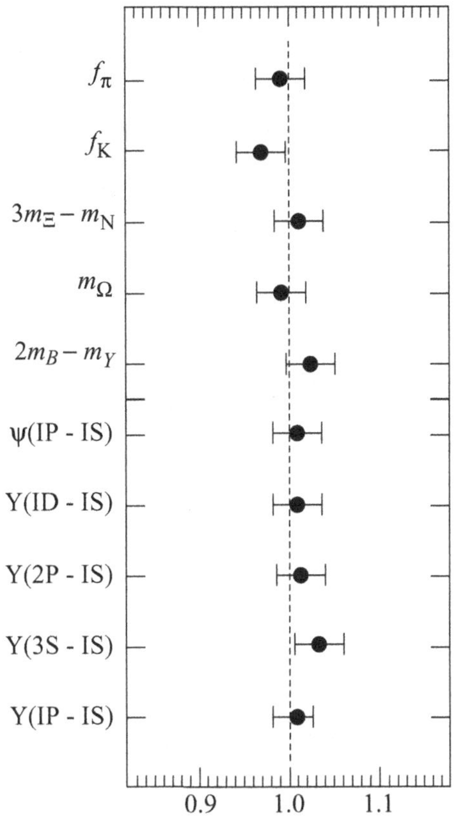

Figure 17.2 Quantities calculated in lattice QCD divided by their experimental values:

$$f_\pi = {}^{\alpha_\pi}/\sqrt{2}\,G_F V_{ud} \quad \text{see Section 9.2,}$$
$$f_K = {}^{\alpha_K}/\sqrt{2}\,G_F V_{us} \quad \text{see Problem 9.10.}$$

m_Ω is the mass of the Ω(sss), the ground state of the baryon with s quark number three.

$3m_\Xi - m_N$ is a combination of ground state baryon masses Ξ(ssu) and the neutron N(ddu).

The other mass differences are between states of the $c\bar{c}$ and $b\bar{b}$ mesons (Davies *et al.*, 2004).

than 4%. There is no reason here to doubt the validity of QCD as the theory of strong interactions.

17.3 Perturbative QCD and deep inelastic scattering

One of the first applications of perturbative QCD was to the Q^2 dependence of the parton distribution functions of the proton. In the parton model of inelastic

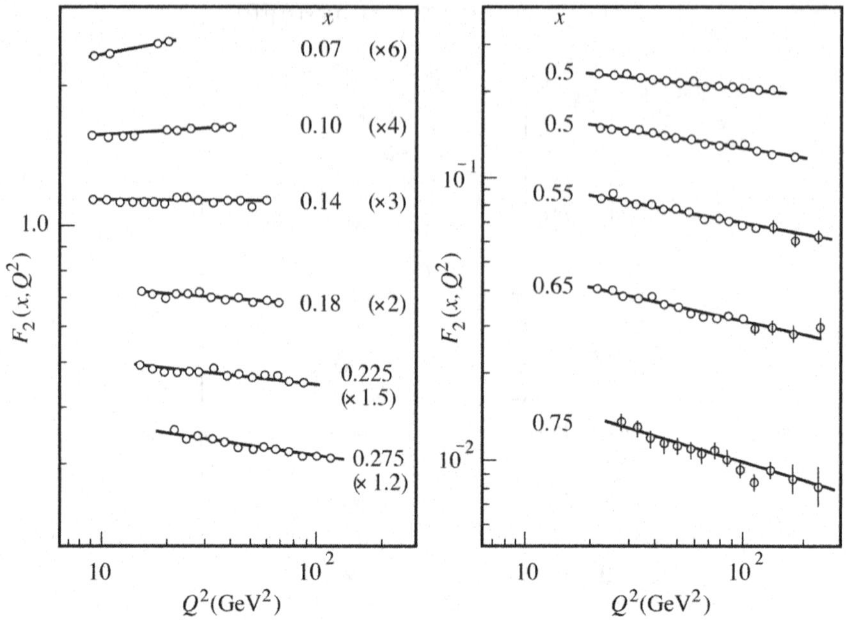

Figure 17.3 The proton structure function $F_2(x, Q^2)$. The experimental points are fitted with curves generated by the evolution equations with $\Lambda = 205$ MeV. To aid reading in the left-hand section, the data have been scaled by the given factors, so for example at $x = 0.18$ the graph is of $2F_2(0.18, Q^2)$. (Taken from *Physics Letters* **B223**, Benvenuti, A. C. *et al.* Test of QCD and a measurement of Λ from scaling violations in the proton structure factor $F_2(x, Q^2)$ at high Q^2 (Benvenuti *et al.*, p. 490), with kind permission of Elsevier Science-NL, Sara Burgerhartstraat 25, 1005 kv Amsterdam, The Netherlands.)

electron–proton scattering (Appendix D), the proton is described by parton distribution functions $p_i(x, Q^2)$, where

$$Q^2 = -q_\mu q^\mu = (\mathbf{p} - \mathbf{p}')^2 - (E - E')^2,$$

$q^\mu = (E - E', \mathbf{p} - \mathbf{p}')$ is the energy and momentum transferred in the inelastic electron scattering, and $x = Q^2/[2M(E - E')]$ where M is the proton mass. The partons are identified as quarks, antiquarks and gluons. Typically, at a fixed value of Q^2, say Q_0^2, distribution functions $p_i(x, Q_0^2)$ are extracted from the data, the number of distribution functions being determined by the number of distinct data sets. At this stage the extraction of the distribution functions is merely a matter of curve fitting: although the functions $p_i(x, Q_0^2)$ should be a consequence of QCD, the problem of establishing their form theoretically is immensely difficult. However, given these distribution functions, and provided Q_0^2 is large enough, perturbative QCD can be used to predict how they evolve with changing Q^2. This evolution

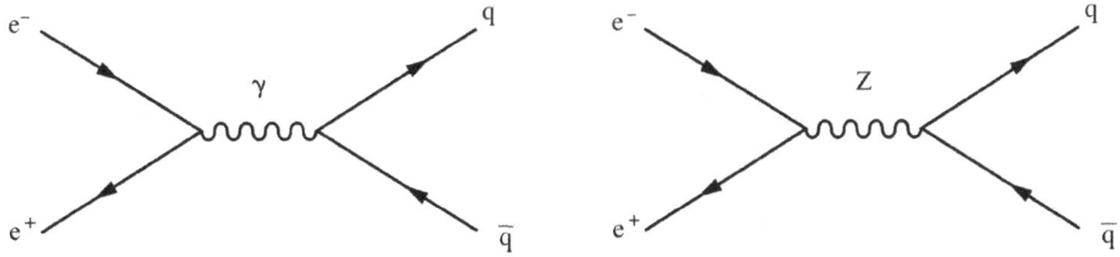

Figure 17.4 e^+e^- annihilation to a quark–antiquark pair with no gluon radiative corrections.

is described by the equations of Altarelli and Parisi (1977), which take account perturbatively of the quark–gluon interactions.

As an example, Fig. 17.3 shows experimental data on the related structure function $F_2(x, Q^2)$ defined in Appendix D, taken by the BCDMS collaboration (Benvenuti *et al.*, 1989). Also shown are the theoretical predictions, at fixed values of x, of the QCD evolution as a function of Q^2. The data are precise and the shapes of all the curves are given by the single parameter Λ. Fits to the data determine $\Lambda = 205 \pm 80\,\text{MeV}$, from which one can infer, using (16.25) with $n_f = 5$, that $\alpha_s(M_z{}^2) = 0.115 \pm 0.007$.

17.4 Perturbative QCD and e^+e^- collider physics

The basic Feynman diagrams for hadron production in e^+e^- colliding beam experiments are shown in Fig. 17.4. In the range 10 GeV to 40 GeV, electromagnetic processes dominate. The data were discussed in Section 1.7.

Around 90 GeV, close to the centre of mass energy for Z production, the weak interaction dominates. The hadronic decays of the Z were discussed in Chapter 15, using perturbation theory. However, there are additional contributions to the cross-section arising from gluon radiation, for example the processes illustrated in Fig. 17.5.

The modification is simply expressed (see Particle Data Group, 1996). If the hadron production cross-section without gluon radiative corrections is denoted by σ_0 then (to order α_s^3) the cross-section σ with corrections is

$$\sigma = f\sigma_0,$$

with

$$f = 1 + \frac{\alpha_s}{\pi} + 1.411\left(\frac{\alpha_s}{\pi}\right)^2 - 12.8\left(\frac{\alpha_s}{\pi}\right)^3, \tag{17.5}$$

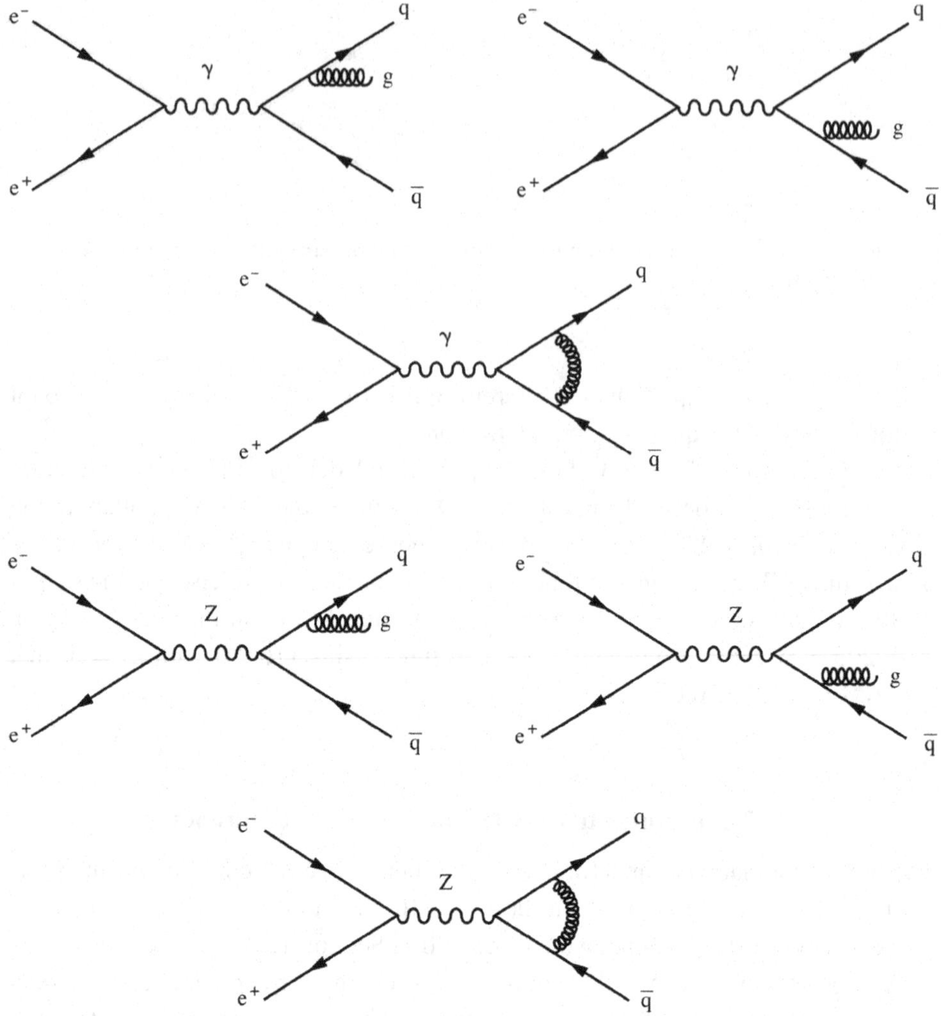

Figure 17.5 The lowest order gluon radiative corrections to quark–antiquark pair production by e^+e^- annihilation.

and $\alpha_s(Q^2)$ taken at Q^2 equal to the square of the centre of mass energy. For example, taking $\alpha_s(M_z^2) = 0.115 \pm 0.007$ from Section 17.3 gives $f = 1.038 \pm 0.003$. This is the value of f used in Chapter 15. Alternatively, the best fit to the hadronic decays of the Z would suggest $f = 1.041 \pm 0.003$, which gives $\alpha_s(M_z^2) = 0.123 \pm 0.007$ and $\Lambda = 310 \pm 90\,\text{MeV}$. The consistency of the theory between the two very different experimental regimes: electron–proton scattering and Z decays, from which these estimates are obtained, is impressive.

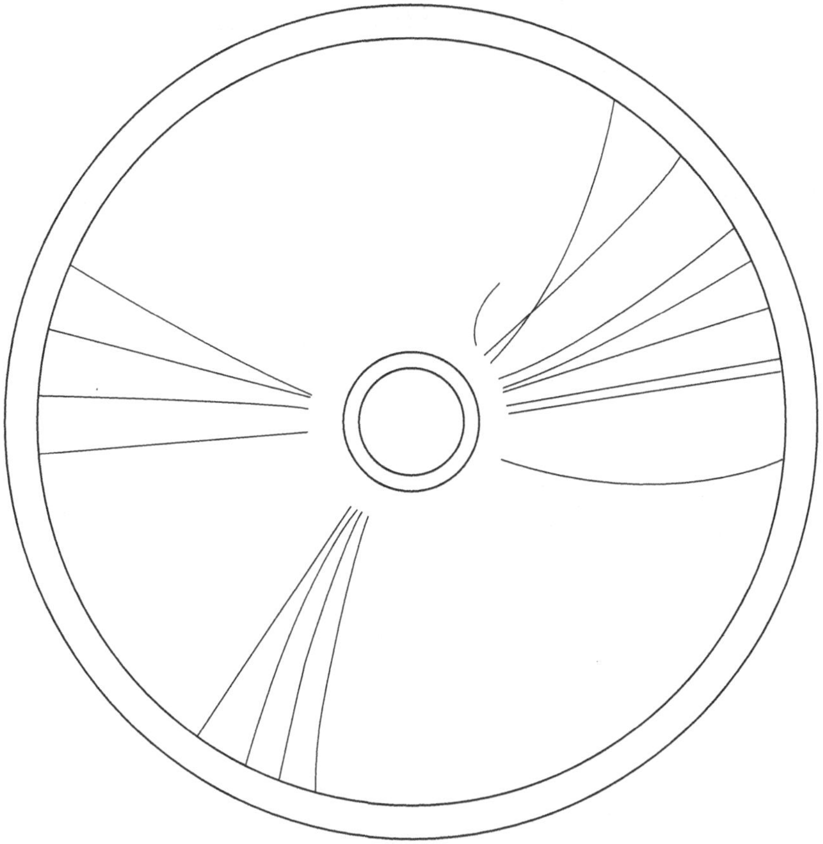

Figure 17.6 A three-jet event recorded by the JADE detector at the PETRA e^+e^- collider, DESY.

The hadrons produced in most e^+e^- annihilations at high energies appear in two back to back jets associated with the originating $q\bar{q}$ pair. Gluon radiation contributing to the f factor is mostly confined to be within the associated quark or antiquark jet. However, according to perturbative QCD it is also possible for a gluon to be radiated into a distinct region of phase space and appear as a third distinct jet. Figure 17.6 is an example of such a three-jet event. Measurements of these three- and even four-jet events gives further strong support to the theory of QCD.

18

The Kobayashi–Maskawa matrix

In Chapter 14, in the theory of the weak interaction of quarks, there appeared the Kobayashi–Maskawa matrix:

$$\mathbf{V} = \begin{pmatrix} V_{\mathrm{ud}} & V_{\mathrm{us}} & V_{\mathrm{ub}} \\ V_{\mathrm{cd}} & V_{\mathrm{cs}} & V_{\mathrm{cb}} \\ V_{\mathrm{td}} & V_{\mathrm{ts}} & V_{\mathrm{tb}} \end{pmatrix} \tag{18.1}$$

and its parameterisation:

$$\mathbf{V} = \begin{pmatrix} c_{12}c_{13} & s_{12}c_{13} & s_{13}\mathrm{e}^{-\mathrm{i}\delta} \\ -s_{12}c_{23} - c_{12}s_{23}s_{13}\mathrm{e}^{\mathrm{i}\delta} & c_{12}c_{23} - s_{12}s_{23}s_{13}\mathrm{e}^{\mathrm{i}\delta} & s_{23}c_{13} \\ s_{12}s_{23} - c_{12}c_{23}s_{13}\mathrm{e}^{\mathrm{i}\delta} & -c_{12}s_{23} - s_{12}c_{23}s_{13}\mathrm{e}^{\mathrm{i}\delta} & c_{23}c_{13} \end{pmatrix} \tag{18.2}$$

where $c_{12} = \cos\theta_{12} > 0$, $s_{12} = \sin\theta_{12} > 0$, etc. The KM matrix couples quark fields of different flavours. It contains four physically significant parameters, which can be taken to be the three rotation angles θ_{12}, θ_{13}, θ_{23}, each lying in the first quadrant, and the phase angle δ.

There is no theory relating these parameters, just as there is no theory relating quark masses. Indeed, the quark sector of the Standard Model may appear to the reader to be lacking in aesthetic appeal. The parameters of the KM matrix must be determined from experiment, and in this chapter we indicate how experimental information has been obtained.

18.1 Leptonic weak decays of hadrons

We have seen in Section 15.3 two unitarity sum rules that support the validity of the Standard Model, and there are many independent measurements that both test for consistency and given consistency determine the parameters. So far no definitive inconsistencies have been established, and a large body of data is well described

(a)

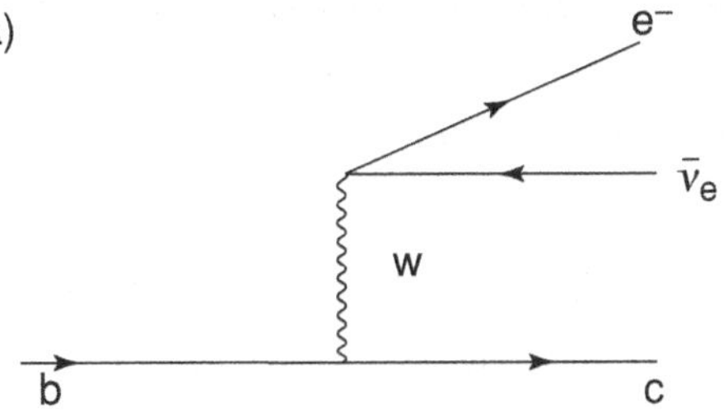

Figure 18.1(a) A Feynman diagram for the leptonic decay $\mathrm{b} \rightarrow \mathrm{c} + \mathrm{e}^- + \bar{\nu}_\mathrm{e}$

(b)

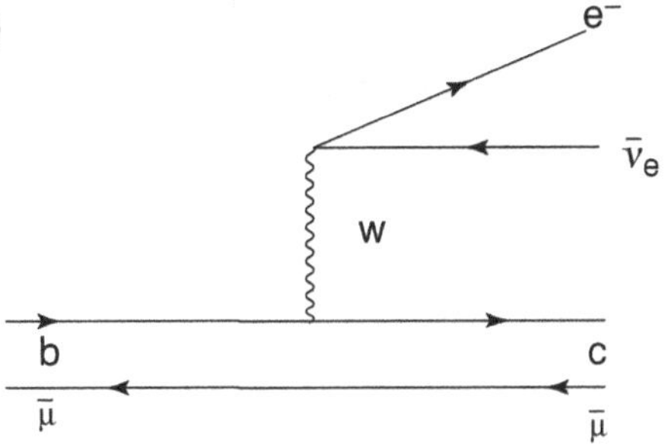

(b) A quark model diagram for the decay $\mathrm{B}^- \rightarrow$ *charmed hadron system* $+$ $\mathrm{e}^- + \bar{\nu}_\mathrm{e}$

with the parameter values $s_{12} = 0.2243 \pm 0.0016$, $s_{23} = 0.0413 \pm 0.0015$, $s_{13} = 0.0037 \pm 0.0005$ and $\delta = 57° \pm 14°$.

A suitable starting point for the consideration of hadronic weak decays is first-order perturbation theory in the effective Lagrangian density of equation (14.21): $L = -2\sqrt{2}G_\mathrm{F} j^\dagger_\mu j^\mu$, where j^μ is given by (14.20). Leptonic decays are the most simple for theoretical analysis because the leptonic parts of a transition matrix element can be calculated with some confidence. If quarks were available as isolated particles, the three rotation angles of the KM matrix could be determined by the measurement of the decay rates of leptonic decays such as

$$\mathrm{b} \rightarrow \mathrm{c} + \mathrm{e} + \bar{\nu}_\mathrm{e}.$$

In lowest order perturbation theory (see Fig. 18.1a) the decay rate for this process is given by

$$\frac{1}{\tau(\mathrm{b} \rightarrow \mathrm{c})} = \frac{G_\mathrm{F}^2 m_\mathrm{b}^5}{192\pi^3} |V_\mathrm{cb}|^2 f\left(\frac{m_\mathrm{c}}{m_\mathrm{b}}\right) \tag{18.3}$$

where $f(x) = 1 - 8x^2 + 8x^6 - x^8 - 24x^4 \ln(x)$ is a factor associated with the available phase space. This programme cannot be carried out directly since the b and c quarks are accompanied by other spectator quarks and gluons (see the quark model diagram of Fig. 18.1b), which involve the calculation of strong interaction matrix elements. To the extent that the hadronic matrix elements can be calculated, a measurement of the decay rate will determine $|V_{cb}|^2$.

18.2 $|V_{ud}|$ and nuclear β decay

Isospin symmetry (see Section 16.6) is important for the determination of the hadronic matrix elements of all nuclear β decays. Such decays involved the quark current

$$j_q^\mu = d_L^\dagger \tilde{\sigma}^\mu u_L = \bar{d}\gamma^\mu (1/2)(1-\gamma^5)u. \tag{18.4}$$

Here we have expressed the current in terms of the Dirac four-component spinors u and d, with the help of the projection operator $(1/2)(1-\gamma^5)$ introduced in (5.32) and noting $\bar{d} = d^\dagger\gamma^o$.

As in Chapter 16, we now take the u and d quarks together in an isotopic doublet:

$$\mathbf{D}(x) = \begin{pmatrix} u(x) \\ d(x) \end{pmatrix}.$$

The isospin operator $(1/2)(\tau^1 - i\tau^2)$ has the property

$$\frac{1}{2}\left(\tau^1 - i\tau^2\right)\begin{pmatrix} u \\ d \end{pmatrix} = \begin{pmatrix} 0 \\ u \end{pmatrix},$$

so that we may write (see (16.31))

$$\begin{aligned} j_q^u &= (1/4)\overline{\mathbf{D}}(x)\gamma^\mu(1-\gamma^5)(\tau^1 - \mathrm{i}\tau^2)\mathbf{D}(x) \\ &= (1/2)\left[v^\mu(x) - a^\mu(x)\right]. \end{aligned} \tag{18.5}$$

We have split the current into the part $v^\mu(x)$, which transforms like a vector under space inversion and the part $a^\mu(x)$, which transforms like an axial vector (see Section 5.5):

$$v^\mu(x) = (1/2)\overline{\mathbf{D}}\gamma^\mu(\tau^1 - \mathrm{i}\tau^2)\mathbf{D}, \tag{18.6}$$

$$a^\mu(x) = (1/2)\overline{\mathbf{D}}\gamma^\mu\gamma^5(\tau^1 - \mathrm{i}\tau^2)\mathbf{D}. \tag{18.7}$$

We saw in Section 16.6 that exact isospin symmetry leads to conserved currents:

$$v_i^\mu = (1/2)\overline{\mathbf{D}}\gamma^\mu\tau^{\mathrm{i}}\mathbf{D}, \tag{18.8}$$

so that the vector part of the β decay current of the u and d quarks is a conserved isospin current.

In the case of nucleons, we denote the isospin doublet of the effective Dirac fields $p(x)$ and $n(x)$ of the proton and neutron by

$$\mathbf{D}_{\mathrm{N}}(x) = \begin{pmatrix} p(x) \\ n(x) \end{pmatrix}. \tag{18.9}$$

An effective Lagrangian density that at the low energies of nuclear physics describes the β decay of a nucleon is

$$L_{\mathrm{eff}} = -2\sqrt{2}G_{\mathrm{F}}C|j_{\mathrm{e}}^{\dagger} j_{\mathrm{N}}^{\mu} + j_{\mathrm{e}}^{\mu\dagger} j_{\mathrm{e}\mu}|, \tag{18.10}$$

with

$$j_{\mathrm{N}}^{\mu} = \frac{1}{4}\overline{\mathbf{D}}_{\mathrm{N}}\gamma^{\mu}(1 - g_{\mathrm{A}}\gamma^{5})(\tau_1 - \mathrm{i}\tau^2)\mathbf{D}_{\mathrm{N}}. \tag{18.11}$$

Experimentally, it is found from a range of nuclear data that

$$C = 0.9713 \pm 0.0013 \quad \text{and} \quad g_{\mathrm{A}} = 1.2739 \pm 0.0019.$$

(See Particle Data Group.)

The vector part of the current j_{N}^{μ} is the conserved isospin current of nuclear physics and corresponds to the more fundamental conserved isospin current at the quark level. Exact isospin symmetry would require that the contribution of the conserved nucleon isospin current to the effective interaction (18.8, 18.9) be the same as that of the quarks in (18.5, 18.6), so that we identify $C = V_{\mathrm{ud}} = 0.9713 \pm 0.0013$.

18.3 More leptonic decays

The most precise estimates of $|V_{\mathrm{us}}|$ have come from observations of leptonic K decays, for example $\mathrm{K}^-(\mathrm{s\bar{u}}) \to \pi^{\mathrm{o}}(\mathrm{u\bar{u}} - \mathrm{d\bar{d}})/\sqrt{2} + \mathrm{e}^- + \bar{\nu}_{\mathrm{e}}$. Analyses of these decays by lattice QCD, quark model calculations, and calculations based on chiral symmetry (see Section 16.7) all converge on the value $|V_{\mathrm{us}}| = 0.224 \pm 0.003$.

Estimates of $|V_{\mathrm{cs}}|$ and $|V_{\mathrm{cd}}|$ can be extracted from D decays, for example $\mathrm{D}^-(\mathrm{\bar{c}d}) \to \mathrm{K}^{\mathrm{o}}(\mathrm{\bar{s}d}) + \mathrm{e}^- + \bar{\nu}_{\mathrm{e}}$ or $\mathrm{D}^-(\mathrm{\bar{c}d}) \to \pi^{\mathrm{o}}(\mathrm{u\bar{u}} - \mathrm{d\bar{d}})/\sqrt{2} + \mathrm{e}^- + \bar{\nu}_{\mathrm{e}}$. These decay rates are proportional to $|V_{\mathrm{cs}}|^2$ and $|V_{\mathrm{cd}}|^2$ respectively.

More experimental information on $|V_{\mathrm{cd}}|^2$ comes from the deep inelastic scattering of neutrinos by atomic nuclei through processes such as

$$\nu_{\mu} + \mathrm{d} \to \mu^- + \mathrm{c}. \qquad \text{(See Appendix D.)}$$

Atomic nuclei provide an abundant source of d quark targets. The cross-section for producing a c quark rather than a u quark can be inferred by identifying those c quarks that decay as $\mathrm{c} \to \mathrm{d} + \mu^+ + \nu_{\mu}$. Overall, a characteristic $\mu^+\mu^-$ pair is produced.

The conclusions, after much work along the lines indicated, and without imposing the unitarity condition, are

$$|V_{cd}| = 0.224 \pm 0.014, \ |V_{cs}| = 1.04 \pm 0.16.$$

Leptonic decays of B mesons ($b\bar{u}$, $b\bar{d}$, $\bar{b}u$ and $\bar{b}d$) provide the best data on $|V_{cb}|$ and $|V_{ub}|$, Three experimental facilities have been constructed to measure B decays: in the USA at Cornell (Cleo) and Stanford (Babar), and in Japan (Belle). At these 'B meson factories' many million B mesons have been produced for analysis.

In the case of $|V_{cb}|$, the hadronic matrix elements for decays like $B^- \to D^\circ + e^- + \bar{\nu}_e$ can be calculated taking the heavy b quark in the $B^-(b, \bar{u})$ meson as static in first approximation. Analysis of the data gives

$$|V_{cb}| = 0.0413 \pm 0.0015, \ |V_{ub}| = 0.00367 \pm 0.00047.$$

The remaining three elements of the KM matrix involve the top quark. The mean life of the top quark is so short it is likely to decay before it has time to settle into a top quark hadron. The methods described above are unavailable for $|V_{ti}|$ (i = d, s or b).

18.4 *CP* symmetry violation in neutral kaon decays

In Section 14.4 we obtained the important result that the quark sector of the Standard Model is not invariant under the charge conjugation, parity, operation unless all the elements of the KM matrix can be made real. With the parameterisation (18.2), this requires that the phase angle $\delta = 0$.

CP violation was first observed in 1964 in the decay of neutral K mesons. The states of definite quark number are the $K^\circ(d\bar{s})$ and $\bar{K}^\circ(\bar{d}s)$. These mesons are readily produced in strong interactions, for example $\pi^-(\bar{u}d) + p(uud) \to K^o(d\bar{s}) + \Lambda(uds)$. Without the weak interaction the K° and $\bar{K}^\circ$ would have equal mass and be stable. The weak interaction is responsible for their instability and *CP* violation would be manifest if for example it were seen that the decay rates $K^o \to \pi^+\pi^-$ and $\bar{K}^\circ \to \pi^+\pi^-$ were different. Such a difference can occur in second-order perturbation theory in the weak interaction (first order in G_F. See (14.21)). This is known as direct *CP* violation.

The weak interaction also gives rise to the phenomenon of mixing (Appendix E, Fig. E1). Although mixing occurs only at second order in G_F it has the dramatic effect of splitting the mass degeneracy: it results in two mixed states of different mass. If *CP* were conserved the mixed states would be

$$|K_1^o\rangle = \left(1/\sqrt{2}\right)\left(|K^o\rangle + |\bar{K}^o\rangle\right) \quad \text{and} \quad |K_2^o\rangle = \left(1/\sqrt{2}\right)\left(|K^o\rangle - |\bar{K}^o\rangle\right).$$

Acting on K^o and $\bar{K}^o$, the *CP* operator may be taken to give

$$CP\left|K^o\right\rangle = \left|\bar{K}^o\right\rangle \quad \text{and} \quad CP\left|\bar{K}^o\right\rangle = \left|K^o\right\rangle.$$

Then $\left|K_1^o\right\rangle$ and $\left|K_2^o\right\rangle$ are eigenstates of *CP* with eigenvalues $+1$ and -1 respectively. Experimentally two states with a mass difference 3.5×10^{-12} MeV are indeed observed; they also have very different mean lives

$$\tau_s = 8.9 \times 10^{-11}\text{s}, \ \tau_L = 5.17 \times 10^{-8}\text{s}.$$

The K_s^o decays predominantly into two pions, $\pi^+\pi^-$ or $\pi^o\pi^o$. Each of these two-pion states is an eigenstate of CP, with eigenvalue $+1$ (Problem 18.2). In its mesonic decay modes, the K_L^o decays predominantly into $\pi^o\pi^o\,\pi^o$, and these three-pion states are eigenstates of *CP* with eigenvalue -1 (Problem 18.3). However, in about three decays in a thousand K_L^o decays into two pions, with *CP* eigenvalue $+1$. If CP were conserved K_L^o would be either K_1^o or K_2^o and could not have both two pion and three pion decay modes. *CP* violation is also seen in leptonic K decays. These show that direct *CP* violation is not responsible for the anomalous K_L^o decays but they are predominantly due to *CP* violation in mixing.

It is shown in Appendix E that neither $\left|K_s^o\right\rangle$ nor $\left|K_L^o\right\rangle$ is an eigenstate of *CP*, but each can be written in terms of $|K^o\rangle$ and $\left|\bar{K}^o\right\rangle$:

$$\begin{aligned} \left|K_s^o\right\rangle &= N\left[p\,|K^o\rangle + q\,\left|\bar{K}^o\right\rangle\right], \\ \left|K_L^o\right\rangle &= N\left[p\,|K^o\rangle - q\,\left|\bar{K}^o\right\rangle\right]. \end{aligned} \tag{18.12}$$

N is the normalisation factor: $(|p|^2 + |q|^2)^{-1/2}$. Note that q is not equal to p. In Appendix E we indicate how p and q can be calculated in the Standard Model.

We can similarly express $\left|K_s^o\right\rangle$ and $\left|K_L^o\right\rangle$ in terms of $\left|K_1^o\right\rangle$ and $\left|K_2^o\right\rangle$:

$$\begin{aligned} \left|K_s^o\right\rangle &= \left(N/\sqrt{2}\right)\left[(p+q)\left|K_1^o\right\rangle + (p-q)\left|K_2^o\right\rangle\right], \\ \left|K_L^o\right\rangle &= \left(N/\sqrt{2}\right)\left[(p-q)\left|K_1^o\right\rangle + (p+q)\left|K_2^o\right\rangle\right]. \end{aligned} \tag{18.13}$$

Neglecting direct *CP* violation only K_1^o can decay into $\pi\pi$ so that the ratio of the decay rates

$$\frac{\Gamma(K_L) \to \pi\pi}{\Gamma(K_S) \to \pi\pi} = \frac{|p/q - 1|^2}{|p/q + 1|^2} = (5.25 \pm 0.05) \times 10^{-6}\,(\text{from experiment}).$$

Defining $p/q = 1 + 2\varepsilon_K$ we infer that $|\varepsilon_K| = 2.3 \times 10^{-3}$; ε_K is a measure of *CP* violation.

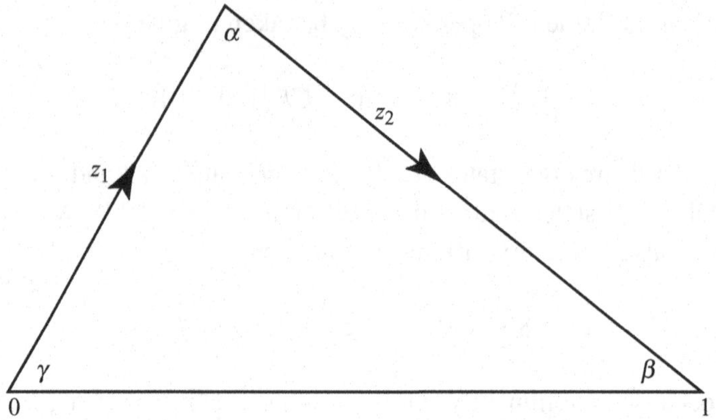

Figure 18.2 The unitarity triangle.

18.5 B meson decays and B°, $\bar{\mathrm{B}}$° mixing

At the B meson factories the 4s ($\mathrm{b}\bar{\mathrm{b}}$) meson is copiously produced by e^+e^- collisions with beam energies turned to the meson mass. The meson decays almost exclusively into B^+, B^- or B°, $\bar{\mathrm{B}}$° pairs and so provides a rich source of B mesons. With a mass of 5.28 GeV, B mesons decay into many different final states and many exhibit *CP* violation. An indication of why this is so can be seen by a consideration of the unitarity condition

$$V_{\mathrm{ud}}V^*_{\mathrm{ub}} + V_{\mathrm{cd}}V^*_{\mathrm{cb}} + V_{\mathrm{td}}V^*_{\mathrm{tb}} = 0,$$

which can be written as

$$z_1 + z_2 = 1 \tag{18.14}$$

where we have defined $z_1 = -\dfrac{V_{\mathrm{ud}}V^*_{\mathrm{ub}}}{V_{\mathrm{cd}}V^*_{\mathrm{cb}}}$ and $z_2 = -\dfrac{V_{\mathrm{td}}V^*_{\mathrm{tb}}}{V_{\mathrm{cd}}V^*_{\mathrm{cb}}}$.

z_1 and z_2 are complex numbers that, in the complex plane form a triangle, the unitarity triangle illustrated in Fig. 18.2. Also it can be seen from the parameters given in Section 18.1 that $V_{\mathrm{cd}}V^*_{\mathrm{cb}}$ is almost real and negative. Neglecting its very small imaginary part, the angle $\gamma = \delta$, the phase of V^*_{ub}, and β is the phase of V^*_{td}. Of all the unitarity triangles, this is the only one with direct access to the two KM matrix elements with large phases; it also involves the b quark and hence B mesons.

Of particular importance has been the measurement of the angle α through both charged and neutral decays $\mathrm{B} \to \pi\pi$, $\mathrm{B} \to \pi\rho$ and $\mathrm{B} \to \rho\rho$ and of the angle β through B°, $\bar{\mathrm{B}}^0$ mixing. As one example it is shown in Appendix E how $\sin(2\beta)$ is measured at the B factories.

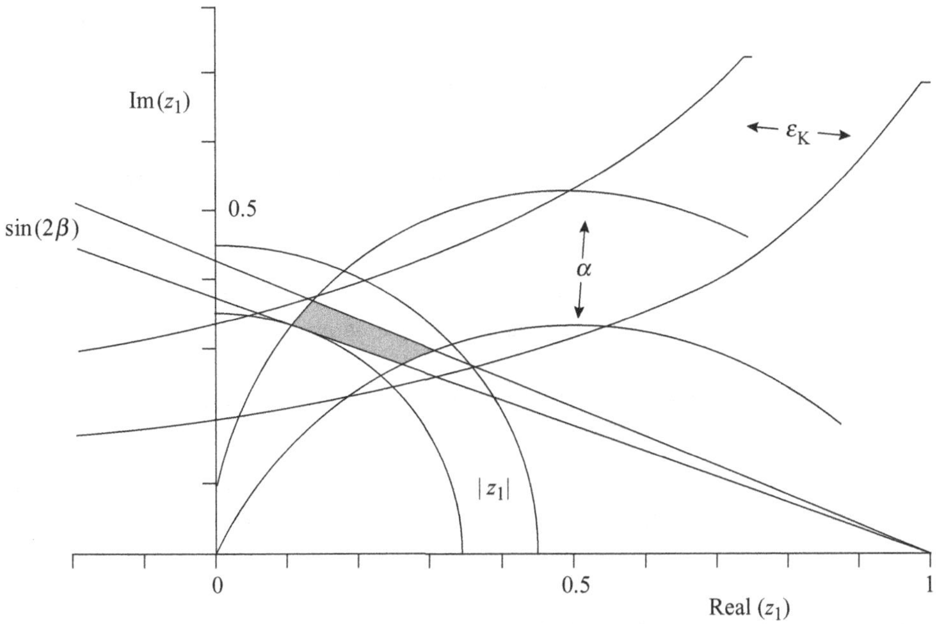

Figure 18.3 The apex of the unitarity triangle is in, or near, the shaded region of the plot.

The unitary triangle is specified by the position of its apex. This requires two parameters, say the real and imaginary parts of z_1. A single parameter defines a line on the complex plane and a parameter with errors defines a band. Four such bands inferred from experiment are shown in Fig. 18.3. The most important point illustrated by the figure is the consistency between four independent measurements. There is no indication of the Standard Model failing. The KM phase δ ($\approx\gamma$) can be seen to be in the region $\delta = 57^{\circ} \pm 14^{\circ}$. The apex of the unitarity triangle is in, or near, the shaded region of the figure.

18.6 The *CPT* theorem

We denote by T the operation of time reversal, $t \to t' = -t$. The *CPT theorem* states that, under very general conditions, a Lorentz invariant quantum field theory is invariant under the combined operations of charge conjugation, space inversion, and time reversal. The theorem was discovered by Pauli in 1955.

For the Standard Model, the *CPT* theorem implies that, since *CP* is not a symmetry of the Model, then neither is time reversal *T*. One may contemplate the implications for the 'Arrow of Time'.

Problems

18.1 Draw quark model diagrams for the decays

$$\pi^- \to \mu^- + \bar{\nu}_\mu, \quad K^- \to \mu^- + \bar{\nu}_\mu.$$

Show that the decay amplitudes are proportional to V_{ud} and V_{us} respectively, and $V_{us}/V_{ud} = \tan\theta_{12}$.

Neglecting the effects of the different quark masses, the ratio α_K/α_π calculated in Problem 9.10 would equal V_{us}/V_{ud}. Use this observation to estimate $\sin\theta_{12}$.

18.2 A π^o meson is even under the charge conjugation operation C, i.e. $C|\pi^0\rangle = |\pi^0\rangle$. Also, $C|\pi^+\rangle = |\pi^-\rangle$ and $C|\pi^-\rangle = |\pi^+\rangle$.

Show that two pions $|\pi^o, \pi^o\rangle$ or $|\pi^+, \pi^-\rangle$ in a relative S state and with their centre of mass at rest satisfy $CP|\pi, \pi\rangle = |\pi, \pi\rangle$.

18.3 Show that a state of three π^o mesons $|\pi^o, \pi^o, \pi^o\rangle$ with angular momentum zero and centre of mass at rest satisfies $CP|\pi^o, \pi^o, \pi^o\rangle = -|\pi^o, \pi^o, \pi^o\rangle$. (See Problem 18.2.)

18.4 Show that the area of the unitary triangle of Fig. 18.5 is $J/2$.

18.5 Show that if the quark fields are subject to a change of phase

$$d \to e^{i\theta_d} d, \quad b \to e^{i\theta_b} b,$$

then the unitary triangle of Fig. 18.5 is rotated through an angle $(\theta_d - \theta_b)$.

19

Neutrino masses and mixing

In this chapter we introduce the phenomenology of neutrino masses and mixing, and show how the phenomenology can be made to be consistent with the $SU(2) \times U(1)$ broken gauge symmetry of the Standard Model. We take it that neutrinos and antineutrinos are distinct Dirac fermions, setting aside, until Chapter 21, the suggestions that neutrinos are Majorana fermions.

The phenomenology arose from the observations that the number of electron neutrinos arriving at the Earth from the Sun is only about half of the number expected from our knowledge of the nuclear reactions that occur in the Sun, and the physics of the Sun's interior. These observations are now explained as the result of some electron neutrinos turning into muon neutrinos and tau neutrinos during their transit between their creation in the interior of the Sun and their observation on Earth. These transitions violate the conservation laws of Section 9.3. We will show that they occur because the e, μ and τ neutrinos are not massless but, as conceived by Pontecorvo (1968) they do not have a definite mass, i.e., they are not eigenstates of the mass operator.

19.1 Neutrino masses

The most general Lorentz invariant neutrino mass term that can be introduced into the Lagrangian density of the Standard Model is

$$\mathcal{L}^{\nu}_{\text{mass}}(x) = -\sum_{\alpha,\beta} \nu^{\dagger}_{\alpha\text{L}}(x)\, m_{\alpha\beta}\, \nu_{\beta\text{R}}(x) \;+\; \text{Hermitian conjugate}, \tag{19.1}$$

where $m_{\alpha\beta}$ is an arbitrary 3×3 complex matrix, α and β run over the three neutrino types e, μ, τ, and $\nu_{\alpha\text{L}}(x)$, $\nu_{\alpha\text{R}}(x)$ are left-handed and right-handed two-component spinor fields. (Spinor indices are omitted here.)

An arbitrary complex matrix can be put into real diagonal form with the help of two unitary matrices (see Problem A.4). We can write

$$m_{\alpha\beta} = \sum_i U^{L*}_{\alpha i} m_i U^{R}_{\beta i}, \tag{19.2}$$

where m_i are three real and positive masses; $\mathbf{U}^{L}$ and $\mathbf{U}^{R}$ are unitary matrices. It is evident that $U^{L}_{\alpha i}$ and $U^{R}_{\beta i}$ can be replaced by $U^{L}_{\alpha i}e^{-i\delta_i}$ and $U^{R}_{\beta i}e^{-i\delta_i}$, where the δ_i are three arbitrary phases.

If we now define the fields

$$\begin{aligned} \nu_{iL}(x) &= \sum_\alpha U^{L}_{\alpha i} \nu_{\alpha L}(x), \\ \nu_{iR}(x) &= \sum_\alpha U^{R}_{\alpha i} \nu_{\alpha R}(x), \end{aligned} \tag{19.3}$$

the mass term takes the standard Dirac form (5.12)

$$\mathcal{L}^{\nu}_{\text{mass}}(x) = -\sum_i \mathrm{m}_i \left(\nu^{\dagger}_{iL}\nu_{iR} + \nu^{\dagger}_{iR}\nu_{iL}\right). \tag{19.4}$$

It is easy to show that the transformations given by equations (19.3) retain the Dirac form of the dynamical terms:

$$\begin{aligned} \mathcal{L}^{\nu}_{\text{dyn}} &= \sum_\alpha \mathrm{i}\left[\nu^{\dagger}_{\alpha L}\tilde{\sigma}^{\mu}\partial_\mu \nu_{\alpha L} + \nu^{\dagger}_{\alpha R}\sigma^{\mu}\partial_\mu \nu_{\alpha R}\right] \\ &= \sum_i \mathrm{i}\left[\nu^{\dagger}_{iL}\tilde{\sigma}^{\mu}\partial_\mu \nu_{iL} + \nu^{\dagger}_{iR}\sigma^{\mu}\partial_\mu \nu_{Ri}\right]. \end{aligned} \tag{19.5}$$

$(\mathcal{L}^{\nu}_{\text{dyn}} + \mathcal{L}^{\nu}_{\text{mass}})$ is the Lagrangian density of free neutrinos of masses m_1, m_2, m_3. Since $\mathbf{U}^{L}$ and $\mathbf{U}^{R}$ are unitary matrices, and a unitary matrix $\mathbf{U}$ satisfies $\mathbf{U}\mathbf{U}^{\dagger} = \mathbf{U}^{\dagger}\mathbf{U} = \mathbf{I}$, we can invert equations (19.3) to give

$$\begin{aligned} \nu_{\alpha L}(x) &= \sum_i U^{L*}_{\alpha i} \nu_{iL}(x), \\ \nu_{\alpha R}(x) &= \sum_i U^{R*}_{\alpha i} \nu_{iR}(x). \end{aligned} \tag{19.6}$$

The e, μ and τ neutrinos are mixtures of the neutrinos having definite mass. We shall see that this leads to the phenomenon of neutrino oscillations.

19.2 The weak currents

Neutrinos interact with each other and with other particles through the weak currents. The charged weak current (9.2), expressed in terms of the neutrino mass eigenfields using (19.6), becomes

$$j^{\mu} = \sum_\alpha \alpha^{\dagger}_{L}\tilde{\sigma}^{\mu}\nu_{\alpha L} = \sum_{\alpha,i} \alpha^{\dagger}_{L}\tilde{\sigma}^{\mu}U^{L*}_{\alpha i}\nu_{iL} \tag{19.7}$$

α_L are the charged lepton fields $\alpha = e,\ \mu,\ \tau$.

The neutral weak current (9.17) keeps the same form: since $\mathbf{U}^{\mathrm{L}}$ is unitary, we have

$$\sum_{\alpha} (\nu_{\alpha \mathrm{L}})^{\dagger} \tilde{\sigma}^{\mu} \nu_{\alpha \mathrm{L}} = \sum_{i} (\nu_{i\mathrm{L}})^{\dagger} \tilde{\sigma}^{\mu} \nu_{i\mathrm{L}}. \tag{19.8}$$

As an example of how these modifications influence the physics discussed in earlier chapters, consider our effective pion interaction (9.1):

$$\mathcal{L}_{\mathrm{int}} = \alpha_{\pi} \left[j^{\mu} \partial_{\mu} \Phi_{\pi} + j^{\mu\dagger} \partial_{\mu} \Phi_{\pi}^{\dagger} \right].$$

The β decay rate formula (9.3) for $\pi^- \to \mathrm{e}^- + \bar{\nu}_{\mathrm{e}}$ becomes three decay rates:

$$\frac{1}{\tau\left(\pi^- \to \mathrm{e}^- \bar{\nu}_i\right)} = \frac{\alpha_{\pi}^2}{4\pi} \left(1 - \frac{v_{\mathrm{e}}}{c}\right) p_{\mathrm{e}}^2 E_{\mathrm{e}} \left|U_{ei}^{\mathrm{L}}\right|^2, \; i = 1, 2, 3.$$

In the derivation of this result the effects of small neutrino masses have been neglected. Because neutrino masses are small (see Table 1.2), it is not possible with present technology to discern differences in energy between these decay modes. The total decay rate is measured, and since $\sum_i U_{ei}^{\mathrm{L}} U_{ei}^{\mathrm{L}*} = 1$ we recover the expression (9.3) for this. A similar conclusion can be drawn about the processes $\pi^- \to \mu^- + \bar{\nu}_{\mu}$ and $\tau^- \to \pi^- + \nu_{\tau}$, described in Section 9.2 by the same effective Lagrangian, and about the results on muon decay of Section 9.4.

19.3 Neutrino oscillations

The Lagrangian density (19.1) with (19.5) for a free neutrino yields the equations

$$\begin{aligned} \mathrm{i}\tilde{\sigma}^{\mu} \partial_{\mu} \nu_{\alpha} \mathrm{L} - m_{\alpha\beta} \nu_{\beta \mathrm{R}} &= 0, \\ \mathrm{i}\sigma^{\mu} \partial_{\mu} \nu_{\alpha \mathrm{R}} - m_{\beta\alpha}^{*} \nu_{\beta \mathrm{L}} &= 0. \end{aligned} \tag{19.9}$$

These equations are a generalisation of the Dirac equations (5.11), and in this section we shall interpret their solutions as neutrino wave functions for the three types $\alpha = e,\ \mu,\ \tau$, not as neutrino fields. We shall look for energy eigenfunctions with time dependence $\mathrm{e}^{-\mathrm{i}Et}$.

Zero mass neutrinos would have plane wave solutions of negative helicity (see Section 6.6). For a wave in the z direction

$$\nu_{\alpha \mathrm{L}}(z, t) = \mathrm{e}^{-\mathrm{i}E(t-z)} f_{\alpha} \begin{pmatrix} 0 \\ 1 \end{pmatrix}, \quad \nu_{\alpha \mathrm{R}} = 0,$$

where the f_{α} are constants.

The introduction of neutrino masses modifies these solutions by allowing the f_α to depend on z:

$$\begin{aligned} \nu_{\alpha L}(z,t) &= e^{-iE(t-z)} f_\alpha(z) \begin{pmatrix} 0 \\ 1 \end{pmatrix}, \\ \nu_{\alpha R}(z,t) &= e^{-iE(t-z)} g_\alpha(z) \begin{pmatrix} 0 \\ 1 \end{pmatrix}. \end{aligned} \tag{19.10}$$

Substituting in the Dirac equations gives

$$\begin{aligned} & i\frac{d}{dz} f_\alpha(z) - m_{\alpha\beta} g_\beta(z) = 0, \\ & \left(2E - i\frac{d}{dz}\right) g_\gamma(z) - m^*_{\alpha\gamma} f_\alpha(z) = 0. \end{aligned} \tag{19.11}$$

$$\left(\text{Note that } \tilde{\sigma}^3 \begin{pmatrix} 0 \\ 1 \end{pmatrix} = \begin{pmatrix} 0 \\ 1 \end{pmatrix},\ \sigma^3 \begin{pmatrix} 0 \\ 1 \end{pmatrix} = -\begin{pmatrix} 0 \\ 1 \end{pmatrix}.\right)$$

For neutrino energies much greater that their mass we can neglect $-i\,dg_\gamma/dz$ compared with $2Eg_\gamma$ (see Problem 19.1) to obtain

$$g_\gamma(z) = m^*_{\alpha\gamma} f_\alpha(z)/2E, \tag{19.12}$$

and hence by substitution three coupled equations for $f_\alpha(z)$:

$$i\frac{d}{dz} f_\beta(z) = m_{\beta\gamma} m^*_{\alpha\gamma} f_\alpha(z)/2E.$$

Diagonalising the mass matrices $m_{\beta\gamma}$ and $m_{\alpha\gamma}$ gives

$$i\frac{d}{dz} f_\beta(z) = U^{L*}_{\beta i} U^{L}_{\alpha i} f_\alpha(z) m_i^2/2E. \tag{19.13}$$

The right-handed U^R do not now appear, so that the label L is now redundant and we shall put $U^L_{\alpha i} = U_{\alpha i}$ for the remainder of this section.

To solve these equations we construct linear combinations

$$f_i(z) = U_{\alpha i} f_\alpha(z); \qquad i = 1, 2, 3. \tag{19.14}$$

which satisfy, using (19.13),

$$\begin{aligned} i\frac{d}{dz} f_i(z) = iU_{\alpha i}\frac{d}{dz} f_\alpha(z) &= U_{\alpha i} U^*_{\alpha j} U_{\beta j} m_j^2 f_\beta(z)/2E \\ &= \delta_{ij} U_{\beta j} m_j^2 f_\beta(z)/2E = \left(m_i^2/2E\right) f_i(z). \end{aligned} \tag{19.15}$$

These uncoupled equations have the simple solutions

$$f_i(z) = e^{-i(m_i^2/2E)z} f_i(0).$$

Inserting the factor $e^{-iE(t-z)}$, the ν_i neutrino wave function is

$$\nu_i(z,t) = e^{-iEt+i\left(E-m_i^2/2E\right)z} f_i(0). \tag{19.16}$$

This state has energy E and momentum $p_i = E - m_i^2/2E$. For $m_i^2 \ll E^2$, $p_i^2 = E^2 - m_i^2$, which is the relativistic relationship for a particle of mass m_i. Thus the neutrino ν_i carries mass m_i. $\nu_i(z,t)$ are the left-handed wavefunctions of (19.3).

Suppose that at $z = 0$ a neutrino of type α is born. The ν_α wavefunction is a linear superposition of mass eigenstates ν_i with $f_i(0) = U_{\alpha i} f_\alpha(0)$. Different mass eigenstates propagate with different phases so that the neutrino type changes with z:

$$f_\beta(z) = U_{\beta i}^* f_i(z) = U_{\beta i}^* e^{-i\left(m_i^2/2E\right)z} U_{\alpha i} f_\alpha(0). \tag{19.17}$$

To be exact a neutrino is born as a wave packet in some localised region of space time around some point $z = 0$, $t = 0$. A realistic treatment of its propagation requires the construction of the appropriate wave packet. We take it that the packet travels with almost the speed of light and with little distortion so that having travelled a distance $z = D$ the probability amplitude for finding a neutrino type β will be $e^{-iE(t-D)} f_\beta(D)$.

The probability of a transition $P_D(\nu_\alpha \to \nu_\beta)$ is

$$P_D(\nu_\alpha \to \nu_\beta) = \left| U_{\beta i}^* e^{-i\left(m_i^2/2E\right)z} U_{\alpha i} \right|^2 = \sum_{ij} U_{\beta i}^* U_{\alpha i} U_{\beta j} U_{\alpha j}^* e^{-i\left(\Delta m_{ij}^2 D/2E\right)}. \tag{19.18}$$

$\mathrm{Re}(U_{\beta i}^* U_{\alpha i} U_{\beta j} U_{\alpha j}^*)$ is symmetric and $\mathrm{Im}\,(U_{\beta i}^* U_{\alpha i} U_{\beta j} U_{\alpha j}^*)$ antisymmetric under the interchange of i and j, from this and the unitarity of U we can write

$$\begin{aligned} P_D(\nu_\alpha \to \nu_\beta) = \delta_{\alpha\beta} &- 4 \sum_{i>j} \mathrm{Re}\left(U_{\beta i}^* U_{\alpha i} U_{\beta j} U_{\alpha j}^*\right) \sin^2\left(\frac{\Delta m_{ij}^2 D}{4E}\right) \\ &+ 2 \sum_{i>j} \mathrm{Im}\left(U_{\beta i}^* U_{\alpha i} U_{\beta j} U_{\alpha j}^*\right) \sin\left(\frac{\Delta m_{ij}^2 D}{2E}\right) \end{aligned} \tag{19.19}$$

where $\Delta m_{ij}^2 = m_i^2 - m_j^2$.

These expressions describe the phenomena of *neutrino oscillations*. We note that experiments designed to observe and measure neutrino oscillations (Chapter 20) can only give values for the differences Δm_{ij}^2, and cannot give values for the individual masses m_i. The differences must satisfy the condition

$$\Delta m_{12}^2 + \Delta m_{23}^2 + \Delta m_{31}^2 = 0.$$

Restoring factors of c and $\hbar$, it will be useful to write

$$\frac{\Delta m_{ij}^2 D}{4E} = \Delta m_{ij}^2 c^4 \left(\frac{D}{\hbar c}\right)\frac{1}{4E} = 1.27\left(\frac{\Delta m_{ij}^2 c^4}{\mathrm{leV}^2}\right)\left(\frac{D}{1\ \mathrm{km}}\right)\left(\frac{1\ \mathrm{GeV}}{E}\right). \tag{19.20}$$

By considering the equations for the charge conjugate wave functions ν_α^c (see Section 7.4), similar formulae result, but with $U_{\alpha i}$ replaced by its complex conjugate $U_{\alpha i}^*$. If Im $\{U_{\beta i}^* U_{\alpha i} U_{\beta j} U_{\alpha j}^*\}$ is not zero it changes sign for antineutrinos and $P_D(\bar{\nu}_\alpha \to \bar{\nu}_\beta) \neq P_D(\nu_\alpha \to \nu_\beta)$. The lepton sector joins the quark sector in displaying matter–antimatter asymmetry.

19.4 The MSW effect

In many experiments that investigate oscillations the neutrinos are not completely free, but pass through matter on their journey from source to detector. This modifies the free wave functions discussed in the previous sections. In particular, matter contains electrons that interact with neutrinos through the charged weak currents. The effective interaction Lagrangian for this process is given by (9.8):

$$\mathcal{L}_{\mathrm{int}} = -2\sqrt{2} G_F g_{\mu\nu} j^\mu j^{\nu\dagger},$$

where, from (9.2), $j^\mu = e_L^\dagger \tilde{\sigma}^\mu \nu_{eL}$, $j^{\nu\dagger} = \nu_{eL}^\dagger \tilde{\sigma}^\mu e_L$, giving

$$\begin{aligned}\mathcal{L}_{\mathrm{int}} &= -2\sqrt{2} G_F g_{\mu\nu} \left(e_L^\dagger \tilde{\sigma}^\mu \nu_{eL}\right)\left(\nu_{eL}^\dagger \tilde{\sigma}^\nu e_L\right) \\ &= -2\sqrt{2} G_F g_{\mu\nu} \left(e_L^\dagger \tilde{\sigma}^\mu e_L\right)\left(\nu_{eL}^\dagger \tilde{\sigma}^\nu \nu_{eL}\right).\end{aligned} \tag{19.21}$$

The last step uses a Fierz transformation (Appendix A),

For matter at rest, the expectation value of $e_L^\dagger \tilde{\sigma}^o e_L = e_L^\dagger e_L = \frac{1}{2} N_e(x)$ where $N_e(x)$ is the total electron density at x. The factor of $1/2$ stems from the involvement of the left-handed electron field components only. Also, apart possibly from ferromagnetic effects, we can expect that the expectation value of $e_L^\dagger \tilde{\sigma}^i e_L = 0$. The neutrino Lagrangian density acquires an additional term $-\sqrt{2} G_F N_e(x)\, \nu_{eL}^\dagger \nu_{eL}$. This results in the modified equations for $f(z)$:

$$\mathrm{i}\frac{\mathrm{d} f_\beta(z)}{\mathrm{d}z} - m_{\beta\gamma} m_{\alpha\gamma}^* f_\alpha(z)/2E - V(z)\,\delta_{\beta e} f_e(z) = 0,$$

or equivalently (see equation 19.15)

$$\mathrm{i}\frac{\mathrm{d} f_i(z)}{\mathrm{d}z} = \frac{m_i^2}{2E} f_i(z) + V(z)\, \mathrm{U}_{ei} \mathrm{U}_{ej}^* f_j(z) \tag{19.22}$$

where $V(z) = \sqrt{2} N_e(z)\, G_F$.

The influence of matter on the propagation of neutrinos was pointed out by Wolfenstein (1978), and further elaborated by Mikheyev and Smirnov (1986). It is known as the MSW effect.

The neutral weak currents also contribute to the Lagrangian density of all neutrino types and result in an additional common phase factor on the wave functions of all types, which has no influence on neutrino oscillations.

19.5 Neutrino masses and the Standard Model

In the Weinberg–Salam electroweak theory for leptons of Chapter 12 we introduced three left-handed lepton doublet fields:

$$\mathbf{L}_e = \begin{pmatrix} \mathrm{L}_{eA} \\ \mathrm{L}_{eB} \end{pmatrix} = \begin{pmatrix} \nu_{eL} \\ e_{eL} \end{pmatrix}, \quad \mathbf{L}_\mu = \begin{pmatrix} \nu_{\mu L} \\ \mu_L \end{pmatrix}, \quad \mathbf{L}_\tau = \begin{pmatrix} \nu_{\tau L} \\ \tau_L \end{pmatrix},$$

and three right-handed singlets e_R, μ_R, τ_R. Under an $SU(2)$ transformation,

$$\mathbf{L}_\alpha \to \mathbf{L}'_\alpha = \mathbf{U}\mathbf{L}_\alpha, \quad \alpha_R \to \alpha'_R = \alpha_R.$$

Dirac neutrinos having mass implies the existence of right-handed neutrino fields. In the Standard Model the right-handed neutrino fields, like the right-handed fields of the charged leptons, must be $SU(2)$ singlets. Neutrino masses are introduced into the model in the same way as the u, c and t quarks by coupling to the Higgs field. An $SU(2)$ invariant coupling of the Higgs field to neutrinos is then (equation (14.9) and Problem 14.3.)

$$\mathcal{L}^\nu_{\text{Higgs}} = -\sum_{\alpha\beta}\left[G^\nu_{\alpha\beta}\left(L^\dagger_\alpha \varepsilon \Phi^*\right)\nu_{\beta R} - G^{\nu *}_{\alpha\beta}\nu^\dagger_{\beta R}\left(\Phi^T \varepsilon L_\alpha\right)\right] \tag{19.23}$$

where $G^\nu_{\alpha\beta}$ is a complex 3×3 matrix. On symmetry breaking this gives the neutrino mass term

$$\mathcal{L}^\nu_{\text{mass}} = -\phi_o \sum_{\alpha,\beta}\left[G^\nu_{\alpha\beta}\nu^\dagger_{\alpha L}\nu_{\beta R} + G^{\nu *}_{\alpha\beta}\nu^\dagger_{\beta R}\nu_{\alpha L}\right]. \tag{19.24}$$

This is just the mass term of equation (19.1) if we identify $\phi_o G^\nu_{\alpha\beta}$ with $m_{\alpha\beta}$.

19.6 Parameterisation of U

We have taken the parameters m_e, m_μ, m_τ and g_2 to be real and positive, but this is in fact a phase convention: any phase on these parameters can be absorbed in phase factors multiplying the lepton fields, and such phase factors are of no physical significance. It is also the case that the definition of the mass matrix $m_{\alpha\beta}$ depends on a phase convention.

Define the six neutrino fields $\nu'_{\alpha\mathrm{L}}, \nu'_{\alpha\mathrm{R}} (\alpha = \mathrm{e}, \mu, \tau)$ and the six charged lepton fields $\alpha'_{\mathrm{L}}, \alpha'_{\mathrm{R}}$ by

$$\nu_{\alpha\mathrm{L}} = \mathrm{e}^{\mathrm{i}\theta_\alpha}\nu'_{\alpha\mathrm{L}}, \quad \nu_{\alpha\mathrm{R}} = \mathrm{e}^{\mathrm{i}\gamma_\alpha}\nu'_{\alpha\mathrm{R}}, \quad \begin{pmatrix}\alpha_{\mathrm{L}}\\ \alpha_{\mathrm{R}}\end{pmatrix} = \mathrm{e}^{\mathrm{i}\theta_\alpha}\begin{pmatrix}\alpha'_{\mathrm{L}}\\ \alpha'_{\mathrm{R}}\end{pmatrix}.$$

The leptonic part of the electroweak Lagrangian density described in Chapter 12 (equation (12.12)), and the charged current (equation (12.16)) and neutral current (equation (12.23)) that give the neutrino coupling to the $\mathrm{W}^{\pm}$ and Z fields, are unchanged in form under these transformations. The neutrino mass matrix retains the same form but with $m_{\alpha\beta}$ replaced by

$$m'_{\alpha\beta} = \mathrm{e}^{-\mathrm{i}\theta_\alpha + \mathrm{i}\gamma_\beta} m_{\alpha\beta}.$$

We can redefine $m_{\alpha\beta}$ in this way, keeping the physical content of the theory unchanged.

The unitary matrix $\mathbf{U}^{\mathrm{L}}$ was defined by $m_{\alpha\beta} = \sum_i U^{\mathrm{L}*}_{\alpha i} m_{\mathrm{i}} U^{\mathrm{R}}_{\beta\mathrm{i}}$. Hence we can redefine $U^{\mathrm{L}'}_{\alpha i} = \mathrm{e}^{\mathrm{i}(\theta_\alpha - \delta_i)} U^{\mathrm{L}}_{\alpha i}$, where the phase factors $\mathrm{e}^{\mathrm{i}\delta_i}$ were introduced in Section 19.1. As in our discussion of the KM matrix in Section 14.2, when the non-physical phase factors have been taken out, the resulting matrix depends on four physical parameters. We parameterise it in the same way as the KM matrix but replace θ_{1j} by $\theta_{\mathrm{e}j}$, θ_{2j} by $\theta_{\mu j}$ and θ_{3j} by $\theta_{\tau j}$, etc. It can be called the neutrino mass mixing matrix.

The term exhibiting matter–antimatter asymmetry in $P_D(\nu_\alpha \to \nu_\beta)$ is (see Problem 19.2)

$$2\sum_{i>j} \mathrm{Im}\left(U^*_{\beta i} U_{\alpha i} U_{\beta j} U^*_{\alpha j}\right) \sin\frac{\Delta m^2_{ij} D}{2E}$$
$$= \begin{cases} 0 \ \text{if} \alpha = \beta \\ \pm 8J \sin\left(\dfrac{\Delta m^2_{21} D}{4E}\right) \sin\left(\dfrac{\Delta m^2_{32} D}{4E}\right) \sin\left(\dfrac{\Delta m^2_{31} D}{4E}\right), \ \text{otherwise} \end{cases}$$

where $J = c_{\mathrm{e}2} c^2_{\mathrm{e}3} c_{\mu 3} s_{\mathrm{e}2} s_{\mathrm{e}3} s_{\mu 3} \sin\delta$, cf. (14.18, 14.19), the minus sign is taken for transitions $\mathrm{e} \to \mu$, $\mu \to \tau$, $\tau \to \mathrm{e}$, and the plus sign otherwise.

19.7 Lepton number conservation

Having defined the phase conventions that fix the parameters of the neutrino mixing matrix, the Lagrangian density has only one remaining global $U(1)$ symmetry. It is unchanged if all lepton fields, charged and neutral, left-handed and right-handed, are multiplied by the same phase factor $\mathrm{e}^{\mathrm{i}\delta}$. Following the method of Section 7.1, we consider an arbitrary small space- and time-dependent variation in δ, and conclude that we have one conserved current:

$$j^\mu(x) = \sum_\alpha \Big[\alpha_L^\dagger(x)\tilde{\sigma}^\mu \alpha_L(x) + \alpha_R^\dagger(x)\sigma^\mu \alpha_R(x) + \nu_{\alpha L}^\dagger(x)\tilde{\sigma}^\mu \nu_{\alpha L}(x) + \nu_{\alpha R}^\dagger(x)\sigma^\mu \nu_{\alpha R}(x)\Big]. \qquad (19.25)$$

The quantity $\int j^{\mathrm{o}}(x)\,\mathrm{d}^3\mathbf{x}$ counts the number of leptons minus the number of antileptons, and this number is conserved.

19.8 Sterile neutrinos

We will see in the next chapter that there is some experimental indication that there are more than three neutrino mass eigenstates. If these indications are confirmed then we will be obliged to introduce a fourth neutrino type (perhaps more), say ν_w. Since there is no indication of another charged lepton to partner ν_{wL} in an $SU(2)$ doublet, and since the decays of the Z (Section 13.6) confirm that only three neutrino types participate in the weak interaction, both ν_{wL} and ν_{wR} must be $SU(2)$ singlets and have no electroweak interactions except through the mass eigenstate. Such a neutrino is known as a sterile neutrino.

Problems

19.1 Neglect the term $\mathrm{i}(\mathrm{d}g_\gamma/\mathrm{d}z)$ in (19.11) and show that $g_\gamma(z) = m^*_{\alpha\gamma}\, {}^{f_\alpha(z)}\!/_{2E}$. Show that an estimate of $\mathrm{i}(\mathrm{d}g_\gamma/\mathrm{d}z)$ is then $\mathrm{i}\mathrm{d}g_\gamma(z)/\mathrm{d}z = S_{\gamma\beta}(2Eg_\beta(z))$ with $S_{\gamma\beta} = m^*_{\alpha\gamma} m_{\alpha\beta}/4E^2$, very small for E much greater than the masses.

19.2 Define $F_{\beta\alpha ij} = \mathrm{Im}\,(U^*_{\beta i} U_{\alpha i} U_{\beta j} U^*_{\alpha j})$

(a) Show that $F_{\beta\alpha ij} = -F_{\beta\alpha ji}$ and that $\sum_i F_{\beta\alpha ij} = 0$, and hence that $F_{\beta\alpha 12} = F_{\beta\alpha 23} = F_{\beta\alpha 31}$. Define $J = F_{\mu e12}$ (this conforms with (14.18) and (19.25)). Using the trigonometric identity $\sin(x) + \sin(y) - \sin(x+y) = 4\sin(x/2)\sin(y/2)\sin((x+y)/2)$.

(b) verify the matter–antimatter asymmetry term in (19.25) for $P_D(\nu_\alpha \to \nu_\beta)$.

20

Neutrino masses and mixing: experimental results

The cross-sections for neutrino–lepton and neutrino–quark interactions are exceedingly small: the collection of data from a particular experiment may extend over several years. The aims of neutrino experiments include: establishing the existence of neutrino oscillations, checking the validity of the theory of Chapter 19, measuring the parameters of the mixing matrix **U** and determining the mass eigenstates of the neutrino. In this chapter we shall present some results of recent experimental work, and indicate how they have been obtained.

20.1 Introduction

Setting aside the possible existence of sterile neutrinos, it is thought that there are three neutrino mass eigenstates, which we shall label by $i = 1, 2, 3$. Measurements of neutrino oscillations give $(\text{mass})^2$ differences:

$$\Delta m_{ij}^2 = m_i^2 - m_j^2.$$

It is estimated from experiment that

$$1.3 \times 10^{-3}\ \text{eV}^2 < |\Delta m_{32}^2| < 3 \times 10^{-3}\ \text{eV}^2,$$

and

$$6.5 \times 10^{-5}\ \text{eV}^2 < |\Delta m_{21}^2| < 8.5 \times 10^{-5}\ \text{eV}^2.$$

Then $\Delta m_{31}^2 = \Delta m_{32}^2 + \Delta m_{21}^2$.

For illustrative numerical calculations in this chapter we shall take $|\Delta m_{32}^2| = 2 \times 10^{-3}\ \text{eV}^2$ and $|\Delta m_{21}^2| = 7 \times 10^{-5}\ \text{eV}^2$.

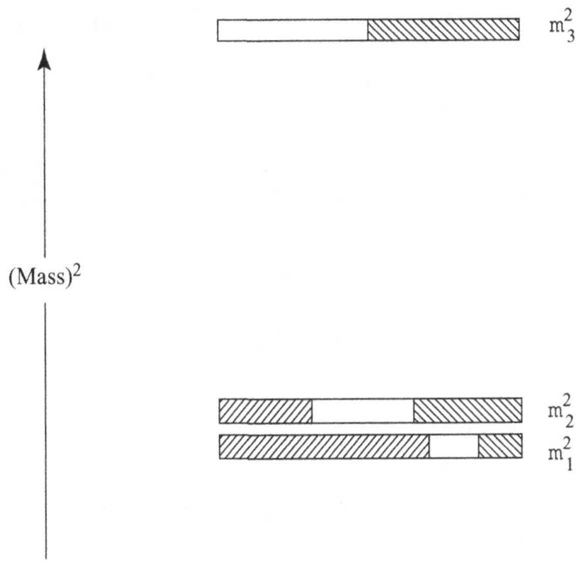

Figure 20.1 A three neutrino mass-squared spectrum. The ν_e fraction of each mass eigenstate is indicated by right-leaning hatching, the ν_μ fraction is blank and the ν_τ fraction by left-leaning hatching (see the report by B. Kayser, Particle Data Group, 2004). The mass-squared base line is not known.

The 3×3 unitary mixing matrix is approximately

$$\mathbf{U} = \begin{bmatrix} c & s & s_{e3}e^{i\delta} \\ -s/\sqrt{2} & c/\sqrt{2} & 1/\sqrt{2} \\ s/\sqrt{2} & -c/\sqrt{2} & 1/\sqrt{2} \end{bmatrix} \tag{20.1}$$

where $c \approx \cos\theta_{e2} \approx 0.84$ and $s \approx \sin\theta_{e2} \approx 0.54$.

It is estimated that $|s_{e3}|^2 < 0.05$. A term $s_{e3}e^{i\delta}$ with $\sin\delta \neq 0$ would violate *CP* conservation and lead to matter–antimatter asymmetry. Such asymmetry has not yet (2006) been discerned. If $s_{e3} \neq 0$ there are small complex corrections to other elements of the matrix. (The matrix (20.1) may be obtained from the unitary KM matrix of Section 14.3 by taking $c_{13}(=c_{e3}) = 1$, $c_{12}\,(=c_{e2}) = c$, $c_{23}(=c_{\mu 3}) = 1/\sqrt{2}$, $s_{13} = s_{e3}$).

The (mass)^2 differences imply either a spectrum of (mass)^2 eigenstates as in Fig. 20.1, with the closest eigenstates having the smallest mass, or the figure might be inverted, with the closest (mass)^2 eigenstates the heaviest. The mixing matrix determines the fractions of ν_e, ν_μ and ν_τ states making up the states 1, 2, 3, and these are indicated on the figure.

In many data analyses the approximation is made of setting $s_{e3} = 0$. We shall see that any particular analysis is then greatly simplified since the number of participating neutrino mass eigenstates is reduced from three to two. Apart from our discussion of the CHOOZ experiment, we shall always make this approximation. However, as the quality of data improves, and in particular when and if s_{e3} is seen to be finite, the approximation will be abandoned. It is important to note that with $s_{e3} = 0$ there is no *CP* violation.

The analysis of data from accelerator and reactor neutrinos is the least complicated, since the MSW effect is negligible at the levels of precision so far obtained, and our formula (19.19) can be directly invoked.

20.2 K2K

The Japanese K2K experiment studies a muon neutrino beam that is engineered at the KEK proton accelerator. 12 GeV protons hit an aluminium target, producing mainly positive pions that decay $\pi^+ \to \mu^+ + \nu_\mu$ (Section 9.2). The beam characteristics are measured by near detectors located 300 m down-stream from the proton target. The mean ν_μ energy is 1.3 GeV. There is then a 250 km flight path to the Super-Kamiokandi detector in the Komioka mine. This detector consists of 22.5 kilotonnes of very pure water (H_2O). Muon neutrinos are observed through their reaction with neutrons in the oxygen nuclei: $\nu_\mu + n \to p + \mu^-$. The neutrino energy E_ν can be determined from measurements of the energy and direction of the muon.

To reach the detector, a neutrino has to pass through the Earth's upper crust. However, we ignore any MSW effect for the moment, and take the values of Δm_{21}^2 given in Section 20.1. $\Delta m_{21}^2 = 7 \times 10^{-5}\,\mathrm{eV}^2$ and $D = 250\,\mathrm{km}$. From (19.20) the oscillating function $\sin^2\left(\frac{\Delta m_{21}^2 D}{4E_\nu}\right) = \sin^2\left(0.022\left(\frac{1\ \mathrm{GeV}}{E_\nu}\right)\right) < 10^{-3}$ for all relevant E_ν. This is so small that with present precision it can be ignored. Also, since $\Delta m_{31}^2 = \Delta m_{32}^2 + \Delta m_{21}^2$ the two other oscillating functions are almost equal, and we will take them both as $\sin^2\left(\frac{\Delta m_{At}^2 D}{4E_\nu}\right)$ with Δm_{At}^2 a mean value of Δm_{32}^2 and Δm_{31}^2. For historical reasons Δm_{At}^2 is called the *atmospheric mass squared difference.*

With these approximations, setting $U_{e3} = 0$ and using the unitarity of U, equations (19.19) give

$$\begin{aligned}
P_D(\nu_\mu \to \nu_\mu) &= 1 - 4|U_{\mu 3}|^2(1 - |U_{\mu 3}|^2)\sin^2\left(\frac{\Delta m_{\mathrm{At}}^2 D}{4E_\nu}\right), \\
P_D(\nu_\mu \to \nu_e) &= 0, \\
P_D(\nu_\mu \to \nu_\tau) &= 4|U_{\mu 3}|^2(1 - |U_{\mu 3}|^2)\sin^2\left(\frac{\Delta m_{\mathrm{At}}^2 D}{4E_\nu}\right).
\end{aligned} \tag{20.2}$$

From these equations, and because of the smallness of $|U_{e3}|^2$, the ∇m_{At}^2 oscillation is almost entirely between ν_μ and ν_τ. Since the MSW effect is for electron

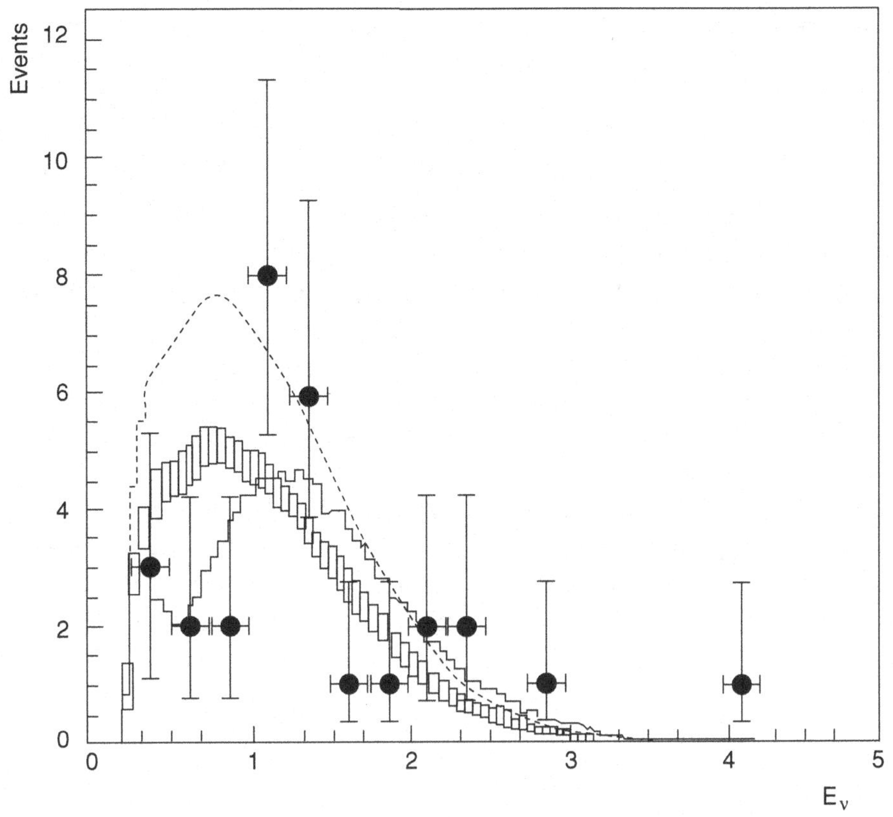

Figure 20.2 K2K data (M. H. Ahn *et al. Phys. Rev. Letts.* **90**, 041801 (2003)). Points with error bars are data. The box histogram is the expected spectrum without oscillations, where the height of the box is the systematic error. The solid line is the best-fit spectrum. These histograms are normalised by the number of events observed (29). In addition, the dashed line shows the expectation with no oscillations normalised to the expected number of events (44).

neutrinos only, it can with present precision be neglected. With $U_{e3} = 0$, we have $|U_{\mu 3}| = \sin\theta_{\mu 3}$ and we arrive at our final formula:

$$P_D(\nu_\mu \to \nu_\mu) = 1 - \sin^2(2\theta_{\mu 3})\sin^2\left(\frac{\Delta m^2_{\mathrm{At}} D}{4E_\nu}\right) \tag{20.3}$$

for fitting the K2K data. This is presented in Fig. 20.2 in which the number of events in the designated energy bins are shown as a function of the mean neutrino energy of each bin. The dashed curve is the expected number distribution dN/dE_ν without oscillation, and when integrated over E_ν is clearly larger than the total number (29) of events accepted. The best fit with equation (20.3), modified to take account of corrections such as energy resolution, is also shown.

It corresponds to $\Delta m^2_{\mathrm{At}} = 2.8 \times 10^{-3}\,\mathrm{eV}^2$ and $\sin^2(2\theta_{\mu 3}) = 1$. The latter allows $\theta_{\mu 3} = \pi/4,\ \cos\theta_{\mu 3} = \sin\theta_{\mu 3} = 1/\sqrt{2}$.

20.3 Chooz

Chooz is a village close to a French nuclear power station. The power station's two reactors are rich sources of electron antineutronos $\bar{\nu}_e$. The fluxes and energy distributions, centred around 3 MeV, of these antineutrinos are very well understood. The detector, shielded from cosmic ray muons by its location deep underground, was positioned about 1 km from the reactors.

The antineutrinos $\bar{\nu}_e$ were detected by their inverse β decay interaction with protons, $\bar{\nu}_e + \mathrm{p} + 1.8\,\mathrm{MeV} \to \mathrm{n} + \mathrm{e}^+$, in a hydrogen rich paraffinic liquid scintillator.

As with the K2K experiment, the oscillatory function $\sin^2\left(\Delta m^2_{21} D/4E_\nu\right)$ is, from (19.20), negligibly small, $< 2 \times 10^{-3}$ (taking $D = 1$ km, $E_\nu > 1.8$ MeV). The MSW effect can also be neglected, since for material in the Earth's crust $V(z) \sim 10^{-13}\,\mathrm{eV} \ll \Delta m^2_{21}/2E_\nu < \Delta m^2_{32}/2E_\nu$. We can, again, to a good approximation, put $\Delta m^2_{32} D/4E_\nu = \Delta m^2_{31} D/4E_\nu = \Delta m^2_{\mathrm{At}} D/4E_\nu$ to obtain

$$P_D(\bar{\nu}_e \to \bar{\nu}_e) = 1 - 4|U_{e3}|^2\left(1 - |U_{e3}|^2\right)\sin^2\left(\Delta m^2_{\mathrm{At}} D/4E_\nu\right). \tag{20.4}$$

Setting $|U_{e3}| = \sin\theta_{e3},\ D = 1\,\mathrm{km},\ \Delta m^2_{\mathrm{At}} = 2 \times 10^{-3}\,\mathrm{eV}^2$, we find from (19.20)

$$P_D(\bar{\nu}_e \to \bar{\nu}_e) = 1 - \sin^2(2\theta_{e3})\sin^2[2.54(3\,\mathrm{MeV}/E_\nu)].$$

To the experimental precision obtained, there was no reduction in flux at the detector and no oscillation, and it was concluded (Apollonio *et al.*, 2003) that $\sin^2(2\theta_{e3}) < 0.18$, which implies $|U_{e3}|^2 =< 0.05$, the result we quote in Section 20.1 of this chapter.

20.4 KamLAND

Like Chooz, the Kamioka Liquid scintillator AntiNeutrino Detector (KamLAND) experiment uses reactor antineutrinos. The sources are a group of nuclear power stations in Japan situated at various distances $\sim$ 100 km to 200 km from the detector. As at Chooz, the detector makes use of the inverse β decay $\bar{\nu}_e + \mathrm{p} \to \mathrm{n} + \mathrm{e}^+$.

The experiment was designed to explore the $\Delta m^2_{21} \sim 7 \times 10^{-5}\,\mathrm{eV}^2$ mass region. For a particular reactor at distance D from the detector, we have from (19.19) and setting $|U_{e3}|^2 = 0$, that the survival probability is given by

$$P_D(\bar{\nu}_e \to \bar{\nu}_e) = 1 - 4|U_{e1}|^2|U_{e2}|^2\sin^2\left(\frac{\Delta m^2_{21} D}{4E}\right) \tag{20.5}$$

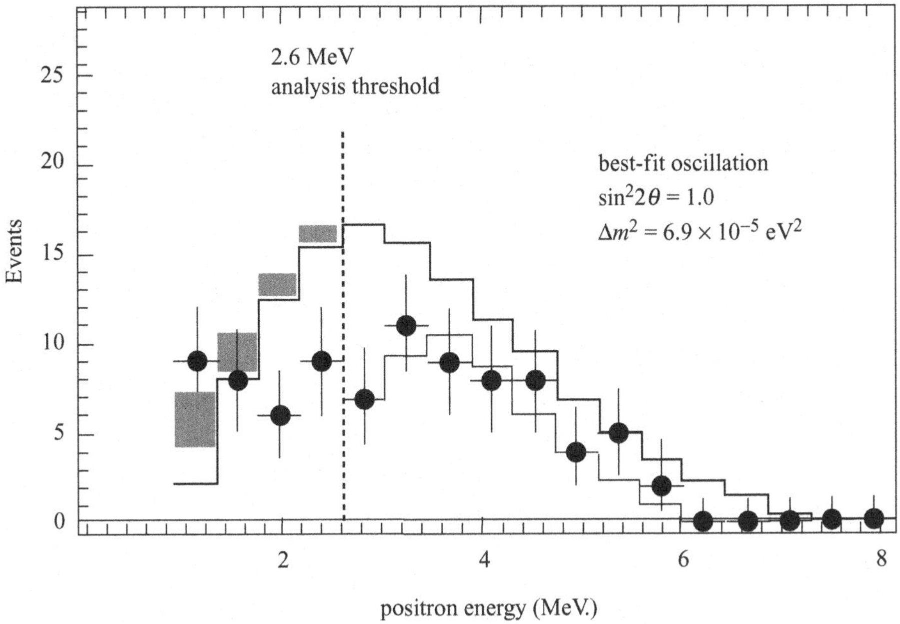

Figure 20.3 KamLAND data (K. Eguchi *et al. Phys. Rev. Lett* **90**, 021802 (2003)). The energy distribution of the observed positrons in bins of 0.425 MeV (solid circles with error bars), along with the expected no oscillations distribution (upper histogram) and the best fit including oscillations using (20.5) (lower histogram). The shaded bands indicate the systematic error in the best fit distribution. The vertical dashed line corresponds to the analysis threshold at 2.6 MeV.

and from the parameterization (14.16)

$$4|U_{e1}|^2|U_{e2}|^2 = \cos^4\theta_{e3}\sin^2 2\theta_{e2} \approx \sin^2 2\theta_{e2}.$$

As at Chooz, MSW effects are negligible. The measured positron energy spectrum is compared with the positron energy spectrum that would be expected if there were no antineutrino oscillations. This spectrum can be very well estimated from knowledge of the various reactor characteristics.

Some results from KamLAND are shown in Fig. 20.3. The energy spectrum of the positrons is clearly below what it would be without oscillation. The best fit to the data using an expression based on (20.5) has

$$|m_{21}^2| = 6.9\times 10^{-5}\,\text{eV}^2,$$
$$0.84 < \sin^2 2\theta_{e2} < 1.$$

The KamLAND analysis took some account of systematic errors arising from the simplifying assumption $|U_{e3}|^2 = 0$.

20.5 Atmospheric neutrinos

The Earth is continually bombarded by cosmic rays, which consist for the most part of high energy protons and electrons. The protons, in their collisions with nuclei in the upper atmosphere, produce π mesons. The π mesons decay by the chains (Section 9.2, Section 9.4):

$$\pi^+ \to \mu^+ + \nu_\mu \qquad , \qquad \pi^- \to \mu^- + \bar{\nu}_\mu \quad .$$
$$\quad \llcorner\!\!\to e^+ + \nu_e + \bar{\nu}_\mu \qquad\qquad \llcorner\!\!\to e^- + \bar{\nu}_e + \nu_\mu$$

The neutrinos and antineutrinos are produced at a mean height $\sim 20\,\text{km}$, with energies extending to the multi-GeV region. The ratio of the flux $\nu_\mu + \bar{\nu}_\mu$ to the flux of $\nu_e + \bar{\nu}_e$ is evidently about 2.

In water detectors, such as Super Kamiokandi, charged leptons are produced through reactions essentially of the form

$$\begin{aligned} \nu_e + n &\to e^- + p, \qquad & \bar{\nu}_e + p &\to e^+ + n; \\ \nu_\mu + n &\to \mu^- + p, \qquad & \bar{\nu}_\mu + p &\to \mu^+ + n. \end{aligned}$$

The charged leptons emit Cerenkov radiation, which provides information on the energy, direction and identity of the incident neutrino.

Figure 20.4 shows some results from the Super-Kamiokandi detector. The plots show the ratio of observed ν_e- and ν_μ-like events to Monte Carlo calculations in the absence of oscillations, as a function of D/E_ν. E_ν is the neutrino energy and D the distance from the point of production $\sim$ 20 km above the Earth's surface, to the detector. D is then inferred from the measured neutrino direction. For multi-GeV electron neutrinos, the MSW modification to the equations has to be included for those neutrinos passing through the Earth on their way to the detector.

The ν_e data show no sign of oscillation, but there is a clear deficit of muon neutrinos. The best fit to the data has $\Delta m^2_{\text{At}} = 2.2 \times 10^{-3}\ \text{eV}^2$, and like K2K has $\sin^2 2\theta_{\mu 3} = 1$, where for $D/E_\nu < 10^3\ \text{km/GeV}$ the Δm^2_{32} and Δm^2_{31} oscillations are combined into one Δm^2_{At} oscillation. The absence of discernible $\nu_e \to \nu_e$ oscillations in the data was the first indication of the smallness of $|U_{e3}|^2$, which again implies that the Δm^2_{At} oscillations are predominantly between ν_μ and ν_τ.

20.6 Solar neutrinos

The nuclear and thermal physics of the Sun is well understood. The solar neutrino spectra predicted by the Standard Solar Model and shown in Fig. 20.5 may be assumed with confidence.

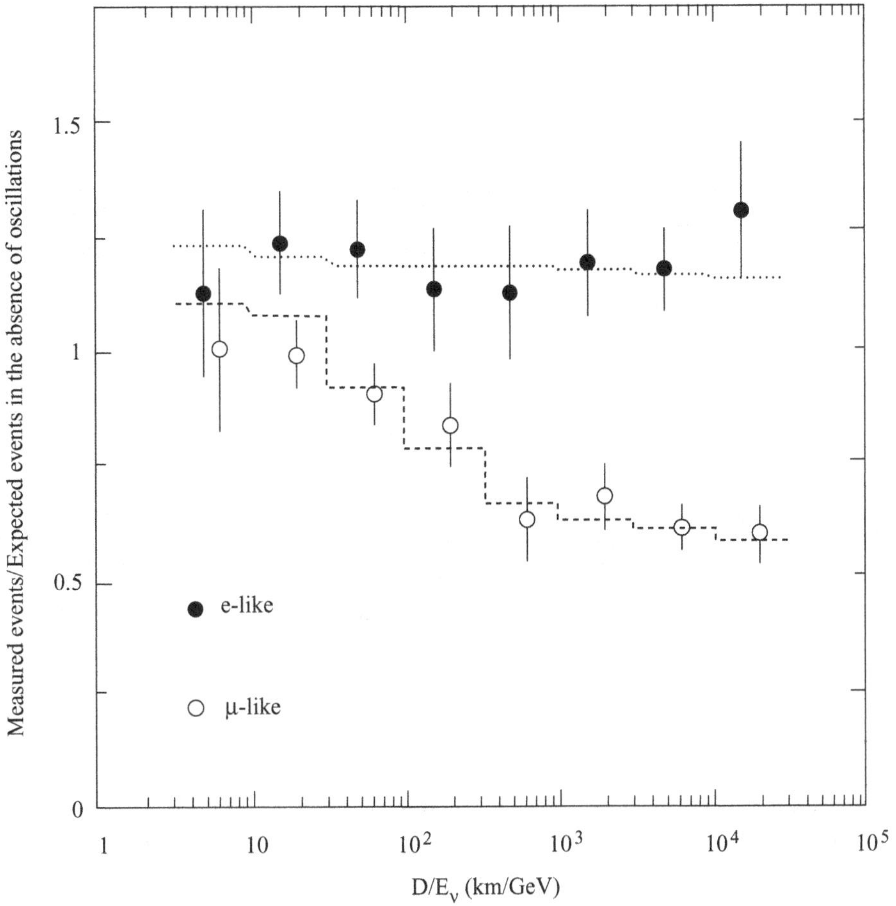

Figure 20.4 Data from Super Kamiokande (Y. Fukunda *et al. Phys. Rev. Lett.* **82**, 1562 (1998). The ratio of measured events to expected events in the absence of oscillations. The lines show the expected shape for $\nu_\mu \leftrightarrow \nu_\tau$ with $\Delta m^2_{\mathrm{At}} = 2.2 \times 10^{-3}\,\mathrm{eV}^2$ and $\sin^2(2\theta_{\mu 3}) = 1$. There is no significant $\nu_e \leftrightarrow \nu_e$ oscillation observed.

The first measurements of the spectra were made by R. Davis and his collaborators in the deep Homestake mine in the U.S.A, (Davis 1964). The detection of the neutrinos was made through the reaction

$$\nu_e + {}^{37}_{17}\mathrm{Cl} + 0.81\ \mathrm{MeV} \rightarrow e^- + {}^{37}_{18}\mathrm{Ar}.$$

The Super-Kamiokande detector also made measurements of the solar neutrino flux with E_ν greater than about 6 Mev (Fukuda *et al.* 1996).

Because of the high energy threshold these measurements were blind to the principal flux from the 'pp' reaction. The GALLEX (Italy) and SAGE (Russia)

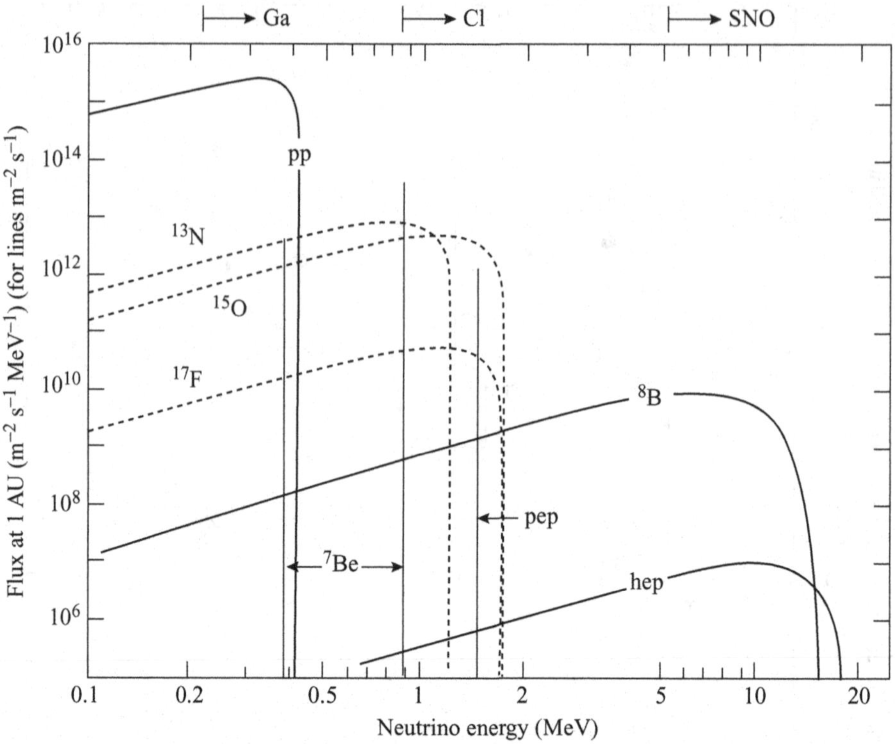

Figure 20.5 The solar neutrino spectra predicted by the standard solar model. Spectra for the pp chain are shown by solid lines and those for the CNO chain by dashed lines. (See Bahcall, J. N. and Ulrich, R. K. (1988), *Rev. Mod. Phys.* **60**, 297.)

experiments were designed to remedy this, and examine the pp flux through the reaction (Hampel *et al.*, 1999; Gavrin *et al.*, 2003)

$$\nu_e + {}^{17}_{31}\mathrm{Ga} + 0.23\ \mathrm{MeV} \to {}^{71}_{32}\mathrm{Ge} + e^-.$$

The SNO (Sudbury Neutrino Observatory, Canada) is a heavy water detector. Neutrinos, with E_ν greater than about 5 Mev, are detected through the reactions, (Ahmad *et al.* 2002)

$$\nu_e + D_2 + 1.44\ \mathrm{MeV} \to e^- + p + p,$$
$$\nu_e + D_2 + 2.22\ \mathrm{MeV} \to p + n + \nu.$$

The first of these reactions is a charged current interaction and can be initiated only by an electron neutrino. The second is a neutral current interaction, initiated with equal probability by an electron, muon or tau neutrino. The SNO experiment also measured the reaction rate of elastic neutrino scattering from electrons,

$$\nu + e^- \rightarrow \nu + e^-.$$

Again, this reaction can be triggered by a neutrino of any type. Measurements can be used to infer both the ν_e flux $\phi(\nu_e)$) and the total flux $\phi(\nu_e + \nu_\mu + \nu_\tau)$.

The early results from the Homestake detector gave a measured flux of only about one third of that expected from the standard Solar Model without oscillation. Super Kamiokande, GALLEX and SAGE gave about half the expected rate. SNO found that

$$\frac{\phi(\nu_e)}{\phi(\nu_e + \nu_\mu + \nu_\tau)} = 0.306 \pm 0.05.$$

The measured total neutrino flux was consistent with that expected from the Standard Solar Model and clearly, since the Sun produces only electron neutrinos, many have made the transition to ν_μ and ν_τ.

20.7 Solar MSW effects

We showed in Section 19.4 that plane wave neutrino mass eigenstates depended on functions $f_i(z)$, that satisfied

$$\mathrm{i}\frac{\mathrm{d}f_i}{\mathrm{d}z} = \frac{m_i^2}{2E} f_i + V(z) U_{\mathrm{ej}}^* U_{\mathrm{ei}} f_j. \tag{20.6}$$

The source of solar neutrinos is the central region of the Sun, where the Standard Solar Model gives $V(\mathrm{o}) = 7.6 \times 10^{-12}$ eV. Comparing this with $\Delta m_{21}^2/2E$, which with the 'reference parameters' of Section 20.1 equals $3.5 \times 10^{-12}(10\,\mathrm{MeV}/E_\nu)$, it is clear that the interpretation of the data from solar neutrino experiments requires a serious consideration of the MSW effect.

As a starting approximation we again neglect the small term $U_{\mathrm{e}3}$. With $U_{\mathrm{e}3} = 0$ the solution of (20.6) for $f_3(z)$ is

$$f_3(z) = \mathrm{e}^{\mathrm{i}m_3^2 z/2E} f_3(0),$$

independent of $V(z)$. With $U_{\mathrm{e}3} = 0$, and since the initial neutrino is an electron neutrino, $f_3(0) = 0$ and it follows that $f_3(z)$ is zero for all z: it plays no part in the oscillations. The approximation again reduces the analysis to a two-neutrino phenomenon in $f_1(z)$ and $f_2(z)$. After some algebra it can be shown that the solar neutrino data can be analysed with the equations

$$\begin{aligned} \mathrm{i}\frac{\mathrm{d}f_\mathrm{e}}{\mathrm{d}z} &= \frac{\Delta m_{21}^2}{2E}(-\cos(2\theta_{\mathrm{e}2})f_\mathrm{e} + \sin(2\theta_{\mathrm{e}2})f_x) + V(z)f_\mathrm{e} \\ \mathrm{i}\frac{df_x}{dz} &= \frac{\Delta m_{21}^2}{2E}(\sin(2\theta_{e2})f_e + (\cos 2\theta_{e2})f_x). \end{aligned} \tag{20.7}$$

$f_x = c_{\mu 3} f_\mu - s_{\mu 3} f_\tau$ is a combination of f_μ and f_τ, $V(z)$ is known from the Standard Solar Model. The equations have to be integrated numerically.

All the solar neutrino data is consistent with the oscillation interpretation, and analysis of the data gives $3 \times 10^{-5}\,\mathrm{eV}^2 < \Delta m_{21}^2 < 1.9 \times 10^{-4}\,\mathrm{eV}^2$, $30.2^\circ < \theta_{\mathrm{e}2} < 34.9^\circ$ with high probability (95% confidence level). The best fit is with $\Delta m_{21}^2 = 6.9 \times 10^{-5}\,\mathrm{eV}^2$, $\theta_{\mathrm{e}2} = 32^\circ$.

The solar neutrino data give a tighter constraint on $\theta_{\mathrm{e}2}$ than KamLAND. Also, with the MSW effect, the solution of equations (20.7) depends on the sign of Δm_{21}^2. It is found to be positive, as is indicated in Figure 20.1

20.8 Future prospects

There are several planned experiments that will make a more thorough investigation of neutrino masses and mixing phenomena. Apart from the possibility of sterile neutrinos, indications of which have not been confirmed, there is no evidence to contradict the three-neutrino theory of Chapter 19. However, it can be seen from the quality of the data presented in this chapter that the neutrino mass theory is not as well established as other branches of The Standard Model. Within the theory experiments are planned to make more precise measurements of the Δm^2 and the parameters of the neutrino mixing matrix.

The principal focus of experimental activity is on the construction of muon neutrino beams as in the K2K experiment. An advantage of accelerator-generated neutrinos is the control that one has on the flux and energy distribution. K2K is an ongoing experiment but by late 2006 the muon neutrino experiments CNGS and MINOS (Main Injector Neutrino Oscillation Search) will be in operation. The CNGS neutrinos are generated at CERN and detected at the GRAN SASSO underground laboratory in Italy. The MINOS beam is generated at Fermilab and detected in the Soudan mine in Minnesota. Both experiments will look for evidence of the rare $\nu_\mu \to \nu_\mathrm{e}$ transition and for the expected $\nu_\mu \to \nu_\tau$ oscillations. If the theory of Chapter 19 is not challenged it is expected that by 2010 we will have much tighter bounds on both $\sin^2(2\theta_{\mathrm{e}3})$ and $|\Delta m_{\mathrm{At}}^2|$.

In the more distant future a new very high intensity proton accelerator will be built at Tokai, Japan. The experiment T2K will take over from K2K with a neutrino beam of much higher intensity. Detection at Super Kamiokande will give a base line

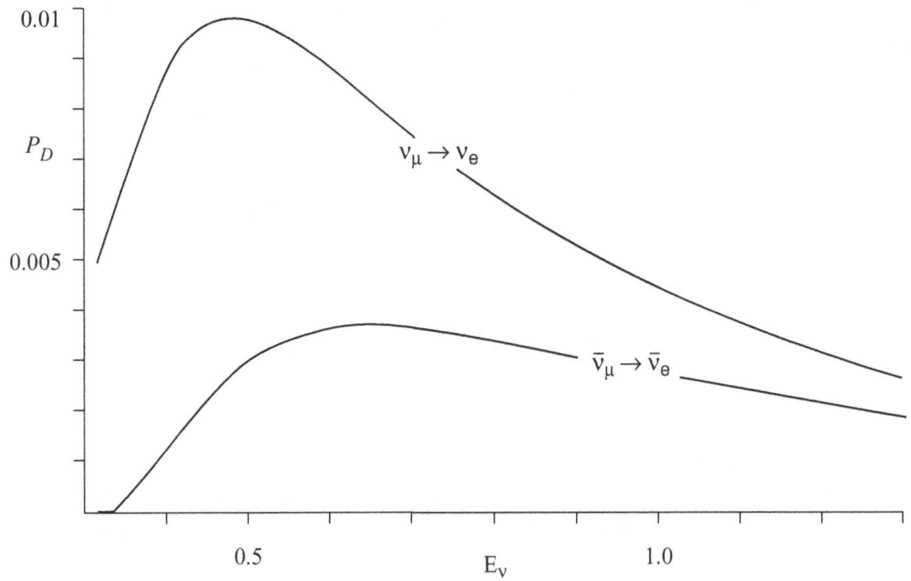

Figure 20.6 The upper curve is $P_D(\nu_\mu \rightarrow \nu_e)$.
The lower curve $P_D(\bar{\nu}_\mu \rightarrow \bar{\nu}_e)$.
The parameters are $\Delta m^2_{32} = 2 \times 10^{-3}\,\mathrm{eV}^2$, $\Delta m^2_{21} = 7 \times 10^{-5}\,\mathrm{eV}^2$, $\cos\theta_{e2} = 0.84$, $\cos\theta_{\mu 3} = 1/\sqrt{2}$, $\sin\theta_{e3} = 0.05$, $\delta = \pi/4$, $D = 295$ km.
The MSW effect, which depends on the local geology will be significant but calculable. It is not included here.

$D \approx 295$ km. An upgrade to higher intensity for MINOS is also planned with a new experiment NOνA. By 2015 with T2K and NOνA it is expected that if $\sin^2(2\theta_{e3}) > 0.01$ then it will be detected. The MSW effect will influence these measurements and the sign of Δm^2_{32} could be established, and hence the mass ordering. If $\sin^2(2\theta_{e3})$ can be measured then it is also possible to have a measurement of the CP violating phase δ. Figure 20.6 shows the transition probabilities $P_D(\nu_\mu \rightarrow \nu_e)$ and $P_D(\bar{\nu}_\mu \rightarrow \bar{\nu}_e)$ as a function of E_ν with the T2K baseline. δ is taken as 45° and the other parameters are a plausible set. Although the probabilities are small, the particle and antiparticle probabilities differ considerably (see Section 19.3).

21

Majorana neutrinos

Majorana fields were introduced in Section 6.6. If neutrino fields are Majorana, then there is no distinction to be made between neutrinos and antineutrinos. As explained in Section 6.7, the smallness of neutrino masses makes the differences between Dirac and Majorana neutrinos difficult to discern experimentally.

In this chapter we elaborate on the theory of Majorana neutrinos and show how they can be accommodated within the Standard Model. Finally we describe experiments on 'double β decay' that may determine the nature of neutrinos.

21.1 Majorana neutrino fields

We shall denote left-handed and right-handed Majorana neutrino fields by $\nu_{\mathrm{L}}(x)$ and $\nu_{\mathrm{R}}(x)$. From (6.28 and 6.29), making the identifications

$$b_{\mathbf{p}+} = d_{\mathbf{p}+}, \ \ b_{\mathbf{p}-} = d_{\mathbf{p}-}$$

we have for a Majorana neutrino field carrying mass m

$$\nu_{\mathrm{L}} = \frac{1}{\sqrt{V}} \sum_{\mathbf{p}} \sqrt{\frac{m}{2E_p}} \ \left[\left(b_{\mathbf{p}+}\mathrm{e}^{-\theta/2}\,|+\rangle\right) + b_{\mathbf{p}-}\mathrm{e}^{\theta/2}\,|-\rangle\,\mathrm{e}^{\mathrm{i}(\mathbf{p}\cdot\mathbf{r}-Et)} \right.$$
$$\left. + \left(b^*_{\mathbf{p}+}\mathrm{e}^{\theta/2}\,|-\rangle - b^*_{\mathbf{p}-}\mathrm{e}^{-\theta/2}\,|+\rangle\right)\mathrm{e}^{\mathrm{i}(-\mathbf{p}\cdot\mathbf{r}+Et)}\right], \tag{21.1}$$

$$\nu_{\mathrm{R}} = \frac{1}{\sqrt{V}} \sum_{\mathbf{p}} \sqrt{\frac{m}{2E_p}} \ \left[\left(b_{\mathbf{p}+}\mathrm{e}^{\theta/2}\,|+\rangle\right) + b_{\mathbf{p}-}\mathrm{e}^{-\theta/2}\,|-\rangle\,\mathrm{e}^{\mathrm{i}(\mathbf{p}\cdot\mathbf{r}-Et)} \right.$$
$$\left. + \left(-b^*_{\mathbf{p}+}\mathrm{e}^{-\theta/2}\,|-\rangle + b^*_{\mathbf{p}-}\mathrm{e}^{\theta/2}\,|+\rangle\right)\mathrm{e}^{\mathrm{i}(-\mathbf{p}\cdot\mathbf{r}+Et)}\right]. \tag{21.2}$$

The fields $\nu_{\mathrm{L}}(x)$ and $\nu_{\mathrm{R}}(x)$ are not independent. It is easily shown, using Problem 6.5, that

$$\left(\mathrm{i}\sigma^2\right)|-\rangle^* = |+\rangle\,, \quad \left(\mathrm{i}\sigma^2\right)|+\rangle^* = -\,|-\rangle\,,$$

and then that

$$\nu_{\mathrm{R}} = (\mathrm{i}\sigma^2)\nu_{\mathrm{L}}^* \quad \text{and} \quad \nu_{\mathrm{L}} = -(\mathrm{i}\sigma^2)\nu_{\mathrm{R}}^*. \tag{21.3}$$

Thus either field may be derived from the other. As a consequence, only left-handed Majorana fields or only right-handed Majorana fields need appear in any theory.

The charge conjugate field ν_{L}^c was defined in (7.11b) by

$$\nu_{\mathrm{L}}^c = -(\mathrm{i}\sigma^2)\nu_{\mathrm{R}}^*.$$

But by the results above $-(\mathrm{i}\sigma^2)\nu_{\mathrm{R}}^* = \nu_{\mathrm{L}}$, so that

$$\nu_{\mathrm{L}}^c = \nu_{\mathrm{L}}. \tag{21.4}$$

Thus the charge conjugate of a Majorana field is identical to the field. There is no room in the theory of Majorana neutrinos for a distinguishable antineutrino. For a given momentum, there are two basic particle states, which we may take to be one with helicity $+1/2$, the other with helicity $-1/2$. (In these respects, Majorana neutrinos are somewhat similar to photons, but with photons having helicities ± 1).

21.2 Majorana Lagrangian density

The Majorana field is constructed from solutions of the Dirac equation. We saw in Section 5.2 that the Lagrangian density for a free Dirac particle of mass m is

$$\mathcal{L}^{\mathrm{Dirac}} = \mathrm{i}\psi_{\mathrm{L}}^\dagger \tilde{\sigma}^\mu \partial_\mu \psi_{\mathrm{L}} + \mathrm{i}\psi_{\mathrm{R}}^\dagger \sigma^\mu \partial_\mu \psi_{\mathrm{R}} - m\left(\psi_{\mathrm{L}}^\dagger \psi_{\mathrm{R}} + \psi_{\mathrm{R}}^\dagger \psi_{\mathrm{L}}\right).$$

In the case of a Majorana field, ν_{R} is determined by ν_{L}, and given by (21.3) above. We choose to work with ν_{L}, and therefore take the Majorana Lagrangian density to be

$$\mathcal{L}^{\mathrm{M}} = \frac{1}{2}\Big[\mathrm{i}\nu^\dagger \tilde{\sigma}^\mu \partial_\mu \nu + \mathrm{i}(\mathrm{i}\sigma^2\nu^*)^\dagger \sigma^\mu \partial_\mu (\mathrm{i}\sigma^2 \nu^*) - m\left\{\nu^\dagger(\mathrm{i}\sigma^2)\nu^* + \nu^{\mathrm{T}}(-\mathrm{i}\sigma^2)\nu\right\}\Big],$$

where $\nu = \nu_{\mathrm{L}}$. For the remainder of this chapter we shall drop the subscript L, for clarity of notation. ν is a two component left-handed neutrino field. We have introduced a factor of $1/2$ to compensate for double counting.

The second dynamical term in $\mathcal{L}^{\mathrm{M}}$ is equivalent to the first (Problem 21.1), so that the Lagrangian density may be written

$$\mathcal{L}^{\mathrm{M}} = \mathrm{i}\nu^\dagger \tilde{\sigma}^\mu \partial_\mu \nu - \frac{m}{2}\left\{\nu^\dagger\left(\mathrm{i}\sigma^2\right)\nu^* + \nu^{\mathrm{T}}\left(-\mathrm{i}\sigma^2\right)\nu\right\}. \tag{21.5}$$

It is interesting and important to note that, with finite mass m and with the Majorana constraints, we lose the $U(1)$ symmetry that gave neutrino number conservation in the Dirac case (Section 7.1). We shall see that with Majorana neutrinos the overall lepton number is no longer conserved.

Noting the factor ½ in the Lagrangian density, the Hamiltonian operator H and momentum operator **P** for Majorana neutrinos are (see Section 6.5)

$$\begin{aligned} \mathrm{H} &= \frac{1}{2}\sum_{\mathbf{p},\varepsilon} \left(b^*_{\mathbf{p}\varepsilon} b_{\mathbf{p}\varepsilon} - b_{\mathbf{p}\varepsilon} b^*_{\mathbf{p}\varepsilon}\right) E_{\mathbf{p}} = \sum_{\mathbf{p},\varepsilon} \left(b^*_{\mathbf{p}\varepsilon} b_{\mathbf{p}\varepsilon}\right) E_{\mathbf{p}}, \\ \mathbf{P} &= \frac{1}{2}\sum_{\mathbf{p},\varepsilon} \left(b^*_{\mathbf{p}\varepsilon} b_{\mathbf{p}\varepsilon} - b_{\mathbf{p}\varepsilon} b^*_{\mathbf{p}\varepsilon}\right) \mathbf{p} = \sum_{\mathbf{p},\varepsilon} \left(b^*_{\mathbf{p}\varepsilon} b_{\mathbf{p}\varepsilon}\right) \mathbf{p}, \end{aligned} \tag{21.6}$$

where $\varepsilon = \pm 1$ is the helicity index.

21.3 Majorana field equations

A variation $\delta\nu^*$ in the Majorana action yields the field equation

$$\mathrm{i}\tilde{\sigma}^\mu \partial_\mu \nu = m\left(\mathrm{i}\sigma^2\right)\nu^*.$$

(Note that there are two contributions from the mass term in the Lagrangian density.)

In a frame K' in which the Majorana neutrino is at rest, $p'_i \nu' = -\mathrm{i}\partial'_i \nu' = 0\,(i = 1, 2, 3)$, and the field equation reduces to

$$\mathrm{i}\frac{\partial \nu'}{\partial t'} = m\left(\mathrm{i}\sigma^2\right)\nu'^* \tag{21.7}$$

It is easy to verify that this equation has two solutions of the form

$$\nu'_1 = b\mathrm{e}^{-\mathrm{i}Et'}\begin{pmatrix}1\\0\end{pmatrix} + b^*\mathrm{e}^{\mathrm{i}Et'}\begin{pmatrix}0\\1\end{pmatrix} \quad \text{and} \quad \nu'_2 = b\mathrm{e}^{-\mathrm{i}Et'}\begin{pmatrix}0\\1\end{pmatrix} - b^*\mathrm{e}^{\mathrm{i}Et'}\begin{pmatrix}1\\0\end{pmatrix},$$

with $E = m$. (21.8)

We may then, as in Section 6.3, transform to a frame K in which the Majorana neutrino is moving with velocity $v > 0$ in the Oz direction:

$$\begin{aligned} \nu_1 = \mathrm{M}^{-1}\nu'_1 &= \begin{pmatrix} e^{-\theta/2} & 0 \\ 0 & \mathrm{e}^{\theta/2} \end{pmatrix}\left[b\mathrm{e}^{-\mathrm{i}mt'}\begin{pmatrix}1\\0\end{pmatrix} + b^*\mathrm{e}^{\mathrm{i}mt'}\begin{pmatrix}0\\1\end{pmatrix}\right] \\ &= b\mathrm{e}^{-mt'}\mathrm{e}^{-\theta/2}\begin{pmatrix}1\\0\end{pmatrix} + b^*\mathrm{e}^{\mathrm{i}mt'}\mathrm{e}^{\theta/2}\begin{pmatrix}0\\1\end{pmatrix}. \end{aligned}$$

Substituting $t' = t\cosh\theta - z\sinh\theta$,

$$\nu_1 = b\mathrm{e}^{-\theta/2}\begin{pmatrix}1\\0\end{pmatrix}\mathrm{e}^{\mathrm{i}(pz-Et)} + b^*\mathrm{e}^{\theta/2}\begin{pmatrix}0\\1\end{pmatrix}\mathrm{e}^{\mathrm{i}(-pz+Et)}. \tag{21.9}$$

Similarly there are solutions of the form

$$\nu_2 = b\mathrm{e}^{\theta/2}\begin{pmatrix}0\\1\end{pmatrix}\mathrm{e}^{\mathrm{i}(pz-Et)} - b^*\mathrm{e}^{-\theta/2}\begin{pmatrix}1\\0\end{pmatrix}\mathrm{e}^{\mathrm{i}(-pz+Et)}. \tag{21.10}$$

All other plane wave solutions may be generated from these by rotations, and we recover the general field (21.1).

21.4 Majorana neutrinos: mixing and oscillations

The most general Lorentz invariant Majorana mass term that can be introduced into a Lagrangian density is

$$\mathcal{L}_{\mathrm{mass}}(x) = -\frac{1}{2}\sum_{\alpha,\beta}\nu_\alpha^{\mathrm{T}}\left(-\mathrm{i}\sigma^2\right)\nu_\beta m_{\alpha\beta} + \text{Hermitian conjugate}. \tag{21.11}$$

α and β run over the three neutrino types, e, μ and τ; ν_α, ν_β are left-handed Majorana fields; $m_{\alpha\beta}$ is an arbitrary complex matrix. In contrast to the case of Dirac neutrinos, $m_{\alpha\beta}$ can be taken to be symmetric. This is because fermion fields anti-commute, so that $\nu_\alpha^{\mathrm{T}}\left(-\mathrm{i}\sigma^2\right)\nu_\beta$ is symmetric on the interchange of α and β (see Problem 21.2).

A general symmetric complex matrix can be transformed into a real diagonal matrix with positive diagonal elements by means of a single unitary matrix **U** (see, for example, Horn and Johnson (1985)). If $m_{\alpha\beta} = m_{\beta\alpha}$, we can write

$$m_{\alpha\beta} = \sum_{i=1}^{3} U_{\alpha i}\, m_i\, U_{\beta i}, \tag{21.12}$$

where the m_i are three positive masses. Note that **U** has no phase ambiguities, whereas Dirac neutrinos have phase ambiguities (see (19.2)).

If we now define the fields

$$\nu_i(x) = \sum_\alpha U_{\alpha i}\nu_\alpha(x), \tag{21.13}$$

the mass term takes the standard Majorana form:

$$\mathcal{L}_{\mathrm{mass}} = -\frac{1}{2}\sum_i m_i\nu_i^{\mathrm{T}}\left(-\mathrm{i}\sigma^2\right)\nu_i + \text{Hermitian conjugate}.$$

The dynamical terms in the Lagrangian density keep the same form under the transformation:

$$\mathcal{L}_{\mathrm{dyn}} = \sum_\alpha \mathrm{i}\nu_\alpha^\dagger\tilde{\sigma}^\mu\partial_\mu\nu_\alpha = \sum_\mathrm{i}\mathrm{i}\nu_\mathrm{i}^\dagger\tilde{\sigma}^\mu\partial_\mu\nu_i.$$

$(\mathcal{L}_{\text{dyn}} + \mathcal{L}_{\text{mass}})$ is the Lagrangian density of free Majorana neutrinos of masses m_1, m_2, m_3. Inverting equation (21.13), the neutrino fields $\nu_\alpha(x)$ appear as mixtures of the neutrino fields of definite mass:

$$\nu_\alpha(x) = \sum_i U^*_{\alpha i} \nu_i(x). \tag{21.14}$$

This is of the same form as equation (19.6) for Dirac neutrinos. The consequences for the weak currents and neutrino oscillations are the same as in Section 19.2 and Section 19.3 for Dirac neutrinos but antineutrinos are interpreted as the neutrinos that accompany a negative charge lepton in weak interaction decays.

21.5 Parameterisation of U

A 3×3 unitary matrix **U** is specified by nine real parameters, but by absorbing phase factors into the definition of the lepton fields, as in Section 19.6, $U_{\alpha \text{i}}$ can be redefined as

$$U'_{\alpha i} = \mathrm{e}^{\mathrm{i}\theta_\alpha} U_{\alpha i},$$

without changing the physical content of the theory. Thus U can be characterised by $9 - 3 = 6$ parameters. The Dirac neutrino mixing matrix (Section 19.6) is determined by four parameters, and requires extension, to include two more parameters. One may take

$$U_{\text{Majorana}} = U_{\text{Dirac}} \times \begin{pmatrix} \mathrm{e}^{\mathrm{i}\Delta 1} & 0 & 0 \\ 0 & \mathrm{e}^{\mathrm{i}\Delta 2} & 0 \\ 0 & 0 & 1 \end{pmatrix}. \tag{21.15}$$

Potentially we have two more *CP* violating parameters. However Δ_1 and Δ_2 make no contribution to the *CP* violation of the oscillation phenomena of Chapters 19 and 20 (see (19.19) and Problem 21.3)

21.6 Majorana neutrinos in the Standard Model

To bring Majorana neutrinos carrying mass into the Standard Model, we must maintain the *SU*(2) symmetry of the weak interaction. As in the case of Dirac neutrinos, a suitable *SU*(2) invariant expressions that we can construct from the Higgs doublet field Φ and a lepton doublet L_α is $(\Phi^{\mathrm{T}}\, \varepsilon\, L_\alpha)$ (See Section 19.5). On symmetry breaking, this becomes $(\Phi^{\mathrm{T}}\, \varepsilon\, L_\alpha) = -(\phi_0 + h/\sqrt{2})\nu_\alpha$.

$\phi_0 \approx 180$ GeV is the Higgs field vacuum expectation value and $h(x)$ is the Higgs boson field.

From these SU(2) invariant expressions we can construct an SU(2) invariant Lagrangian density that on symmetry breaking becomes

$$\mathcal{L}_{\text{mass}} = -\frac{1}{2}(\phi_0 + h/\sqrt{2})^2\, \nu_\alpha^{\mathrm{T}}(-\mathrm{i}\sigma^2)\nu_\beta K_{\alpha\beta} + \text{Hermitian conjugate.} \tag{21.16}$$

The matrix $K_{\alpha\beta}$ couples the neutrino fields to the Higgs field, and we can identify the mass term

$$m_{\alpha\beta} = \phi_0^2 K_{\alpha\beta}. \tag{21.17}$$

Hence the coupling matrix K has dimension $(\text{mass})^{-1}$, which implies (see Section 8.4) that it is an 'effective' Lagrange density. Coupling terms such as this render the theory unrenormalisable.

21.7 The seesaw mechanism

To address the question of renormalisability consider the Lagrangian density

$$\mathcal{L} = \mathrm{i}\nu_{\mathrm{L}}^\dagger\tilde{\sigma}^\mu\partial_\mu\nu_{\mathrm{L}} + \mathrm{i}R^\dagger\sigma^\mu\partial_\mu R - \frac{M}{2}\left(\mathrm{i}R^{\mathrm{T}}\sigma^2 R - \mathrm{i}R^\dagger\sigma^2 R^*\right) - \mu\nu_L^\dagger R - \mu R^\dagger\nu_{\mathrm{L}}. \tag{21.18}$$

M and μ are mass parameters; ν_{L} and R are two component left-handed and right-handed spinor fields respectively. Discarding the terms coupling ν_{L} and R, the Lagrangian density is that of a massless left-handed neutrino field ν_{L}, and a right-handed Majorana neutrino field carrying mass M.

We now suppose that M is so large that the dynamical term $\mathrm{i}R^\dagger\sigma^\mu\partial_\mu R$ may be neglected, to leave

$$\mathcal{L} = \mathrm{i}\nu_{\mathrm{L}}^\dagger\tilde{\sigma}^\mu\partial_\mu\nu_{\mathrm{L}} - \frac{M}{2}(R^{\mathrm{T}}(\mathrm{i}\sigma^2)R - R^\dagger(\mathrm{i}\sigma^2)R^*) - \mu\nu_{\mathrm{L}}^\dagger \mathrm{R} - \mu R^\dagger\nu_{\mathrm{L}}. \tag{21.19}$$

A variation δR^* in the action gives the field equation for R:

$$M\mathrm{i}\sigma^2 R^* - \mu\nu_{\mathrm{L}} = 0.$$

And multiplying by $\mathrm{i}\sigma^2/M$ we obtain

$$R = -(\mu/M)\mathrm{i}\sigma^2\nu_{\mathrm{L}}^*. \tag{21.20}$$

Substituting back into (21.19) gives the effective Lagrangian density

$$\mathcal{L} = \mathrm{i}\nu_{\mathrm{L}}^\dagger\tilde{\sigma}^\mu\partial_\mu\nu_{\mathrm{L}} + (\mu^2/2M)\left(\nu_{\mathrm{L}}^\dagger\mathrm{i}\sigma^2\nu_{\mathrm{L}}^* + \nu_{\mathrm{L}}^{\mathrm{T}}(-\mathrm{i}\sigma^2)\nu_L\right). \tag{21.21}$$

The sign of the mass term can be changed by making the phase change $\nu_{\mathrm{L}} \to \nu'_{\mathrm{L}} = \mathrm{i}\nu_{\mathrm{L}}$. The effective $\mathcal{L}$ is then a free neutrino field of mass $m = \mu^2/M$. Taking for μ a typical lepton mass, say the mass of the muon (10^2 MeV), we can make m the magnitude of a neutrino mass by taking M sufficiently large, $>10^7$ GeV. The generalisation of the seesaw mechanism to include three neutrino types is straightforward.

Taking R to be an $SU(2)$ singlet, the Lagrangian density (21.19) can be made compatible with the Standard Model by replacing $\mu\nu_{\mathrm{L}}^{\dagger}R$ with the $SU(2)$ invariant $C(L_{\mathrm{L}}^{\dagger}\phi)R$, and similarly replacing $\mu R^{\dagger}\nu$, where C is a dimensionless coupling constant. After symmetry breaking, $\mu\nu_{\mathrm{L}}^{\dagger}R$ becomes $C\left(\phi_{\mathrm{O}} + h(x)/\sqrt{2}\right)\nu_{\mathrm{L}}^{\dagger}R$ and setting aside the coupling to the Higgs boson, the mass $\mu = C\phi_0$. It should be noted though that although there are no dimensioned coupling constants the mass M is not generated by the Higgs mechanism.

21.8 Are neutrinos Dirac or Majorana?

The principal feature that distinguishes massive Majorana neutrinos from massive Dirac neutrinos is that Majorana neutrinos do not conserve lepton number. As pointed out in Section 21.2, in the Majorana case the $U(1)$ symmetry that gives lepton number conservation in the Dirac case is lost. The experimental observation of a lepton number violating process would therefore be of great interest. ‘Double β decay’ is the most promising phenomenon for investigation.

The first direct laboratory observation of double β decay was made in 1987, with the decay

$$^{82}_{34}\mathrm{Se} \to {}^{82}_{36}\mathrm{Kr} + \mathrm{e}^- + \mathrm{e}^- + \bar{\nu}_e + \bar{\nu}_e + 3.03\,\mathrm{MeV}.$$

The mean lifetime for this decay has been measured to be $(9.2 \pm 1)\ 10^{19}$ yrs.

If neutrinos are Dirac particles, $\bar{\nu}_{\mathrm{e}}$ is the appropriate symbol in this decay. If neutrinos are Majorana particles, ν and $\bar{\nu}$ are identical. The observed decay does not distinguish between the two interpretations. The process is illustrated in Fig. 21.1a. An electron and a $\bar{\nu}$ in the Dirac case, or a ν in the Majorana case, are created at each interaction point at which a d quark is transformed into a u quark. The nucleus becomes $^{82}_{35}\mathrm{Br}$, possibly in an excited state, between the interaction points.

If neutrinos are Majorana, the decay might be a neutrinoless double β decay, as envisaged in Fig. 21.1b. The neutrino created at X_1 is annihilated at X_2, giving a change of 2 in lepton number. This process is not available if neutrinos are Dirac particles. In the absence of neutrinos to share the energy, the sum of the energies of

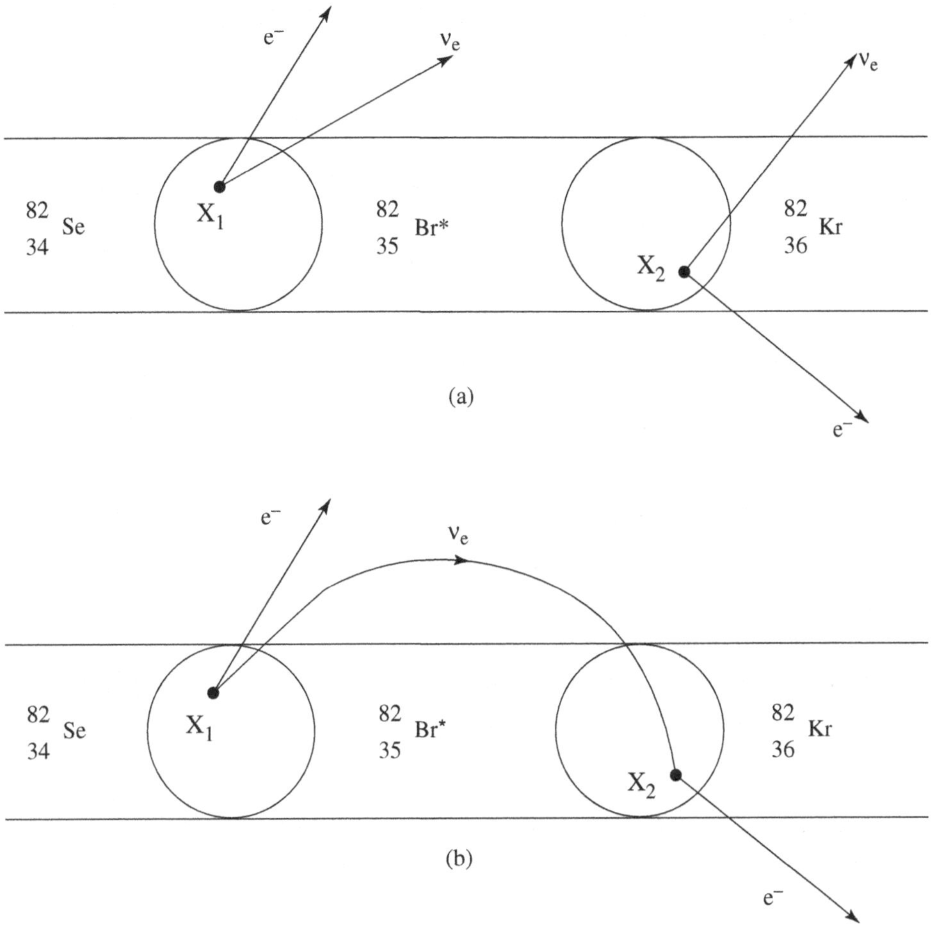

Figure 21.1 (a) Illustrates the two neutrino double β decay of ${}^{82}_{34}\text{Se}$. The decay occurs at the second order of perturbation theory in the weak interaction and involves a sum over many states of ${}^{82}_{35}\text{Br}$ (denoted by ${}^{82}_{35}\text{Br}^*$).
(b) Illustrates the neutrinoless double β decay, a Majorana neutrino created in the transition ${}^{82}_{34}\text{Se} \to {}^{82}_{35}\text{Br}^*$ is annihilated in the transition ${}^{82}_{35}\text{Br}^* \to {}^{82}_{36}\text{Kr}$. In perturbation theory this involves a sum over all momentum states of the neutrino as well as many states of ${}^{82}_{35}\text{Br}$.

the two electrons emitted would be sharply peaked at the decay energy. (The recoil energy of the nucleus would be small.)

Double β decay and neutrinoless double β decay occur at the second order of perturbation theory in the effective weak interaction of equation (14.22). For Majorana neutrinos, double β decay and neutrinoless double β decay are competing processes. Neutrinoless decays are heavily suppressed. From the field equation.

Table 21.1. *From Elliot and Vogel hep/ph/0202264 Feb 2002*

Nucleus	$T^{2\nu}_{1/2}$ (years)	Estimate $T^{0\nu}_{1/2}$ (years)	Measured 0υ half life Lower limit (years)
^{48}Ca	$(4.2 \pm 1.2)\,10^{19}$	$(2.2 \pm 1.3)\,10^{25}$	$> 9.5 \times 10^{21}$
^{76}Ge	$(1.3 \pm 0.1)\,10^{21}$	$(3.2 \pm 2.4)\,10^{25}$	$> 1.9 \times 10^{25}$
^{82}Se	$(9.2 \pm 1.0)\,10^{19}$	$(1.3 \pm 1.0)\,10^{25}$	$> 2.7 \times 10^{22}$
^{100}Mo	$(8.0 \pm 0.6)\,10^{18}$	$(8.4 \pm 7.2)\,10^{26}$	$> 5.5 \times 10^{22}$
^{116}Cd	$(3.2 \pm 0.3)\,10^{19}$	$(1.0 \pm 0.9)\,10^{25}$	$> 7.0 \times 10^{22}$

(21.1), the decay amplitude for the neutrinoless mode, with an intermediate neutrino of mass m_i and energy E_ν, is proportional to

$$(m_i/2E_\nu)\left\lfloor e^{-\theta/2}\,e^{\theta/2} + e^{\vartheta/2}\,e^{-\theta/2}\right\rfloor = (m_i/E_\nu)\,.$$

The two terms come from the two helicity states. The corresponding factors in two neutrino β decay are dominated by the term $(m_i/2E_\nu)\,e^\theta$, and $e^\theta \approx 2\cosh\theta = (2E_\nu/m_i)$, giving unity.

With three neutrino mass eigenstates the decay rate will be proportional to $(1/\bar{E}^2_\nu)|\sum_i m_i U_{ei}|^2$ where $\bar{E}_\nu$ is some mean neutrino energy that can be expected to be a nuclear excitation energy.

Table 21.1. gives some measured two neutrino β decay half lives, and corresponding estimates of the half lives of the neutrinoless decays. These theoretical estimates are sensitive to the nuclear model used.

Problems

21.1 Show that $(i\sigma^2\nu^*)^\dagger\sigma^\mu\partial_\mu(i\sigma^2\nu^*) = \nu^\dagger\tilde{\sigma}^\mu\partial_\mu\nu$.

21.2 Show that, taking account of the anticommuting spinor fields,

$$\nu^T_\alpha\sigma^2\nu_\beta = \nu^T_\beta\sigma^2\nu_\alpha.$$

21.3 Denoting the Majorana and Dirac mixing matrices by U^M and U^D, show that $U^M_{\beta j}U^{M*}_{\alpha j} = U^D_{\beta j}U^{D*}_{\alpha j}$ and hence that the phenomenology of mixing is the same for both Majorana and Dirac neutrinos.

22
Anomalies

In the Standard Model, the fermion fields of the leptons and quarks interact through the mediation of vector bosons. As we remarked in Chapter 10, the renormalisability of the Model requires the vector boson fields to be introduced through the mechanisms of local gauge symmetry. Renormalisation requires the insertion of counter terms in the Lagrangian (Chapter 8). It is important that the counter terms maintain the local gauge symmetries, along with their corresponding conserved currents. As a consequence, one of the global current conservation laws of the Standard Model, that we have obtained by treating the fields as classical fields, has to be modified when the classical fields are quantised. This is an example of an *anomaly*. We shall see that baryon number and lepton number are not strictly conserved quantities in quantum field theory.

22.1 The Adler–Bell–Jackiw anomaly

Bell and Jackiw and, independently, Adler were the first to find an anomaly in a field theory (see Treiman *et al.*, 1985). They were concerned with the axial vector current associated with the chiral symmetries introduced in Section 16.7. To appreciate the nature of this anomaly, consider the model Lagrangian density

$$\mathcal{L} = \bar{\psi}[\gamma^\mu(\mathrm{i}\partial_\mu - qA_\mu) - m]\psi - \frac{1}{4}F_{\mu\nu}F^{\mu\nu}. \tag{22.1}$$

This has the local gauge symmetry of electromagnetism; it is invariant under the transformation

$$\begin{aligned} \psi(x) &\to \psi'(x) = \mathrm{e}^{-\mathrm{i}q\chi(x)}\psi(x), \\ A_\mu(x) &\to A'_\mu(x) = A_\mu(x) + \partial_\mu\chi(x). \end{aligned} \tag{22.2}$$

If $m = 0$, $\mathcal{L}$ also has a global chiral symmetry: it is then invariant under the transformation

$$\psi(x) \to \psi'(x) = \mathrm{e}^{\mathrm{i}\alpha\gamma^5}\psi(x), \tag{22.3}$$

as may easily be verified using the properties of the γ matrices (Section 5.5).

Applying the transformation (22.3) to the Lagrangian density (22.1), with α taken to be infinitesimal and space and time dependent, gives an infinitesimal change $\delta\mathcal{L}$ in $\mathcal{L}$ which (after an integration by parts in the action) may be taken to be

$$\delta\mathcal{L} = \alpha(x)[\partial_\mu j_A^\mu - 2\mathrm{i}m\bar{\psi}\gamma^5\psi],$$

where

$$j_A^\mu = \bar{\psi}\gamma^\mu\gamma^5\psi \tag{22.4}$$

is the *axial current.* (See Problem 5.6.)

It follows from Hamilton's principle that, for fields that obey the field equations,

$$\partial_\mu j_A^\mu = 2\mathrm{i}m\bar{\psi}\gamma^5\psi. \tag{22.5}$$

If $m = 0$, the axial current is conserved:

$$\partial_\mu j_A^\mu = 0 \quad \text{if} \quad m = 0. \tag{22.6}$$

The results (22.5) and (22.6) have been obtained treating the fields as classical fields. In quantum field theory the fields become quantum operators, and the currents can be calculated in perturbation theory. It is found that in order to keep the electric charge conserved and maintain electromagnetism as a local gauge symmetry, perturbation theory requires

$$\partial_\mu j_A^\mu = 2\mathrm{i}m\bar{\psi}\gamma^5\psi - \frac{e^2}{2\pi^2}\varepsilon^{\mu\nu\lambda\rho}\partial_\mu A_\nu \partial_\lambda A_\rho. \tag{22.7}$$

With $m = 0$ the axial current is not conserved, but instead

$$\partial_\mu j_A^\mu = -\frac{e^2}{2\pi^2}\varepsilon^{\mu\nu\lambda\rho}\partial_\mu A_\nu \partial_\lambda A_\rho. \tag{22.8}$$

This is the Adler–Bell–Jackiw axial anomaly. It is found to be the only anomalous term in $\partial_\mu j_A^\mu$. Using Problem 4.3, we can write (22.8) in the explicitly gauge invariant form

$$\partial_\mu j_A^\mu = -\frac{e^2}{\pi^2}\mathbf{E}\cdot\mathbf{B}. \tag{22.9}$$

It is interesting to note that from (22.8) we can construct a current

$$j_{\text{total}}^\mu = j_A^\mu + \frac{e^2}{4\pi^2}\varepsilon^{\mu\nu\lambda\rho}A_\nu F_{\lambda\rho}, \tag{22.10}$$

which evidently is conserved:

$$\partial_\mu j^\mu_{\text{total}} = 0. \tag{22.11}$$

j^μ_{total} is gauge dependent (it contains A_ν) and hence lacks immediate physical significance. Nevertheless it follows from (22.11) that the charge

$$Q(t) = \int j^{\text{o}}_{\text{total}}\, \text{d}^3x \tag{22.12}$$

is constant in time. $Q(t)$ is a gauge invariant quantity.

22.2 Cancellation of anomalies in electroweak currents

In the Standard Model, there are anomalies that have an origin and structure similar to the axial anomaly described in Section 22.1. In particular in the electroweak sector the gauge bosons couple to currents that have both vector and axial vector components, as, for example, in (12.15) where

$$j^\mu_e = e^\dagger_{\text{L}}\tilde{\sigma}^\mu \nu_{\text{L}} = \bar{e}\gamma^\mu(1/2)(1-\gamma^5)\nu_{\text{e}}. \tag{22.13}$$

It is the mix of vector and axial vector that gives rise to anomalies that threaten the renormalisability of the electroweak sector. Detailed calculations show that, in a theory that has only leptons and no quarks, anomalies do spoil the conservation laws of the currents that couple to the bosons. Conversely, in a theory with only quarks and no leptons there are again anomalies. Remarkably, in a theory which includes both leptons and quarks the anomalies cancel exactly, provided that the number of lepton families is equal to the number of quark families, and then the electroweak gauge currents are strictly conserved (t'Hooft, 1976). Thus equality in the number of lepton families and quark families is of fundamental importance to the renormalisability of the Standard Model.

There are no serious anomalies associated with the gluon fields of the strong interaction.

22.3 Lepton and baryon anomalies

We now turn to the currents that, classically, arise from global symmetries and conserve the number of leptons and the number of quarks. We will first consider the situation if neutrinos are shown to be Dirac fermions. For Dirac neutrinos there is a conserved lepton current given by (22.25)

$$J^\mu_{\text{lepton}}(x) = \sum_{\alpha=\text{e},\mu,\text{r}} \left[\alpha^\dagger_{\text{L}}(x)\tilde{\sigma}^\mu\alpha_{\text{L}}(x) + \alpha^\dagger_{\text{R}}(x)\sigma^\mu\alpha_{\text{R}}(x) + \nu^\dagger_{\alpha\text{L}}(x)\tilde{\sigma}^\mu\nu_{\alpha\text{L}}(x) + \nu^\dagger_{\alpha\text{R}}(x)\sigma^\mu\nu_{\alpha\text{R}}(x)\right]. \tag{22.14}$$

and classically

$$\partial_\mu (J^\mu_{\text{lepton}}) = 0. \tag{22.15}$$

On quantisation, this current is not conserved. The divergence equation has to be modified in a way reminiscent of (22.8) and becomes

$$\partial_\mu \left(J^\mu_{\text{lepton}} \right) = \frac{3}{64\pi^2} \varepsilon^{\mu\nu\lambda\rho} \left[\frac{1}{2} g_2^2 \mathrm{Tr} \left(\mathbf{W}_{\mu\nu} \mathbf{W}_{\lambda\rho} \right) - g_1^2 \mathbf{B}_{\mu\nu} \mathbf{B}_{\lambda\rho} \right]. \tag{22.16}$$

The fields $\mathbf{W}_{\mu\nu}$, $\mathbf{B}_{\mu\nu}$, and the coupling constants g_1 and g_2, were introduced in Chapter 11.

The total quark number is also classically conserved but the same anomalous term as in (22.15) arises when the quark fields are quantised for each colour. Summing over the three colours we have

$$\partial_\mu J^\mu_{\text{quark}} = 3 \partial_\mu J^\mu_{\text{lepton}}. \tag{22.17}$$

Since baryon number is one third of the quark number, this can also be written

$$\partial_\mu J^\mu_{\text{baryon}} = \partial_\mu J^\mu_{\text{lepton}}, \tag{22.18}$$

where $J^\mu_{\text{lepton}} = J^\mu_{\text{e}} + J^\mu_{\text{muon}} + J^\mu_{\text{tau}}$.

Thus if neutrinos are Dirac particles, anomalies reduce the two classically conserved currents of the Standard Model to one that can be taken as $J^\mu_{\text{baryon}} - J^\mu_{\text{lepton}}$. The independent current $J^\mu_{\text{baryon}} + J^\mu_{\text{lepton}}$ is not conserved.

Let us now consider the lepton number current. This is not conserved but, as we found with the chiral anomaly, there is nevertheless an associated current that is conserved, and we may write

$$\partial_\mu \left(J^\mu_{\text{lepton}} - J^\mu_{\text{T}} \right) = 0, \tag{22.19}$$

where

$$J^\mu_{\text{T}} = \frac{3}{32\pi^2} \varepsilon^{\mu\nu\lambda\rho} \left[\frac{1}{2} g_2^2 \mathrm{Tr} \left(\mathbf{W}_\nu \mathbf{W}_{\lambda\rho} - (ig_2/3) \mathbf{W}_\nu \mathbf{W}_\lambda \mathbf{W}_\rho \right) - g_1^2 B_\nu B_{\lambda\rho} \right]. \tag{22.20}$$

J^μ_{T} is called the *topological current*, and

$$N_{\text{T}} = \int J^0_{\text{T}} \, \mathrm{d}^3\mathbf{x} \tag{22.21}$$

is the *topological number*.

The lepton number is defined to be

$$N_{\text{lepton}} = \int J^0_{\text{lepton}} \mathrm{d}^3\mathbf{x}, \tag{22.22}$$

and it follows from (22.19) that $N_{\text{lepton}} - N_{\text{T}}$ is constant in time. If N_{T} changes by ΔN_{T}, then N_{lepton} changes by ΔN_{lepton}, and $\Delta N_{\text{lepton}} = \Delta N_T$.

22.4 Gauge transformations and the topological number

Is the topological number a gauge invariant? For simplicity we shall restrict our discussion to fields that are gauge transforms of the vacuum field configuration. Then from (11.4b) and (11.6)

$$B_\mu = (2/g_1)\,\partial_\mu\theta, \tag{22.23}$$
$$\mathbf{W}_\mu = (2i/g_2)\left(\partial_\mu\mathbf{U}\right)\mathbf{U}^\dagger. \tag{22.24}$$

The field strengths $B_{\mu\nu}$ and $\mathbf{W}_{\mu\nu}$ are of course zero everywhere. Also we shall only consider gauge transformations in a local region of space, so that $\theta \to 0$ and $\mathbf{U} \to \mathbf{I}$ as $r \to \infty$. The topological number for this vacuum configuration is

$$N_{\text{T}} = -\frac{1}{8\pi^2}\int \varepsilon^{0ijk}\text{Tr}\left\{(\partial_i\text{U})\,\text{U}^\dagger\left(\partial_j\text{U}\right)\text{U}^\dagger\,(\partial_k\text{U})\,\text{U}^\dagger\right\}\text{d}^3\mathbf{x}, \tag{22.25}$$

using (22.24) in (22.20).

It can be shown that N_{T} is an integer multiple of 3, 0, ±3, ±6, ... We can illustrate this by considering unitary transformations of the form

$$\mathbf{U}(x) = \cos f(r)\mathbf{I} + \text{i}\sin f(r)(\hat{\mathbf{r}}\cdot\tau), \tag{22.26}$$

taking $\alpha = f(r)\hat{\mathbf{r}}$ in (B.9). Here $f(r)$ is a function with the property that $f(r) \to 0$ as $r \to \infty$, so that $\mathbf{U} \to \mathbf{I}$ as $r \to \infty$. If $\mathbf{U}(\mathbf{x})$ is to be defined at $r = 0$, then $\sin f(r)$ must vanish there (since $\hat{\mathbf{r}}$ is not defined at $r = 0$). Thus we require $f(0) = n\pi$ where n is an integer. Subject only to the boundary conditions at $r = 0$ and $r \to \infty$, $f(r)$ can be any continuous and differentiable function.

If $n = 0$, $f(r)$ can be deformed continuously to give $f(r) = 0$, $\mathbf{U} = \mathbf{I}$, for all r; transformations like this are called ‘small’ unitary transformations. If $n \neq 0$ there is no way in which $f(r)$ can be deformed continuously to give $\mathbf{U} = \mathbf{I}$ for all r; these are ‘large’ unitary transformations. Direct computation of (22.25) with $\mathbf{U}$ of the form (22.26) gives

$$N_{\text{T}} = \frac{6}{\pi}\int_0^{n\pi}\sin^2 f\,\text{d}f = 3n. \tag{22.27}$$

It appears that in a theory with no fermions there would be many inequivalent representations of the vacuum state, characterised by a topological number N_{T}. Neglecting the fermions, and treating the $SU(2) \times U(1)$ gauge fields and the Higgs field classically, it is found that to change N_{T} continuously by one unit involves field distortions that require energy. Estimates suggest the energy barrier in field configurations is of height a few times $(4\pi/g_2^2)\ M_{\mathrm{w}} \sim 100\ M_{\mathrm{w}}$. Treating the fields as quantum fields, t'Hooft (1976) found that quantum tunnelling can take place through the barrier, but the probability per unit volume in space-time of a change in N_{T} is very small because of a very small tunnelling factor $\exp(-16\pi^2/g_2^2) \approx 10^{-173}$.

22.5 The instability of matter, and matter genesis

Including the fermions in the Standard Model, if the Higgs and gauge fields pass over the energy barrier separating different topological sectors, the fermion fields must also evolve. Suppose, for example, that $\Delta N_{\mathrm{lepton}} = -3$ and, from (22.18), $\Delta N_{\mathrm{baryon}} = -3$. These conditions are satisfied by, for example, the decay ${}^3_2\mathrm{He} \to \mathrm{e}^+ + \mu^+ + \bar{\nu}_\tau$.

With suppression factors like 10^{-173}, it is unlikely that any helium nucleus in our galaxy has ever decayed in this way since helium nuclei were formed.

It is nevertheless an intriguing possibility that the matter content of the Universe could have been generated by an anomaly mechanism. In the Big Bang model of cosmology, at the very early stage in its evolution the Universe was intensely hot, at a temperature high compared even with the barrier height separating the different topological sectors. Thermal fluctuations over the barrier would produce matter or antimatter depending on the sign of ΔN_{T}. In the beginning the net baryon and lepton numbers might both have taken the symmetrical value zero. To generate the observed preponderance of matter over antimatter requires *CP* violation, and this is an attribute of the Standard Model.

The modifications are straightforward if neutrinos are Majorana fermions. For example, with the Majorana Lagrange density of (21.11), (22.19) becomes

$$\partial_\mu \left(J^\mu_{\mathrm{lepton}} - J^\mu_{\mathrm{T}} \right) = m_{\alpha\beta} \left(\nu_\alpha^{\mathrm{T}} \sigma^2 \nu_\beta + \nu_\beta^{+} \sigma^2 \nu_\alpha^{*} \right) \tag{22.28}$$

as can be shown by making an infinitesimal, space time dependent, phase change on all the lepton fields (see the method of section (22.1)). If neutrinos are Majorana particles then, with the anomalies, no global conservation laws remain.

Epilogue

Reductionism complete?

The Standard Model, extended to include neutrinos carrying mass, gives a remarkably successful account of the experimental data of particle physics obtained up to 2006. Any subsequent theory must, in some sense, correspond to the Standard Model in the energy range that has so far been explored.

Many questions remain to be answered. Why is there the internal electroweak and strong group structure $U(1) \times SU(2) \times SU(3)$, with the three coupling constants g_1, g_2, g_3? Is the origin of mass really to be found in the Higgs field with its two parameters: the Higgs mass and the expectation value of the Higgs field? In the electroweak sector, why are the masses of the charged leptons as they are? There are three parameters here. Another set of parameters comes with allowing neutrinos to have mass: three neutrino masses and four parameters of the mass mixing matrix (or six if it appears that neutrinos correspond to Majorana fields rather than Dirac fields). In the quark sector ten more parameters are introduced: six quark masses, and four parameters in the Kobayashi–Maskawa matrix.

Are these twenty five or twenty six parameters really independent?

Some of these questions may be answered when experimentalists have the LHC (Large Hadron Collider) at CERN, probing to higher energies and thereby to smaller distances to make progress into finding common origins of what are now diverse elements of the Standard Model. The task is to reduce twenty six parameters to one or two, say, before closing the book on the theory of matter and radiation.

Appendix A

An *aide-mémoire* on matrices

A.1 Definitions and notation

An $m \times n$ *matrix* $\mathbf{A} = (A_{ij})$; $i = 1, \ldots, m$; $j = 1, \ldots, n$; is an ordered array of mn numbers, which may be complex:

$$\mathbf{A} = \begin{pmatrix} A_{11} A_{12} \ldots A_{1n} \\ A_{21} A_{22} \ldots \\ \ldots\ldots\ldots\ldots\ldots \\ A_{m1} \ldots \quad A_{mn} \end{pmatrix}.$$

A_{ij} is the *element* of the ith row and jth column.

The *complex conjugate* of $\mathbf{A}$, written $\mathbf{A}^*$, is defined by

$$\mathbf{A}^* = (A^*_{ij}).$$

The *transpose* of $\mathbf{A}$, written $\mathbf{A}^{\mathrm{T}}$, is the $n \times m$ matrix defined by

$$A^{\mathrm{T}}_{ji} = A_{ij}.$$

The *Hermitian conjugate*, or *adjoint*, of $\mathbf{A}$, written $\mathbf{A}^\dagger$, is defined by

$$A^\dagger_{jt} = A^*_{ij} = A^{\mathrm{T}\,*}_{ji}, \text{ or equivalently by} \mathbf{A}^\dagger = (\mathbf{A}^{\mathrm{T}})^*.$$

If λ, μ are complex numbers and $\mathbf{A}$, $\mathbf{B}$ are $m \times n$ matrices, $\mathrm{C} = \lambda\mathbf{A} + \mu\mathbf{B}$ is defined by

$$C_{ij} = \lambda A_{ij} + \mu B_{ij}.$$

Multiplication of the $m \times n$ matrix $\mathbf{A}$ by an $n \times l$ matrix $\mathbf{B}$ is defined by $\mathbf{AB} = \mathbf{C}$, where $\mathbf{C}$ is the $m \times l$ matrix given by

$$C_{ik} = A_{ij} B_{jk}.$$

We use the Einstein convention, that a repeated 'dummy' suffix is understood to be summed over, so that

$$A_{ij} B_{jk} \text{ means} \sum_{j=1}^{n} A_{ij} B_{jk}.$$

Multiplication is associative: (**AB**)**C** = **A**(**BC**). If follows immediately from the definitions that

$$(\mathbf{AB})^* = \mathbf{A}^*\mathbf{B}^*,\ (\mathbf{AB})^{\mathrm{T}} = \mathbf{B}^{\mathrm{T}}\mathbf{A}^{\mathrm{T}},\ (\mathbf{AB})^\dagger = \mathbf{B}^\dagger\mathbf{A}^\dagger.$$

Block multiplication: matrices may be subdivided into blocks and multiplied by a rule similar to that for multiplication of elements, provided that the blocks are compatible. For example,

$$\begin{pmatrix} \mathbf{A}\ \mathbf{B} \\ \mathbf{C}\ \mathbf{D} \end{pmatrix} \begin{pmatrix} \mathbf{E} \\ \mathbf{F} \end{pmatrix} = \begin{pmatrix} \mathbf{AE}+\mathbf{BF} \\ \mathbf{CE}+\mathbf{DF} \end{pmatrix}$$

provided that the l_1 columns of **A** and l_2 columns of **B** are matched by l_1 rows of **E** and l_2 rows of **F**. The proof follows from writing out the appropriate sums.

A.2 Properties of $n \times n$ matrices

We now focus on 'square' $n \times n$ matrices. If **A** and **B** are $n \times n$ matrices, we can construct both **AB** and **BA**. In general, matrix multiplication is non-commutative, i.e. in general, $\mathbf{AB} \neq \mathbf{BA}$.

The $n \times n$ *identity matrix* or *unit matrix* I is defined by $I_{ij} = \delta_{ij}$, where δ_{ij} is the Kronecker δ:

$$\delta_{ij} = \begin{cases} 1 & \text{if } i = j, \\ 0 & \text{if } i \neq j. \end{cases}$$

From the rule for multiplication,

$$\mathbf{IA} = \mathbf{AI} = \mathbf{A}$$

for any **A**. **A** is said to be *diagonal* if $A_{ij} = 0$ for $i \neq j$.

Determinants: with a square matrix **A** we can associate the *determinant* of **A**, denoted by det **A** or $|A_{ij}|$, and defined by

$$\det \mathbf{A} = \varepsilon_{ij\ldots t} A_{1i} A_{2j} \ldots A_{nt}$$

(remember the summation convention) where

$$\varepsilon_{ij\ldots t} = \begin{cases} 1 & \text{if } i, j, \ldots, t \text{ is an even permutation of } 1, 2, \ldots, n, \\ -1 & \text{if } i, j, \ldots, t \text{ is an odd permutation of } 1, 2, \ldots, n, \\ 0 & \text{otherwise.} \end{cases}$$

An important result is

$$\det(\mathbf{AB}) = \det \mathbf{A} \det \mathbf{B}.$$

Note also

$$\det \mathbf{A}^{\mathrm{T}} = \det \mathbf{A}, \quad \det \mathbf{I} = 1.$$

If $\det \mathbf{A} \neq 0$ the matrix **A** is said to be *non-singular*, and $\det \mathbf{A} \neq \mathbf{0}$ is a necessary and sufficient condition for a unique inverse $\mathbf{A}^{-1}$ to exist, such that

$$\mathbf{AA}^{-1} = \mathbf{A}^{-1}\mathbf{A} = \mathbf{I}.$$

Evidently,

$$(\mathbf{AB})^{-1} = \mathbf{B}^{-1}\mathbf{A}^{-1}.$$

The trace of a matrix **A**, written Tr**A**, is the sum of its diagonal elements:

$$\mathrm{Tr}\mathbf{A} = A_{ii}.$$

It follows from the definition that

$$\mathrm{Tr}(\mathbf{AB}) = A_{ij}B_{ji} = B_{ji}A_{ij} = \mathrm{Tr}(\mathbf{BA}),$$

and hence

$$\mathrm{Tr}(\mathbf{ABC}) = \mathrm{Tr}(\mathbf{BCA}) = \mathrm{Tr}(\mathbf{CAB}).$$

A.3 Hermitian and unitary matrices

Hermitian and unitary matrices are square matrices of particular importance in quantum mechanics. In a matrix formulation of quantum mechanics, dynamical observables are represented by Hermitian matrices, while the time development of a system is determined by a unitary matrix.

A matrix **H** is *Hermitian* if it is equal to its Hermitian conjugate:

$$\mathbf{H} = \mathbf{H}^\dagger, \quad \text{or} \quad H_{ij} = H_{ji}^*.$$

The diagonal elements of a Hermitian matrix are therefore real, and an $n \times n$ Hermitian matrix is specified by $n + 2n(n-1)/2 = n^2$ real numbers.

A matrix **U** is *unitary* if

$$\mathbf{U}^{-1} = \mathbf{U}^\dagger, \quad \text{or} \quad \mathbf{UU}^\dagger = \mathbf{U}^\dagger\mathbf{U} = \mathbf{I}.$$

The product of two unitary matrices is also unitary.

A *unitary transformation* of a matrix **A** is a transformation of the form

$$\mathbf{A} \to \mathbf{A}' = \mathbf{UAU}^{-1} = \mathbf{UAU}^\dagger,$$

where **U** is a unitary matrix. The transformation preserves algebraic relationships:

$$(\mathbf{AB})' = \mathbf{A}'\mathbf{B}',$$

and Hermitian conjugation

$$(\mathbf{A}')^\dagger = \mathbf{UA}^\dagger\mathbf{U}^\dagger.$$

Also

$$\mathrm{Tr}\mathbf{A}' = \mathrm{Tr}\mathbf{A}, \quad \det\mathbf{A}' = \det\mathbf{A}.$$

An important theorem of matrix algebra is that, for each Hermitian matrix **H**, there exists a unitary matrix **U** such that

$$\mathbf{H}' = \mathbf{UHU}^{-1} = \mathbf{UHU}^\dagger = \mathbf{H}_D$$

is a real diagonal matrix.

A necessary and sufficient condition that Hermitian matrices $\mathbf{H}_1$ and $\mathbf{H}_2$ can be brought into the diagonal form by the same unitary transformation is

$$\mathbf{H}_1\mathbf{H}_2 - \mathbf{H}_2\mathbf{H}_1 = 0.$$

It follows from this (see Problem A.3) that a matrix **M** can be brought into diagonal form by a unitary transformation if and only if

$$\mathbf{MM}^\dagger - \mathbf{M}^\dagger\mathbf{M} = 0.$$

Note that unitary matrices satisfy this condition.

An arbitrary matrix **M** which does not satisfy this condition can be brought into real diagonal form by a generalised transformation involving two unitary matrices, $\mathbf{U}_1$ and $\mathbf{U}_2$ say, which may be chosen so that

$$\mathbf{U}_1\mathbf{M}\mathbf{U}_2^\dagger = \mathbf{M}_D$$

is diagonal (see Problem A.4).

If **H** is a Hermitian matrix, the matrix

$$\mathbf{U} = \exp(\mathrm{i}\mathbf{H})$$

is unitary. The right-hand side of this equation is to be understood as defined by the series expansion

$$\mathbf{U} = \mathbf{I} + (\mathrm{i}\mathbf{H}) + (\mathrm{i}\mathbf{H})^2/2! + \cdots$$

Then

$$\begin{aligned}\mathbf{U}^\dagger &= \mathbf{I} + (-\mathrm{i}\mathbf{H}^\dagger) + (-\mathrm{i}\mathbf{H}^\dagger)^2/2! + \cdots \\ &= \exp(-\mathrm{i}\mathbf{H}^\dagger) = \exp(-\mathrm{i}\mathbf{H}) = \mathbf{U}^{-1}\end{aligned}$$

(the operation of Hermitian conjugation being carried out term by term). Conversely, any unitary matrix **U** can be expressed in this form. Since an $n \times n$ Hermitian matrix is specified by n^2 real numbers, it follows that a unitary matrix is specified by n^2 real numbers.

A.4 A Fierz transformation

It is easy to show that any 2×2 matrix **M** with complex elements may be expressed as a linear combination of the matrices $\widetilde{\sigma}^\mu$.

$$\mathbf{M} = Z_\mu \widetilde{\sigma}^\mu,$$

and $Z_\mu = \frac{1}{2}\mathrm{Tr}\,(\widetilde{\sigma}^\mu \mathbf{M})$, since $\mathrm{Tr}\,(\widetilde{\sigma}^\mu \widetilde{\sigma}^\nu) = 2\delta_{\mu\nu}$.

Consider the expression
$g_{\mu\nu}\langle a^*|\widetilde{\sigma}^\mu|b\rangle\langle c^*|\widetilde{\sigma}^\nu|d\rangle$, where $|a\rangle$, $|b\rangle$, $|c\rangle$, $|d\rangle$ are two-component spinor fields. Using the result above, we can replace the matrix $|b\rangle\langle c^*|$ by

$$\begin{aligned}|b\rangle\langle c^*| &= \frac{1}{2}Tr(\widetilde{\sigma}^\lambda|b\rangle\langle c^*|)\widetilde{\sigma}^\lambda \\ &= -\frac{1}{2}\langle c^*|\widetilde{\sigma}^\lambda|b\rangle\widetilde{\sigma}^\lambda.\end{aligned}$$

The last step is evident on putting in the spinors indices, and the minus sign arises from the interchange of anticommuting spinor fields.

We now have

$$g_{\mu\nu}\langle a^*|\widetilde{\sigma}^\mu|b\rangle\langle c^*|\widetilde{\sigma}^\nu|d\rangle = -\frac{1}{2}g_{\mu\nu}\langle a^*|\widetilde{\sigma}^\mu\widetilde{\sigma}^\lambda\widetilde{\sigma}^\nu|d >< c^*|\widetilde{\sigma}^\lambda|b\rangle.$$

Using the algebraic identity

$$g_{\mu\nu}\widetilde{\sigma}^\mu\widetilde{\sigma}^\lambda\widetilde{\sigma}^\nu = -2g_{\rho\lambda}\widetilde{\sigma}^\rho,$$

gives $g_{\mu\nu}\langle a^*|\widetilde{\sigma}^\mu|b\rangle\langle c^*|\widetilde{\sigma}^\nu|d\rangle = g_{\rho\lambda}\langle a^*|\widetilde{\sigma}^\rho|d\rangle\langle c^*|\widetilde{\sigma}^\lambda|b\rangle$.

This is an example of a *Fierz transformation.*

Problems

A.1 Show that

$$\varepsilon_{ij\dots t} A_{\alpha i} A_{\beta j} \cdots A_{\nu t} = \varepsilon_{\alpha\beta\dots\nu} \det \mathbf{A}.$$

A.2 Show that if **A, B** are Hermitian, then $\mathrm{i}(\mathbf{AB} - \mathbf{BA})$ is Hermitian.

A.3 Show that an arbitrary square matrix **M** can be written in the form $\mathbf{M} = \mathbf{A} + \mathrm{i}\mathbf{B}$, where **A** and **B** are Hermitian matrices. Find **A** and **B** in terms of **M** and $\mathbf{M}^\dagger$. Hence show that **M** may be put into diagonal form by a unitary transformation if and only if $\mathbf{M}\mathbf{M}^\dagger - \mathbf{M}^\dagger\mathbf{M} = 0$.

A.4 If **M** is an arbitrary square matrix, show that $\mathbf{M}\mathbf{M}^\dagger$ is Hermitian and hence can be diagonalised by a unitary matrix $\mathbf{U}_1$, so that we can write

$$\mathbf{U}_1(\mathbf{M}\mathbf{M}^\dagger)\mathbf{U}_1{}^\dagger = \mathbf{M}_D{}^2$$

where $\mathbf{M}_D$ is diagonal with real diagonal elements ≥ 0. Suppose none are zero. Define the Hermitian matrix $\mathbf{H} = \mathbf{U}_1{}^\dagger\mathbf{M}_D\mathbf{U}_1$. Show that $\mathbf{V} = \mathbf{H}^{-1}\mathbf{M}$ is unitary. Hence show that

$$\mathbf{M} = \mathbf{U}_1{}^\dagger\mathbf{M}_D\mathbf{U}_2,$$

where $\mathbf{U}_2 = \mathbf{U}_1\mathbf{V}$ is a unitary matrix.

Appendix B

The groups of the Standard Model

The Standard Model is constructed by insisting that the equations of the model retain the same form after certain transformations. For instance, we require that the equations take the same form in every inertial frame of reference, so that they are covariant under a Lorentz transformation; this may be a rotation of axes or a boost, or a combination of rotation and boost. The Lagrangian density that describes the Standard Model takes the same form in the new coordinate system, and the Lorentz transformation is said to be a *symmetry transformation*. In the Standard Model, as well as symmetries under coordinate transformations, there are 'internal' symmetries of the particle fields. The corresponding symmetry transformations are conveniently represented by matrices.

It is characteristic of symmetry transformations that they satisfy the mathematical axioms of a *group*, which we set out below. In this appendix we consider some properties of the groups that play a special role in the Standard Model.

B.1 Definition of a group

A group G is a set of elements $a, b, c, \ldots$, together with a rule that combines any two elements a,b of G to form an element ab, which also belongs to G, satisfying the following conditions.

(i) The rule is *associative*: $a(bc) = (ab)c$.

(ii) G contains a unique *identity element* I such that, for every element a of G,

$$aI = Ia = a.$$

(iii) For every element a of G there exists a unique *inverse* element a^{-1} such that

$$aa^{-1} = a^{-1}a = I.$$

If also $ab = ba$ for all a, b the group is said to be *commutative* or *Abelian*.

It is usually easy to determine whether or not a given set of elements and their combination law satisfy these axioms. For example, the set of all integers forms an Abelian group under addition, with 0 the identity element. The set of all non-singular $n \times n$ matrices ($n > 1$) forms a non-Abelian group under matrix multiplication. The permutations of the numbers $1, 2, \ldots, n$ form a group which has $n!$ elements; this is an example of a *finite group*. The group of rotations of the coordinate axes is a three-parameter *continuous group*: an element is specified by three parameters that take on a continuous range of values. We shall be concerned principally with groups of this type.

B.2 Rotations of the coordinate axes, and the group *SO(3)*

Consider a rotation of the coordinate axes about the origin. If the coordinates of a point P are (x^1, x^2, x^3) in a frame of reference K, and (x'^1, x'^2, x'^3) in a frame K', rotated relative to K, the x'^i are related to the x^i by a real linear transformation of the form

$$x'^i = R^i_{\ j} x^j. \tag{B.1}$$

$\mathbf{R} = (R^i_{\ j})$ is the *rotation matrix*. For example, a rotation of the axes through an angle θ about the 03 axis in a right-handed sense is given by

$$\begin{aligned} x'^1 &= x^1 \cos\theta + x^2 \sin\theta, \\ x'^2 &= -x^1 \sin\theta + x^2 \cos\theta, \\ x'^3 &= x^3, \end{aligned}$$

and corresponds to the matrix

$$\mathbf{R}_{03}(\theta) = \begin{pmatrix} \cos\theta & \sin\theta & 0 \\ -\sin\theta & \cos\theta & 0 \\ 0 & 0 & 1 \end{pmatrix}. \tag{B.2}$$

We may regard the x'^i and x^i as 3×1 (column) matrices $\mathbf{x}'$ and $\mathbf{x}$, and write the transformation (B.1) as

$$\mathbf{x}' = \mathbf{R}\mathbf{x}.$$

The transpose $\mathbf{x}^{\mathrm{T}}$ of $\mathbf{x}$ is a 1×3 (row) matrix, and the scalar product of two vectors $\mathbf{x}$ and $\mathbf{y}$ is

$$x'y' = \mathbf{x}^{\mathrm{T}}\mathbf{y} = \mathbf{y}^{\mathrm{T}}\mathbf{x}.$$

In particular, the length OP is given by $\surd(\mathbf{x}^{\mathrm{T}}\mathbf{x})$. Since a rotation of axes preserves scalar products,

$$\mathbf{x}'^{\mathrm{T}}\mathbf{y}' = \mathbf{x}^{\mathrm{T}}\mathbf{R}^{\mathrm{T}}\mathbf{R}\mathbf{y} = \mathbf{x}^{\mathrm{T}}\mathbf{y}.$$

This holds for *all* pairs **x, y**. Hence

$$\mathbf{R}^{\mathrm{T}}\mathbf{R} = \mathbf{I} \tag{B.3}$$

where **I** is the identity matrix: hence the inverse of **R** is the transpose $\mathbf{R}^{\mathrm{T}}$ of **R** and **R** is said to be an *orthogonal matrix*.

$$\text{Since } \det \mathbf{R}^{\mathrm{T}} \det \mathbf{R} = \det(\mathbf{R}^{\mathrm{T}}\mathbf{R}) = \det \mathbf{I} = 1 \text{ and } \det \mathbf{R}^{\mathrm{T}} = \det \mathbf{R}, \tag{B.4}$$

$$(\det \mathbf{R})^2 = 1, \quad \det \mathbf{R} = \pm 1.$$

Matrices corresponding to pure or 'proper' rotations have $\det \mathbf{R} = +1$. We can see this by noting that the identity rotation is a proper rotation, and $\det \mathbf{I} = 1$. Any proper rotation can be constructed as a sequence of infinitesimal rotations starting from **I** and hence by continuity also has determinant $+1$.

The product of two orthogonal matrices is an orthogonal matrix, since

$$(\mathbf{R}_1\mathbf{R}_2)^{\mathrm{T}} = \mathbf{R}_2{}^{\mathrm{T}}\mathbf{R}_1{}^{\mathrm{T}} = \mathbf{R}_2{}^{-1}\mathbf{R}_1{}^{-1} = (\mathbf{R}_1\mathbf{R}_2)^{-1},$$

and if $\det \mathbf{R}_1 = 1$ and $\det \mathbf{R}_2 = 1$,

$$\det(\mathbf{R}_1\mathbf{R}_2) = \det \mathbf{R}_1 \det \mathbf{R}_2 = 1.$$

Hence real orthogonal 3×3 matrices with $\det \mathbf{R} = 1$ form a group under matrix multiplication. This group is called *the special orthogonal group* and is denoted by *SO*(3).

Orthogonal matrices with $\det \mathbf{R} = -1$ also preserve scalar products. It is easy to see that inversion of the coordinate axes in the origin, $x'^i = -x^i$, corresponds to an orthogonal matrix with determinant -1; a general 'improper' rotation corresponds to inversion in the origin together with a proper rotation. Improper rotation matrices do not form a group, since the product of two improper rotations is a proper rotation.

A general proper rotation may be built up as a sequence of rotations about three different axes. For example, consider

$$\mathbf{R}(\psi, \theta, \phi) = \mathbf{R}_{03''}(\psi)\mathbf{R}_{02'}(\theta)\mathbf{R}_{03}(\phi), \tag{B.5}$$

in an obvious notation. The direction of $03''$ is defined by θ and ϕ, and then ψ defines the final orientation of $01''2''$ in the plane perpendicular to $03''$. Thus each element of $SO(3)$ is specified by just three parameters. (ψ, θ, ϕ are known as the Euler angles.)

We can also interpret the transformation (B.1) in an *active* sense. Consider a system described by a wave function $\mathbf{\Phi}(\mathbf{x})$ in the frame K. The system is described by $\mathbf{\Phi}'(\mathbf{x}') = \mathbf{\Phi}(\mathbf{R}^{-1}\mathbf{x}')$ in the frame K'. This is the *passive* interpretation. We might, alternatively, drop the primes on the coordinates and give this equation an active interpretation, supposing that the axes have been held fixed and the system given the inverse rotation $\mathbf{R}^{-1}$. The wave function of the rotated system is $\mathbf{\Phi}'(\mathbf{x}) = \mathbf{\Phi}(\mathbf{R}^{-1}\mathbf{x})$.

B.3 The group *SU(2)*

An $n \times n$ matrix $\mathbf{U}$ is *unitary* if $\mathbf{U}\mathbf{U}^\dagger = \mathbf{U}^\dagger\mathbf{U} = \mathbf{I}$. The product of two unitary matrices is unitary. Hence $n \times n$ unitary matrices form a group under matrix multiplication, denoted by $U(n)$.

Since

$$\det(\mathbf{U}\mathbf{U}^\dagger) = \det\mathbf{U}\det\mathbf{U}^* = \det\mathbf{U}(\det(\mathbf{U})^* = \det\mathbf{I} = 1,$$

we may write $\det\mathbf{U} = \mathrm{e}^{\mathrm{i}n\alpha}$, where α is real.

The *special unitary group SU*(2) is the group of all 2×2 unitary matrices with determinant equal to 1. These form a group, since if $\det\mathbf{U}_1 = 1$ and $\det\mathbf{U}_2 = 1$ then $\det(\mathbf{U}_1\mathbf{U}_2) = \det\mathbf{U}_1\det\mathbf{U}_2 = 1$. $SU(2)$ is a *sub-group* of $U(2)$. Every element of $U(2)$ is the product of a phase factor $\mathrm{e}^{\mathrm{i}\alpha}$, which is an element of $U(1)$, and an element of $SU(2)$.

The group $SU(2)$ is related in a remarkable way to the rotation group $SO(3)$ described in Section B.2. It is central to the electroweak sector of the Standard Model.

Any element of $U(2)$ can be put in the form

$$\mathbf{U} = \exp(\mathbf{iH})$$

where $\mathbf{H}$ is a Hermitian matrix (Appendix A). A general 2×2 Hermitian matrix may be taken as

$$\mathbf{H} = \begin{pmatrix} \alpha^0 + \alpha^3 & a^1 - \mathrm{i}\alpha^2 \\ \alpha^1 + \mathrm{i}\alpha^2 & \alpha^0 - \alpha^3 \end{pmatrix}$$

where the $\alpha^\mu (\mu = 0, 1, 2, 3)$ are four real parameters. This choice enables us to write

$$\mathbf{H} = \alpha^0\mathbf{I} + \alpha^{\mathrm{k}}\sigma^{\mathrm{k}}, \tag{B.6}$$

where the index k runs from 1 to 3, and

$$\sigma^1 = \begin{pmatrix} 0 & 1 \\ 1 & 0 \end{pmatrix}, \quad \sigma^2 = \begin{pmatrix} 0 & -\mathrm{i} \\ \mathrm{i} & 0 \end{pmatrix}, \quad \sigma^3 = \begin{pmatrix} 1 & 0 \\ 0 & -1 \end{pmatrix}.$$

The σ^k are the same as the Pauli spin matrices, and hence they satisfy

$$\begin{aligned} &(\sigma^1)^2 = (\sigma^2)^2 = (\sigma^3)^2 = \mathbf{I};\ \sigma^j\sigma^k + \sigma^k\sigma^j = 0,\ j \neq k; \\ &[\sigma^1, \sigma^2] = \sigma^1\sigma^2 - \sigma^2\sigma^1 = 2\mathrm{i}\sigma_3,\ \text{etc.} \end{aligned} \tag{B.7}$$

Since the unit matrix $\mathbf{I}$ commutes with all matrices, a general member of $U(2)$ can be written as

$$\mathbf{U} = \exp\,\mathrm{i}(\alpha^0\mathbf{I} + \alpha^k\sigma^k) = \exp(\mathrm{i}\alpha^0)\exp(\mathrm{i}\alpha^k\sigma^k).$$

The phase factor $\exp(\mathrm{i}\alpha^0)$ belongs to the group $U(1)$. Hence elements of $SU(2)$ are of the form

$$\mathbf{U}_s = \exp(\mathrm{i}\alpha^k\sigma^k). \tag{B.8}$$

An element may be specified by the three parameters α^k; the matrices σ^k are the corresponding *generators* of the group. Each has zero trace (see Problem B.1).

The algebra of the σ^k matrices enables us to write these elements in closed form. Let us formally consider the α^k to make up a vector $\boldsymbol{\alpha} = \alpha\hat{\boldsymbol{\alpha}}$, where $\hat{\boldsymbol{\alpha}}$ is the corresponding unit vector, and write the 'scalar product' $\alpha^k\sigma^k$ as $\alpha\hat{\boldsymbol{\alpha}}\cdot\boldsymbol{\sigma}$. It is easy to see that

$$(\hat{\boldsymbol{\alpha}}\cdot\boldsymbol{\sigma})^2 = \hat{\alpha}^j\sigma^j\hat{\alpha}^k\sigma^k = \hat{\alpha}^j\hat{\alpha}^j\mathbf{I} = \mathbf{I},$$

since $\sigma^j\sigma^k + \sigma^k\sigma^j = 0$ and $(\sigma^1)^2 = \mathbf{I}$, etc. Then the power series expansion of (B.8) gives

$$\begin{aligned}\mathbf{U}_s &= \mathbf{I} + \mathrm{i}\alpha(\hat{\boldsymbol{\alpha}}\cdot\boldsymbol{\sigma}) + \frac{(\mathrm{i}\alpha)^2}{2!}\mathbf{I} + \cdots \\ &= \cos\alpha\mathbf{I} + \mathrm{i}\,\sin\alpha(\hat{\boldsymbol{\alpha}}\cdot\boldsymbol{\sigma}).\end{aligned} \tag{B.9}$$

To establish the connection between the groups $SU(2)$ and $SO(3)$, we associate with each point $\mathbf{x}$ the Hermitian matrix

$$\mathbf{X}(\mathbf{x}) = \begin{pmatrix} x^3 & x^1 - \mathrm{i}x^2 \\ x^1 + \mathrm{i}x^2 & -x^3 \end{pmatrix}. \tag{B.10}$$

This matrix has $\mathrm{Tr}\,\mathbf{X} = 0$ and $\det\mathbf{X} = -x^kx^k$.

Consider now an element $\mathbf{U}$ of $SU(2)$ and the matrix

$$\mathbf{X}' = \mathbf{U}\mathbf{X}\mathbf{U}^\dagger. \tag{B.11}$$

(We are now dropping the suffix s on $\mathbf{U}$.)

$\mathbf{X}'$ is also Hermitian, and $\mathrm{Tr}\,\mathbf{X}' = \mathrm{Tr}(\mathbf{U}\mathbf{X}\mathbf{U}^\dagger) = \mathrm{Tr}(\mathbf{U}^\dagger\mathbf{U}\mathbf{X}) = \mathrm{Tr}\,\mathbf{X} = 0$. Hence $\mathbf{X}'$ is of the form

$$\mathbf{X}' = \begin{pmatrix} x'^3 & x'^1 - \mathrm{i}x'^2 \\ x'^1 + \mathrm{i}x'^2 & -x'^3 \end{pmatrix}$$

where the x'^k are related to the x^k by a real linear transformation.

Also $\det\mathbf{X}' = \det\mathbf{U}\,\det\mathbf{X}\,\det\mathbf{U}^\dagger = \det(\mathbf{U}\mathbf{U}^\dagger)\det\mathbf{X} = \det\mathbf{X}$, so that $x'^kx'^k = x^kx^k$. Since the length of $\mathbf{x}$ is preserved and the transformation may be continuously generated from the identity matrix (see Problem B.3), the transformation must correspond to a proper rotation of the coordinate axes and hence to a rotation matrix $\mathbf{R}(\mathbf{U})$.

As an example, the $SU(2)$ matrix

$$\mathbf{U} = \exp[\mathrm{i}(\theta/2)\sigma^3] = \cos(\theta/2)\mathbf{I} + \mathrm{i}\,\sin(\theta/2)\sigma^3 = \begin{pmatrix} e^{\mathrm{i}\theta/2} & 0 \\ 0 & e^{-\mathrm{i}\theta/2} \end{pmatrix}, \tag{B.12}$$

where we have used (B.9), corresponds to the rotation matrix $\mathbf{R}_{03}(\theta)$ of equation (B.2). This may be verified by direct matrix multiplication.

The matrices $\mathbf{U}$ and $-\mathbf{U}$ give the same transformation (B.11), and hence correspond to the same rotation matrix: to every element of $SO(3)$ there correspond two elements of $SU(2)$, differing by a factor of -1. In the example (B.12) above, rotations of θ and $\theta + 2\pi$ about the 03 axis correspond to the same rotation matrix, but give matrices $\mathbf{U}$ and $-\mathbf{U}$, respectively in $SU(2)$.

B.4 The group *SL*(2,C) and the proper Lorentz group

The set of all 2×2 matrices with complex elements and with determinant equal to 1 evidently forms a group under matrix multiplication. This group is denoted by *SL*(2,C). It is related to the group of proper Lorentz transformations in much the same way as the group *SU*(2) is related to the group of proper rotations.

We now associate with each point $x = (x^0, \mathbf{x})$ in space-time the general Hermitian matrix

$$\mathbf{X}(x) = \begin{pmatrix} x^0 + x^3 & x^1 - \mathrm{i}x^2 \\ x^1 + \mathrm{i}x^2 & x^0 - x^3 \end{pmatrix} \tag{B.13}$$

which has

$$\det \mathbf{X} = (x^0)^2 - x^k x^k.$$

Consider an element $\mathbf{M}$ of *SL*(2,C) and the matrix $\mathbf{X}'$ given by

$$\mathbf{M}^\dagger \mathbf{X}' \mathbf{M} = \mathbf{X} \text{ or } \mathbf{X}' = (\mathbf{M}^{-1})^\dagger \mathbf{X} \mathbf{M}^{-1}. \tag{B.14}$$

Then $\mathbf{X}'$ is also Hermitian and hence we can write

$$\mathbf{X}' = \begin{pmatrix} x'^0 + x'^3 & x'^1 - \mathrm{i}x'^2 \\ x'^1 + \mathrm{i}x'^2 & x'^0 - x'^3 \end{pmatrix},$$

where the x'^μ are related to the $x^\mu (\mu = 0, 1, 2, 3)$ by a real linear transformation. Also

$$\det \mathbf{M}^\dagger \mathbf{X}' \mathbf{M} = \det \mathbf{M}^\dagger \det \mathbf{X}' \det \mathbf{M} = \det \mathbf{X}' = \det \mathbf{X}$$

so that

$$(x'^0)^2 - x'^k x'^k = (x^0)^2 - x^k x^k.$$

Hence the matrix $\mathbf{M}$ corresponds to a Lorentz transformation matrix $\mathbf{L}(\mathbf{M})$. The matrices $\mathbf{L}(\mathbf{M})$ form a group that includes the identity transformation $\mathbf{L}(\mathbf{I}) = \mathbf{I}$, and hence by continuity correspond to proper Lorentz transformations.

A general proper Lorentz transformation between frames K and K' is specified by six parameters: three parameters to give the velocity $\mathbf{v}$ of K' relative to K and three parameters to give the orientation of K' relative to K. A general 2×2 complex matrix is defined by eight real parameters. The condition $\det \mathbf{M} = 1$ reduces this number to six. Hence a matrix $\mathbf{M}$ can be found corresponding to every proper Lorentz transformation. The matrices $\mathbf{M}$ and $-\mathbf{M}$ give the same transformation (B.14): two elements of *SL*(2,C) correspond to each element of the proper Lorentz group.

The matrix

$$\mathbf{P} = \exp[(\theta/2)\sigma^3] = \cosh(\theta/2)\mathbf{I} + \sinh(\theta/2)\sigma^3 = \begin{pmatrix} \mathrm{e}^{\theta/2} & 0 \\ 0 & \mathrm{e}^{-\theta/2} \end{pmatrix} \tag{B.15}$$

corresponds to the Lorentz boost (2.3) of Chapter 2, as may be verified by direct matrix multiplication.

More generally, a Lorentz boost from a frame K to a frame K' moving with velocity $v = \tanh\theta$ in the direction of the unit vector $\hat{\mathbf{v}}$ is given by

$$\mathbf{P} = \exp[(\theta/2)\hat{\mathbf{v}}\cdot\boldsymbol{\sigma}] = \cosh(\theta/2)\mathbf{I} + \sinh(\theta/2)\hat{\mathbf{v}}\cdot\boldsymbol{\sigma}$$

where $\boldsymbol{\sigma} = (\sigma^1, \sigma^2, \sigma^3)$.

Note that, since the matrices σ^k are Hermitian, so also is any matrix $\mathbf{P}$ corresponding to a Lorentz boost.

B.5 Transformations of the Pauli matrices

In discussing Lorentz transformations, it is convenient to write $\mathbf{I} = \sigma^0$ and introduce the notation

$$\sigma^\mu = (\sigma^0, \sigma^1, \sigma^2, \sigma^3), \qquad \tilde{\sigma}^\mu = (\sigma^0, -\sigma^1, -\sigma^2, -\sigma^3). \tag{B.16}$$

Then from (B.13)

$$\mathbf{X}(x) = x^0\sigma^0 + x^k\sigma^k = x_\mu\tilde{\sigma}^\mu, \qquad \mathbf{X}'(x') = x'_\mu\tilde{\sigma}^\mu.$$

The relation

$$\mathbf{M}^\dagger\mathbf{X}'\mathbf{M} = \mathbf{X}$$

gives

$$x'_\mu\mathbf{M}^\dagger\tilde{\sigma}^\mu\mathbf{M} = x_\nu\tilde{\sigma}^\nu = L^\mu{}_\nu\tilde{\sigma}^\nu x'_\mu$$

(see Problem 2.2). Since the x'_μ are arbitrary, we can deduce

$$\mathbf{M}^\dagger\tilde{\sigma}^\mu\mathbf{M} = L^\mu{}_\nu\tilde{\sigma}^\nu. \tag{B.17}$$

Also (Problem B.6)

$$L^\mu{}_\nu = \frac{1}{2}\mathrm{Tr}(\tilde{\sigma}^\nu M^\dagger\tilde{\sigma}^\mu M).$$

Similarly, by considering the matrix

$$\mathbf{X}_1(x) = x^0\sigma^0 - x^k\sigma^k = x_\nu\sigma^\nu,$$

which also has $\det \mathbf{X}_1 = (x^0)^2 - x^k x^k$, we can show that there exists a matrix $\mathbf{N}$ belonging to $SL(2,C)$ such that

$$\mathbf{N}^\dagger\sigma^\mu\mathbf{N} = L^\mu{}_\nu\sigma^\nu. \tag{B.18}$$

The matrices $\mathbf{M}$ and $\mathbf{N}$ are evidently related. The reader may verify directly that when $\mathbf{M} = \mathbf{P}$, where $\mathbf{P}$ is given by (B.15) and corresponds to a Lorentz boost, we can take $\mathbf{N} = \mathbf{P}^{-1}$, and this will be true for a Lorentz boost in any direction. For a pure rotation of axes, we take $\mathbf{M} = \mathbf{N} = \mathbf{U}$, where $\mathbf{U}$ is a unitary matrix. A general $\mathbf{M}$ can be constructed as a product of a rotation followed by a boost: $\mathbf{M} = \mathbf{PU}$. The corresponding $\mathbf{N}$ is given by $\mathbf{N} = \mathbf{P}^{-1}\mathbf{U}$.

Now $\mathbf{U}$ satisfies $\mathbf{UU}^\dagger = \mathbf{I}$, and we noted that $\mathbf{P}$ is Hermitian, $\mathbf{P} = \mathbf{P}^\dagger$. Hence

$$\mathbf{NM}^\dagger = (\mathbf{P}^{-1}\mathbf{U})(\mathbf{U}^\dagger\mathbf{P}) = \mathbf{I}, \tag{B.19}$$

so that $\mathbf{N}$ is the inverse of $\mathbf{M}^\dagger$.

The results (B.17) and (B.18), together with (B.19), are useful in constructing Lorentz scalars, vectors and higher order tensors.

B.6 Spinors

We define a *left-handed spinor*

$$\mathbf{l} = \begin{pmatrix} l_1 \\ l_2 \end{pmatrix}$$

as a complex two-component entity that transforms under a Lorentz transformation with matrix $\mathbf{L}(\mathbf{M})$ by the rule

$$\mathbf{l}' = \mathbf{Ml} \tag{B.20}$$

i.e. $l'_a = M_{ab}l_b$, where a and b take on the values $1, 2$.

We similarly define a right-handed spinor

$$\mathbf{r} = \begin{pmatrix} r_1 \\ r_2 \end{pmatrix} \tag{B.21}$$

as a two-component entity that transforms by

$$\mathbf{r}' = \mathbf{N}\mathbf{r}.$$

Electrons, and all other fermions in the Standard Model, are described by spinor fields. The nomenclature of 'left-handed' and 'right-handed' is elucidated in Section 6.3.

Spinors have the remarkable property that they can be combined in pairs to make Lorentz scalars, Lorentz four-vectors and higher order Lorentz tensors. For example, $\mathbf{l}^\dagger \mathrm{r} = l^*{}_a r_a$ is a (complex) Lorentz scalar, since

$$\mathbf{l}'^\dagger \mathbf{r}' = (\mathbf{M}\mathbf{l})^\dagger \mathbf{N}\mathbf{r} = \mathbf{l}^\dagger \mathbf{M}^\dagger \mathbf{N}\mathbf{r} = \mathbf{l}^\dagger \mathbf{r}, \tag{B.22}$$

where we have used (B.19).

The quantities

$$\mathbf{l}^\dagger \tilde{\sigma} \mathbf{l} = \mathbf{l}^\dagger(\sigma^0, -\sigma^1, -\sigma^2, -\sigma^3)\mathbf{l},$$
$$\mathbf{r}^\dagger \sigma \mathbf{r} = \mathbf{r}^\dagger(\sigma^0, \sigma^1 \sigma^2, \sigma^3)\mathbf{r},$$

transform like (real) contravariant four-vectors, since

$$\mathbf{l}'^\dagger \tilde{\sigma}^\mu \mathbf{l}' = \mathbf{l}^\dagger \mathbf{M}^\dagger \tilde{\sigma}^\mu \mathbf{M}\mathbf{l} = L^\mu{}_\nu (\mathbf{l}^\dagger \tilde{\sigma}^\nu \mathbf{l}), \tag{B.23}$$

using (B.17), and

$$\mathbf{r}'^\dagger \sigma^\mu \mathbf{r}' = \mathbf{r}^\dagger \mathbf{N}^\dagger \sigma^\mu \mathbf{N}\mathbf{r} = L^\mu{}_\nu (\mathbf{r}^\dagger \sigma^\nu \mathbf{r}), \tag{B.24}$$

using (B.18).

B.7 The group *SU(3)*

The special unitary group $SU(3)$ is the group of all 3×3 unitary matrices with determinant equal to 1. Our discussion will parallel our discussion of the group $SU(2)$ in Section B.3. An element of $SU(3)$ can be expressed as

$$\mathbf{U} = \exp(\mathrm{i}\mathbf{H})$$

where $\mathbf{H}$ is a 3×3 Hermitian matrix. A general 3×3 Hermitian matrix is specified by $3^2 = 9$ real parameters (Appendix A). The condition $\det \mathbf{U} = 1$, or equivalently $\mathrm{Tr}\mathbf{H} = 0$ (Problem B.1), reduces this number to 8. In place of the σ^k matrices used in Section B.3, we have the eight traceless Hermitian matrices introduced by Gell-Mann:

$$\begin{gathered}
\lambda_1 = \begin{pmatrix} 0 & 1 & 0 \\ 1 & 0 & 0 \\ 0 & 0 & 0 \end{pmatrix}, \quad \lambda_2 = \begin{pmatrix} 0 & -\mathrm{i} & 0 \\ \mathrm{i} & 0 & 0 \\ 0 & 0 & 0 \end{pmatrix} \quad \lambda_3 = \begin{pmatrix} 1 & 0 & 0 \\ 0 & -1 & 0 \\ 0 & 0 & 0 \end{pmatrix}, \\
\lambda_4 = \begin{pmatrix} 0 & 0 & 1 \\ 0 & 0 & 0 \\ 1 & 0 & 0 \end{pmatrix}, \quad \lambda_5 = \begin{pmatrix} 0 & 0 & -\mathrm{i} \\ 0 & 0 & 0 \\ \mathrm{i} & 0 & 0 \end{pmatrix}, \quad \lambda_6 = \begin{pmatrix} 0 & 0 & 0 \\ 0 & 0 & 1 \\ 0 & 1 & 0 \end{pmatrix}, \\
\lambda_7 = \begin{pmatrix} 0 & 0 & 0 \\ 0 & 0 & -\mathrm{i} \\ 0 & \mathrm{i} & 0 \end{pmatrix}, \quad \lambda_8 = (1/\sqrt{3}) \begin{pmatrix} 1 & 0 & 0 \\ 0 & 1 & 0 \\ 0 & 0 & -2 \end{pmatrix}.
\end{gathered} \tag{B.25}$$

A general traceless Hermitian matrix is of the form

$$\mathbf{H} = \alpha_1\lambda_1 + \alpha_2\lambda_2 + \cdots + \alpha_8\lambda_8$$
$$= \begin{pmatrix} \alpha_3 + \alpha_8/\sqrt{3} & \alpha_1 - i\alpha_2 & \alpha_4 - i\alpha_5 \\ \alpha_1 + i\alpha_2 & -\alpha_3 + \alpha_8/\sqrt{3} & \alpha_6 - i\alpha_7 \\ \alpha_1 + i\alpha_5 & \alpha_6 + i\alpha_7 & -2\alpha_8/\sqrt{3} \end{pmatrix} \tag{B.26}$$

The matrices λ_a satisfy the commutation relations

$$[\lambda_a, \lambda_b] = 2i \sum_{c=1}^{8} f_{abc}\lambda_c \tag{B.27}$$

where the f_{abc} are the structure constants (cf. equations (B.7)). The f_{abc} are odd in the interchange of any pair of indices, and the non-vanishing f_{abc} are given by the permutations of $f_{123} = 1,\ f_{147} = f_{246} = f_{257} = f_{345} = f_{516} = f_{637} = 1/2,\ f_{458} = f_{678} = \sqrt{3}/2$.

The matrices also have the property

$$\mathrm{Tr}(\lambda_a\lambda_b) = 2\delta_{ab}, \tag{B.28}$$

where δ_{ab} is the Kronecker δ.

These results may be verified by direct calculation.

Problems

B.1 Show that if $\mathbf{U} = \exp(i\mathbf{H})$ and $\mathrm{Tr}\,\mathbf{H} = 0$, then $\det \mathbf{U} = 1$. (Make **H** diagonal with a unitary transformation. **U** is then also diagonal.)

B.2 Verify that the *SU*(2) matrices $\exp[i(\theta/2)\sigma^1]$ and $\exp[i(\theta/2)\sigma^2]$ correspond to rotations $R_{01}(\theta)$ and $R_{02}(\theta)$, respectively.

B.3 Show that the *SU*(2) matrix corresponding to the rotation $R(\psi, \theta, \phi)$ (equation (B.5)) is

$$\begin{pmatrix} e^{i\psi/2}\cos(\theta/2)e^{i\phi/2} & e^{i\psi/2}\sin(\theta/2)e^{-i\phi/2} \\ -e^{-i\psi/2}\sin(\theta/2)e^{i\phi/2} & e^{-i\psi/2}\cos(\theta/2)e^{-i\phi/2} \end{pmatrix}.$$

B.4 Show that $\mathbf{l}^\dagger\tilde{\sigma}^\mu\sigma^\nu\mathbf{r}$ transforms as a tensor and $\mathbf{l}^\dagger(\tilde{\sigma}^\mu\sigma^\nu + \tilde{\sigma}^\nu\sigma^\mu)\mathbf{r} = \mathbf{2}g^{\mu\nu}\mathbf{l}^\dagger\mathbf{r}$.

B.5 Show that the rotation matrix R^i_j of equation (B.1) is related to the *SU*(2) matrix **U** of (B.11) by

$$R^i_j = \frac{1}{2}\mathrm{Tr}(\mathbf{U}\sigma^i\mathbf{U}^\dagger\sigma^j).$$

B.6 Show from (B.17) that

$$L^\mu{}_\nu = \frac{1}{2}\mathrm{Tr}(\tilde{\sigma}^\nu M^\dagger\tilde{\sigma}^\mu M).$$

Appendix C

Annihilation and creation operators

C.1 The simple harmonic oscillator

The reader may well have met annihilation and creation operators in treating the quantum mechanics of the simple harmonic oscillator. In this context, an operator a and its Hermitian conjugate $a^\dagger$ are constructed. These satisfy the commutation relations

$$[a, a^\dagger] = aa^\dagger - a^\dagger a = 1 \tag{C.1}$$

and also of course

$$[a, a] = 0, \quad [a^\dagger, a^\dagger] = 0.$$

The operator $N = a^\dagger a$ is Hermitian. We denote by $|n\rangle$ the normalised eigenstate of N with eigenvalue n. Since $n = \langle n|a^\dagger a|n\rangle$ is the modulus squared of the state $a|n\rangle$, n is real and ≥ 0, and equal to 0 only if $a|n\rangle = 0$.

It follows from the commutation relations that the lowest eigenstate of n is $n = 0$, corresponding to the ground state $|0\rangle$. This is because

$$Na|n\rangle = a^\dagger aa|n\rangle = (aa^\dagger - 1)a|n\rangle = (n-1)a|n\rangle.$$

Thus $a|n\rangle$ is, apart from normalisation, an eigenstate of N with eigenvalue $(n-1)$, unless $a|n\rangle = 0$. Similarly $a|n-1\rangle$ is an eigenstate of N with eigenvalue $(n-2)$, and so on. The process must terminate at the eigenstate $|0\rangle$ with eigenvalue 0, and $a|0\rangle = 0$, since otherwise we would be able to violate the condition $n \geq 0$.

Similarly $a^\dagger|n\rangle$ is, apart from normalisation, an eigenstate of N with eigenvalue $(n+1)$. Thus the eigenvalues of the *number operator* N are the integers 0, 1, 2, 3 . . .

Since $\langle n|a^\dagger a|n\rangle = n$, we have

$$a|n\rangle = n^{1/2}|n-1\rangle. \tag{C.2}$$

Also, $\langle n|aa^\dagger|n\rangle = \langle n|a^\dagger a + 1|n\rangle = n + 1$, so that

$$a^\dagger|n\rangle = (n+1)^{1/2}|n+1\rangle. \tag{C.3}$$

We call a an *annihilation operator* and $a^\dagger$ a *creation operator.*

Written in terms of a and $a^\dagger$, the simple harmonic oscillator Hamiltonian becomes

$$H = \left(a^\dagger a + \frac{1}{2}\right)\hbar\omega = \left(N + \frac{1}{2}\right)\hbar\omega, \tag{C.4}$$

where ω is the frequency of the corresponding classical oscillator (Problem C.1). The term $\frac{1}{2}\hbar\omega$ is the zero-point energy. Since in field theory only energy differences are of physical significance, it is usually convenient to redefine H, dropping the zero-point energy and taking $H = a^\dagger a\hbar\omega$. We may then reinterpret the state $|n\rangle$ as a state in which there are n identical 'particles' each of energy $\hbar\omega$, associated with the oscillator, and say that a and $a^\dagger$ annihilate and create particles.

In the Heisenberg representation (Section 8.2),

$$a(t) = \mathrm{e}^{\mathrm{i}Ht} a \mathrm{e}^{-\mathrm{i}Ht} = \mathrm{e}^{\mathrm{i}N\omega t} a \mathrm{e}^{-\mathrm{i}N\omega t} = \mathrm{e}^{-\mathrm{i}\omega t} a. \tag{C.5}$$

This may be seen by considering the effect of $a(t)$ acting on a state $|n\rangle$, and noting that, since

$$\mathrm{e}^{\pm \mathrm{i}N\omega t}|n\rangle = \mathrm{e}^{\pm n\omega t}|n\rangle,$$

the two expressions for $a(t)$ give the same result. Similarly,

$$a^\dagger(t) = \mathrm{e}^{\mathrm{i}\omega t} a^\dagger. \tag{C.6}$$

C.2 An assembly of bosons

A similar operator formalism may be developed for assemblies of identical particles. We set out first the formalism when the particles are bosons.

Let $u_i(\xi)$ be a complete set of single particle states, where ξ stands for the space and spin coordinate of a particle. We define annihilation and creation operators a_i and $a_i^\dagger$ for each state, satisfying the commutation relations

$$[a_i, a_j{}^\dagger] = \delta_{ij}, \qquad [a_i, a_j] = 0, \qquad [a_i{}^\dagger, a_j{}^\dagger] = 0. \tag{C.7}$$

Any state of the system can be constructed by operating on the *vacuum state* $|0\rangle$, in which there are no particles present, and $a_i|0\rangle = 0$ for all i. For example, a three-particle state having two particles in the state u_1 and one particle in the state u_2 is given (apart from normalisation) by $a_1^\dagger a_1^\dagger a_2^\dagger|0\rangle$. Evidently such a state is symmetric in the interchange of any two particles since the creation operators all commute, and the particles will obey Bose–Einstein statistics.

It follows from the commutation relations that the number operator $N_i = a_i^\dagger a_i$ gives the number of particles in the state u_i. In the case of non-interacting bosons, the $u_i(\xi)$ can be taken as the single particle energy eigenstates and the Hamiltonian operator is then

$$H_0 = \sum_i a_i^\dagger a_i \varepsilon_i = \sum_i N_i \varepsilon_i, \tag{C.7}$$

where the ε_i are the single particle energy levels.

In the Heisenberg representation and with the free particle Hamiltonian H_0, the time dependence of the annihilation and creation operators is like that of simple harmonic oscillator operators, and follows by a similar argument:

$$a_i(t) = \mathrm{e}^{-\mathrm{i}\varepsilon_i t} a_i, \qquad a_i^\dagger(t) = \mathrm{e}^{\mathrm{i}\varepsilon_i t} a_i^\dagger. \tag{C.8}$$

C.3 An assembly of fermions

In the case of an assembly of identical fermions, we define annihilation and creation operators b_i and $b_i{}^\dagger$ for each single particle state $u_i(\xi)$, which are *anticommuting:*

$$\{b_i, b_j{}^\dagger\} = b_i b_j{}^\dagger + b_j{}^\dagger b_i = \delta_{ij}, \qquad \{b_i, b_j\} = 0, \qquad \{b_i{}^\dagger, b_j{}^\dagger\} = 0. \tag{C.9}$$

In particular,

$$(b_i)^2 = 0, \ (b_j{}^\dagger)^2 = 0. \tag{C.10}$$

Thus two fermions cannot be annihilated from the same state, or created in the same state, in accord with the Pauli principle.

The number operator $N_i = b_i{}^\dagger b_i$ satisfies

$$N_i^2 = b_i{}^\dagger b_i b_i{}^\dagger b_i = b_i{}^\dagger(1 - b_i{}^\dagger b_i)\, b_i = b_i{}^\dagger b_i = N_i,$$

or

$$N_i(N_i - 1) = 0,$$

so that the eigenvalues of N_i are 0 and 1. This, again, is in accord with the Pauli principle. A many-particle fermion state can be constructed by operating on the vacuum state $|0\rangle$ with creation operators. For example $b_1{}^\dagger b_2{}^\dagger b_5{}^\dagger|0\rangle$ is a state with a fermion in each of the states u_1, u_2, u_5. Such a state is antisymmetric under particle exchange, and the particles obey Fermi–Dirac statistics.

In the case of an assembly of non-interacting fermions, the Hamiltonian operator is

$$H_0 = \sum_i b_i{}^\dagger b_i \varepsilon_i, \tag{C.11}$$

and in the Heisenberg representation

$$b_i(t) = \mathrm{e}^{-\mathrm{i}\varepsilon_i t} b_i, \qquad b_i{}^\dagger(t) = \mathrm{e}^{\mathrm{i}\varepsilon_i t} b_i{}^\dagger. \tag{C.12}$$

Problems

C.1 With rescaling of coordinates,

$$P = p/(m\hbar\omega)^{1/2}, \qquad X = x(m\omega/\hbar)^{1/2},$$

the simple harmonic oscillator Hamiltonian

$$H = (p^2/2m) + (m\omega^2 x^2/2)$$

becomes

$$H = (\hbar\omega/2)(P^2 + X^2),$$

and

$$[X, P] = \mathrm{i}$$

Show that if $a = (1/\sqrt{2})(X + \mathrm{i}P)$, $a^\dagger = (1/\sqrt{2})(X - \mathrm{i}P)$, then

$$[a, a^\dagger] = 1 \quad \text{and} \quad H = (a^\dagger a + \frac{1}{2})\hbar\omega.$$

C.2 Show that the normalised ground state wave function of the simple harmonic oscillator is $(m\omega/\pi\hbar)^{1/4}\exp(-m\omega x^2/2\hbar)$.

C.3 Using the commutation relations for fermions show that the state $b_i{}^\dagger|0\rangle$ is an eigenstate of $N_i = b_i{}^\dagger b_i$ with eigenvalue 1.

C.4 Show that the matrices

$$\mathbf{b} = \begin{pmatrix} 0 & 1 \\ 0 & 0 \end{pmatrix} \quad \text{and} \quad \mathbf{b}^\dagger = \begin{pmatrix} 0 & 0 \\ 1 & 0 \end{pmatrix}$$

satisfy the commutation relations for fermion annihilation and creation operators.

Appendix D

The parton model

D.1 Elastic electron scattering from nucleons

In the 1950s, experiments on elastic scattering of electrons from nucleon targets at rest in the laboratory revealed the electric charge distribution in protons and neutrons, clearly establishing the size of the nucleons.

The differential cross-section for the elastic scattering of electrons at high energies from a Dirac particle of mass M and charge e may be calculated in QED. To leading order in the fine-structure constant $\alpha = e^2/4\pi$, and neglecting the electron's mass compared with its energy, the differential cross-section for scattering from an unpolarised Dirac particle, initially at rest in the laboratory frame, in which the scattered electron emerges at an angle θ with respect to its incident direction, is

$$\frac{\mathrm{d}\sigma}{\mathrm{d}\Omega} = \frac{\alpha^2}{4E^2 \sin^4(\theta/2)} \left(\frac{E'}{E}\right) \left[\cos^2(\theta/2) + \frac{Q^2}{2M^2} \sin^2(\theta/2)\right], \tag{D.1}$$

where

$(E, \mathbf{p}) =$ initial electron energy-momentum four-vector,
$(E', \mathbf{p}') =$ final electron energy-momentum four-vector,
$q^\mu = (E - E', \mathbf{p} - \mathbf{p}') =$ energy-momentum transfer,
$Q^2 = -q_\mu q^\mu = (\mathbf{p} - \mathbf{p}')^2 - (E - E')^2$.

(See, for example, Gross, 1993, p. 294.)

Note that Q^2 is Lorentz invariant. For elastic scattering at a given energy, the angle θ determines, through energy and momentum conservation, all other quantities in the expression. For example,

$$Q^2 = 4EE' \sin^2(\theta/2), \tag{D.2}$$

where the energy E' is given by

$$M(E - E') - 2EE' \sin^2(\theta/2) = 0 \tag{D.3}$$

(Problem D.1).

Taking M to be the proton mass, the formula (D.1) does not fit the experimental data and, indeed, since the proton has an anomalous magnetic moment $\approx 1.79(e\hbar/2M)$, we would not expect a fit. More generally, the elastic scattering from an unpolarised 'extended' proton is of the form

$$\frac{d\sigma}{d\Omega} = \frac{\alpha^2}{4E^2 \sin^4(\theta/2)} \left(\frac{E'}{E}\right) \left[\left\{ f_1^2(Q^2) + \frac{Q^2}{4M^2} f_2^2(Q^2) \right\} \cos^2(\theta/2) + \frac{Q^2}{2M^2} \{f_1(Q^2) + f_2(Q^2)\}^2 \sin^2(\theta/2) \right]. \tag{D.4}$$

The form of this expression is essentially determined given the proton has spin 1/2 and no electric dipole moment. $f_1(Q^2)$ is called the Dirac form factor of the proton, and $f_2(Q^2)$ is the form factor associated with the anomalous magnetic moment. At $Q = 0$, $f_1(0) = 1$ and $f_2(0) \approx 1.79$ (corresponding to the anomalous moment). The electric and magnetic form factors

$$G_E(Q^2) = f_1(Q^2) - \frac{Q^2}{4M^2} f_2(Q^2), \tag{D.5}$$

$$G_M(Q^2) = f_1(Q^2) + f_2(Q^2), \tag{D.6}$$

can be interpreted in the non-relativistic limit as Fourier transforms of the electric charge and magnetic moment distributions in the proton (Problem D.2). It is from their experimental determination that the size of the proton is inferred. Both $f_1(Q^2)$ and $f_2(Q^2)$ fall off rapidly as Q^2 increases (Fig. D.1). Similar form factors can be defined, and determined experimentally, for the neutron (using scattering data from deuterium targets). The analysis is consistent with the quark model. Since the electric charge is carried by the quarks, the charge and magnetic moment distribution should trace the distributions of quark charge and quark magnetic moment.

D.2 Inelastic electron scattering from nucleons: the parton model

The early elastic scattering experiments were performed at electron energies ≤ 500 MeV. Scattering at higher energies has thrown more light on the behaviour of quarks in nucleons, and revealed properties that will continue to be crucial for pursuing particle physics at the even higher energies of the future. Except where Q^2 is small, inelastic scattering, which involves hadron production, becomes the dominant mode at higher energies. In the case of inelastic scattering, θ and E' are independent variables. In general, there are many other independent variables that describe the final hadronic system, but the very important differential cross-section $d^2\sigma/dE'd\Omega$, called the *inclusive cross-section*, includes all the possible final hadronic states.

At the electron–proton collider HERA at Hamburg a beam of 30 GeV electrons meets a beam of 820 GeV protons head on. Many features of the ensuing electron–proton collisions are described by the *parton model.* which was introduced by Feynman in 1969.

In the parton model each proton in the beam is regarded as a system of sub-particles, called *partons.* These are quarks, antiquarks and gluons. Quarks and antiquarks are the partons that carry electric charge. The proton's energy and momentum P^μ is envisaged as being distributed over the different parton types i with certain probability distributions. The mean number of partons of type i in the proton carrying energy and momentum in the range xP^μ, $(x + dx)P^\mu$, $0 < x < 1$, is written $p_i(x)dx$. Here the label i covers all types of quarks, antiquarks and gluons (u, ū, d, d̄, s, s̄, etc.). Scaling both energy and momentum by the same factor ensures that all the partons have the velocity of the proton. Any transverse momentum a parton may have is neglected. Thus, in the model, each proton in the HERA beam is regarded as a sub-beam of partons. The consequences of the model for

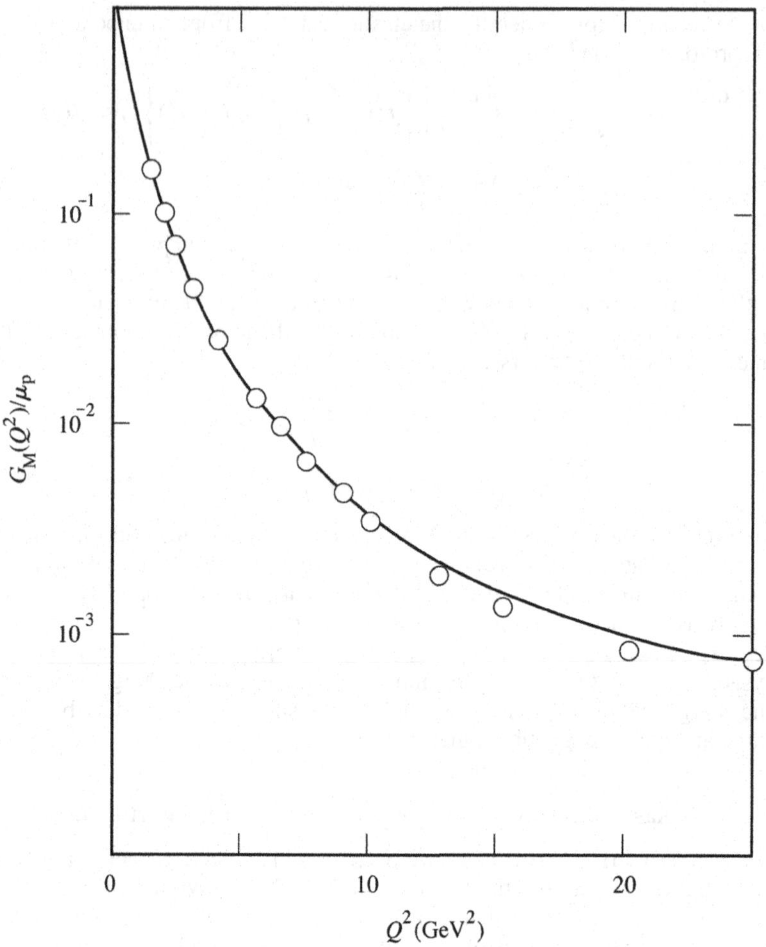

Figure D.1 This figure shows the measured magnetic dipole form factor of the proton. The data are quite well represented by the simple expression

$$G_M(Q^2) = \mu_p \left[\frac{1}{1 + Q^2/\beta^2} \right]^2$$

with $\mu_p = 2.79,\ \beta = 0.84\,\text{GeV}$. This curve is shown.

For $Q^2 < 3\,\text{GeV}^2$, $G_E = (Q^2)G_M(Q^2)/\mu_P$ but for $Q^2 > 5\,\text{GeV}^2$ only $G_M(Q^2)$ can be measured with accuracy (see Coward *et al.*, 1968).

the inclusive cross-section can be most easily demonstrated in the rest frame of the proton. In this frame, a parton with energy–momentum fraction x will behave like a particle of mass xM at rest. For $Q^2 < M_w^2$ the dominant scattering will be electromagnetic scattering from the charged partons: the spin 1/2 quarks and antiquarks. For the elastic scattering from a parton of type i with effective mass xM we have

$$\frac{d^2\sigma^i}{dE'd\Omega} = \frac{(xM)E}{E'}\delta\{(E' - E)(xM) + 2EE' \sin^2(\theta/2)\} \left(\frac{d\sigma^i}{d\Omega} \right)_{\text{elastic}}, \tag{D.7}$$

where $(\mathrm{d}\sigma^i/\mathrm{d}\Omega)_{\text{elastic}}$ is of the form given by (D.4), but with M replaced by (xM), and α^2 by $q_i^2\alpha^2$ where $q_i^2 = (1/3)^2$ or $(2/3)^2$ depending on the type of parton. On integrating over E', the δ-function in (D.7) picks out the energy for elastic scattering through an angle θ, as required by the condition (D.3) with (xM) in place of M.
(Note that $\delta(aE' - b) = (E'/b)\delta(E' - b/a), a > 0)$). If we define

$$\nu = E - E'$$

then

$$\frac{\mathrm{d}^2\sigma^i}{\mathrm{d}E'\,\mathrm{d}\Omega} = \frac{(xM)E}{E'}\delta\{(xM)\nu - Q^2/2\}\left(\frac{\mathrm{d}\sigma^i}{\mathrm{d}\Omega}\right)_{\text{elastic}}. \tag{D.8}$$

Averaging over a large number of collisions, and assuming that the partons scatter incoherently, the inclusive cross-section in the parton model is

$$\begin{aligned}\frac{\mathrm{d}^2\sigma}{\mathrm{d}E'\,\mathrm{d}\Omega} &= \int \frac{(xM)E}{E'}\delta\{(xM)\nu - Q^2/2\}\left(\sum_i p_i(x)\left(\frac{\mathrm{d}\sigma^i}{\mathrm{d}\Omega}\right)_{\text{elastic}}\right)\mathrm{d}x \\ &= \frac{x}{\nu}\frac{E}{E'}\sum_i p_i(x)\left(\frac{\mathrm{d}\sigma^i}{\mathrm{d}\Omega}\right)_{\text{elastic}},\end{aligned} \tag{D.9}$$

where

$$x = Q^2/2M\nu, \tag{D.10}$$

and the sum is over all types of charged partons. Finally, inserting explicitly the general elastic scattering formula (D.4)

$$\frac{\mathrm{d}^2\sigma}{\mathrm{d}E'\mathrm{d}\Omega} = \frac{\alpha^2}{2ME^2\sin^4(\theta/2)}\left[\frac{M}{2\nu}F_2(x, Q^2)\cos^2(\theta/2) + F_1(x, Q^2)\sin^2(\theta/2)\right] \tag{D.11}$$

where

$$F_2(x, Q^2) = x\sum_i p_i(x)q_i^2\left\{\left(f_1^i\right)^2 + \frac{\nu}{2Mx}\left(f_2^i\right)^2\right\}, \tag{D.12}$$

$$F_1(x, Q^2) = \frac{1}{2}\sum_i p_i(x)q_i^2\{\left(f_1^i\right) + \left(f_2^i\right)\}^2 \tag{D.13}$$

(using (D.10), $Q^2/4x^2M^2 = \nu/2Mx$).

In fact the form (D.11) for the inclusive cross-section, in terms of two structure functions $F_1(x, Q^2)$ and $F_2(x, Q^2)$, is quite general, and does not depend on the model we have introduced.

The wavelength $\hbar/Q$ is a measure of the scale on which the structure of the proton is explored in an electron scattering experiment. For low Q, such that $\hbar/Q$ is large compared with the size of the proton, we can anticipate that the electron is scattered coherently from the proton as a whole. It is at high Q that the parton model becomes interesting. For $Q^2 >$ a few GeV^2, incoherent parton scattering seems to dominate, and the quarks and antiquarks in the proton apparently behave almost like free elementary particles: their anomalous moments can be neglected and we can set $f_2^i = 0$. Then from (D.12) and (D.13)

$$F_2(x, Q^2) = 2xF_1(x, Q^2). \tag{D.14}$$

This, the *Callen–Gross relation*, is well satisfied experimentally.

If the charged partons are structureless Dirac particles, $f_1^i = 1$ for all Q^2, so that

$$F_2(x, Q^2) = x\sum_i p_i(x)q_i^2 = F_2(x), \tag{D.15}$$

$$F_1(x, Q^2) = \frac{1}{2}\sum_i p_i(x)q_i^2 = F_1(x), \tag{D.16}$$

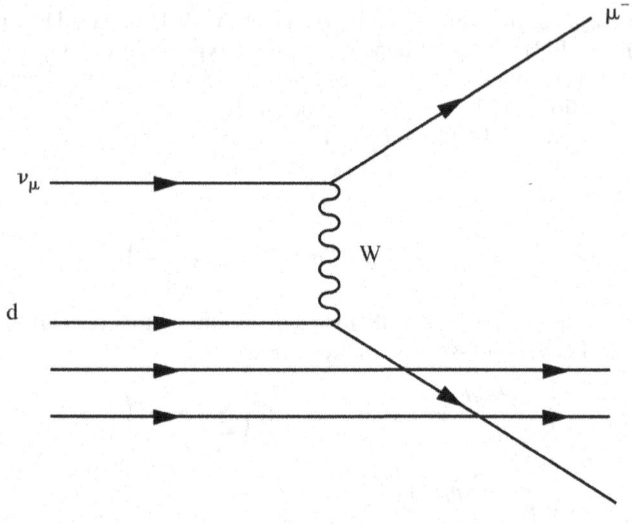

Figure D.2 An illustration of a muon neutrino converting to a muon on scattering from a d quark in a nuclean. The illustration indicates three 'valence quarks'. In fact there is additional scattering from quark–antiquark pairs that are generated by the gluon field.

and both F_2 and F_1 depend only on the dimensionless parameter $x = Q^2/2M\nu$. This is *Bjorken scaling.*

$F_2(x, Q^2)$ is illustrated in Fig. 17.3 over a wide range of values of Q^2 and x. It can be seen that the naïve parton model is not strictly correct, but that the Q^2 dependence is weak compared with that of the elastic form factor of the proton (Fig. D.1). It is usual to rewrite (D.12) as

$$F_2(x, Q^2) = x \sum_i p_i(x, Q^2) q_i^2, \tag{D.17}$$

associating the Q^2 dependence with the parton distribution itself rather than with the parton form factor. (See the discussion of the Altarelli–Parisi equations of QCD in Section 17.3.)

To determine the individual parton distributions $p_i(x, Q^2)$ introduced in equation (D.17) requires more information than is contained in the proton structure functions alone. The neutron has been investigated using deuteron targets, and, using the isospin symmetry between the neutron and proton (u $\leftrightarrow$ d, ū $\leftrightarrow$ d̄), the neutron data give further independent information. The weak interaction between quarks and leptons is described in Chapter 14. Neutrino and antineutrino inclusive cross-sections on proton and deuteron targets (Fig. D.2) give a further four independent relationships, so that, neglecting the contributions of heavier quarks, the individual u, d, s, ū, d̄, s̄ parton distributions can be estimated. In this approximation, (D.17) becomes

$$F_2(x) \approx \frac{4}{9}[xu(x) + x\bar{u}(x)] + \frac{1}{9}[xd(x) + x\bar{d}(x) + xs(x) + x\bar{s}(x)], \tag{D.18}$$

where $u(x) = p_{\mathrm{u}}(x)$, etc.

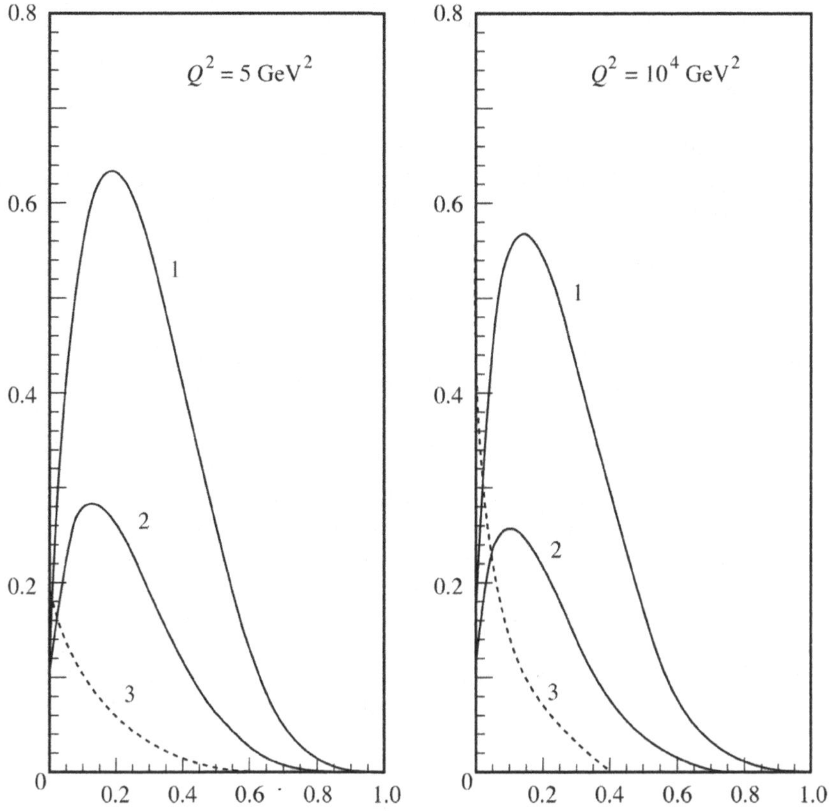

Figure D.3 Curve 1 is of $x(u(x) - \bar{u}(x))$ (see equation (D.18)). $u(x) - \bar{u}(x)$ is called the valence u quark distribution function. Curve 2 is $x(d(x) - \bar{d}(x))$, $(d(x) - \bar{d}(x))$, the valence d quark distribution function.

Curve 3 illustrates the sea quark distribution. Neglecting the generation of $c\bar{c}$, $b\bar{b}$ and $t\bar{t}$ pairs, curve 3 is of $x(\bar{u}(x) + \bar{d}(x) + \bar{s}(x))$.

Figure D.3 shows acceptable sets of parton distributions for the proton at $Q^2 = 5$ GeV^2 and at $Q^2 = 10^4$ GeV^2. With the present precision of the data these curves can be taken only as a fair indication of their forms. They have been constructed to satisfy the condition that the total parton charge is equal to e:

$$\sum_i \int_0^1 q_i p_i(x)\,\mathrm{d}x = 1,$$

but it is important to note that the charged partons carry only about one half of the total proton momentum:

$$\sum_i \int x p_i(x)\,\mathrm{d}x \approx 1/2.$$

The remainder is presumably carried by the electrically neutral gluons.

D.3 Hadronic states

The basic idea of the naïve parton model is that at high Q^2 an electron scatters from a free elementary quark or antiquark, and the scattering process is completed before the recoiling quark has time to interact with its environment of quarks, antiquarks and gluons. Thus in the calculation of the inclusive cross-section the final hadronic states do not appear.

In the model, at large Q^2 both the electron and the struck quark are deflected through large angles. Figure 1.10 shows an example of an event from the ZEUS detector at HERA. The transverse momentum of the scattered electron is balanced by a jet of hadrons that can be associated with the recoiling quark. Another jet, the 'proton remnant' jet is confined to small angles with respect to the proton beam. Events like these give further strong support to the parton model.

The 'deep inelastic' scattering data, when interpreted within the parton model, require the nucleon to have some $\bar{\mathrm{u}}$ and $\bar{\mathrm{d}}$ content, and also to contain $\mathrm{s}\bar{\mathrm{s}}$ quark-antiquark pairs (Fig. D.3). How is this to be reconciled with the simple quark model of nucleons at rest that we used in Chapter 1? A quark of the 'three quark' model of a nucleon, often called a *constituent quark*, is to be regarded as an elementary quark dressed with the strong interaction field, which will itself induce fluctuating quark–antiquark pairs. The quarks in the parton model are to be regarded as more like elementary quarks.

In quantum field theory, it is a non-trivial matter to make a Lorentz transformation on the internal wave function of a complex interacting system like a nucleon. The quark and gluon content of a proton are frame dependent. Because of time dilation, the time scale of the internal dynamics of the nucleon becomes long in a frame in which its momentum is large, and in this frame the parton distribution will be fixed over the time of interaction with an electron in a deep inelastic scattering experiment. The parton distributions in the model are taken to represent the distributions in this 'infinite momentum' frame.

Problems

D.1 Verify equations (D.2) and (D.3).

D.2 In quantum mechanics, the differential cross-section for the elastic scattering of an electron with energy $E \gg m_{\mathrm{e}}$ from a fixed electrostatic potential $\phi(r)$ is given in Born approximation, and neglecting the effects of electron spin, by

$$\frac{\mathrm{d}\sigma}{\mathrm{d}\Omega} = \left(\frac{E}{2\pi}\right)^2 \left(e \int \phi(r)\, \mathrm{e}^{\mathrm{i}\mathbf{q}\cdot\mathbf{r}} \mathrm{d}^3\mathbf{x}\right)^2,$$

where $\mathbf{q}$ is the difference between the initial and final wave vectors of the electron.

a. Show that $q = |\mathbf{q}| = 2E\sin(\theta/2)$, where θ is the scattering angle.

b. Poisson's equation relates the potential $\phi(r)$ to the charge density $\rho(r)$ by $\nabla^2\phi = -\rho$. Noting that $\nabla^2 \mathrm{e}^{\mathrm{i}\mathbf{q}\cdot\mathbf{r}} = -q^2 \mathrm{e}^{\mathrm{i}\mathbf{q}\cdot\mathbf{r}}$, and integrating by parts, show that

$$\frac{\mathrm{d}\sigma}{\mathrm{d}\Omega} = \left(\frac{E}{2\pi}\right)^2 \frac{1}{q^4} \left(e \int \rho(r)\, \mathrm{e}^{\mathrm{i}\mathbf{q}\cdot\mathbf{r}} \mathrm{d}^3\mathbf{x}\right)^2.$$

Thus a measured cross-section can be used to infer the Fourier transform of the charge distribution, as this simple example illustrates.

D.3 Taking Q^2 and ν as independent variables instead of E′ and θ, show that

$$\frac{\mathrm{d}^2\sigma}{\mathrm{d}E'\mathrm{d}\Omega} = \frac{1}{2\pi} \frac{\mathrm{d}^2\sigma}{\mathrm{d}E'\,\mathrm{d}(\cos\theta)} = \frac{EE'}{\pi} \frac{\mathrm{d}^2\sigma}{\mathrm{d}Q^2\mathrm{d}\nu}.$$

Appendix E

Mass matrices and mixing

E.1 K^o and $\bar{K}^o$

A phenomenological description of the time development of an electrically charged meson $|P\rangle$ at rest is given by the equation

$$i\frac{d}{dt}|P\rangle = [m - (i/2)\,\Gamma]\,|P\rangle \tag{E.1}$$

with its solution

$$|P\,(t)\rangle = |P\,(0)\rangle e^{-imt-(1/2)\Gamma t}$$

Here, m is the meson mass, Γ is the decay rate and $1/\Gamma$ is the mean life of the meson.

Electrically neutral mesons, for example $K^o(d\bar{s})$ and $B^o(d\bar{b})$, which have a distinct antimeson, in this example $\bar{K}^o(s\bar{d})$ and $\bar{B}^o(b\bar{d})$, can mix so that (E.1) becomes two coupled equations. For K^o and $\bar{K}^o$ these are

$$i\frac{d}{dt}\begin{pmatrix}|K^o\rangle\\ |\bar{K}^o\rangle\end{pmatrix} = \begin{pmatrix} m-(i/2)\,\Gamma & -p^2\\ -q^2 & m-(i/2)\,\Gamma\end{pmatrix}\begin{pmatrix}|K^o\rangle\\ |\bar{K}^o\rangle\end{pmatrix} \tag{E.2}$$

p^2 and q^2 are two complex numbers. We can regard the 2×2 mass matrix as an 'effective' Hamiltonian H_{weak}. The equality of the diagonal elements of H_{weak} is guaranteed by *CPT* invariance. The weak interaction generates the off-diagonal elements

$$\langle K^o|H_{weak}|\bar{K}^o\rangle = -p^2,\ \langle\bar{K}^o|H_{weak}|K^o\rangle = -q^2.$$

Contributions to p^2 and q^2 are illustrated in Fig. E.1.

By substitution into (E.2) it can be seen that the eigenstates of H_{weak} are

$$|K_S\rangle = N[p|K^o\rangle + q|\bar{K}^o\rangle] \tag{E.3}$$

and

$$|K_L\rangle = N[p|K^o\rangle - q|\bar{K}^o\rangle] \tag{E.4}$$

with eigenvalues $m - i\Gamma/2 - pq$ and $m - i\Gamma/2 + pq$ respectively. $N = (|p|^2 + |q|^2)^{-1/2}$ is a normalising factor. We choose the sign of the square root, $pq = \sqrt{p^2q^2}$, so that $\mathrm{Im}(pq)$ is positive; then K_L has a longer mean life than K_S.

The mass difference $\Delta m = 2\mathrm{Real}(pq)$ (from experiment $\Delta m \approx 3 \times 10^{-12}$ MeV). We shall identify m with the mean mass of K_S and K_L. The mean lives are

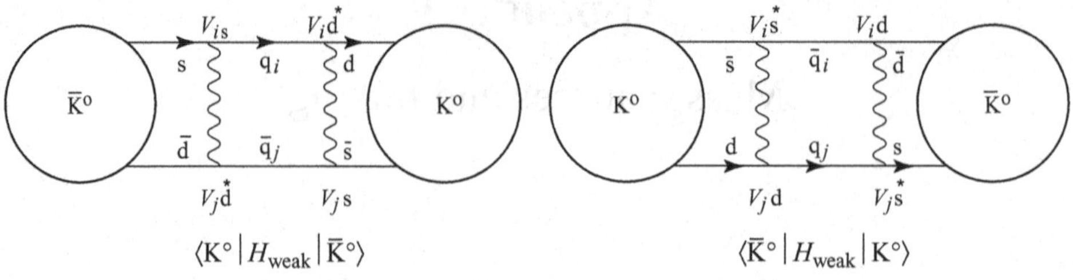

Figure E.1 Quark diagrams illustrating how the weak interaction with W bosons generates mixing. q_i, and q_j are any of the (2/3)e charged quarks u, c or t. The mixing matrix elements are proportional to the products of the four KM factors in the diagrams.

$\tau_L \approx \dfrac{1}{\Gamma - 2\,\mathrm{Im}(pq)}$ and $\tau_S = \dfrac{1}{\Gamma + 2\,\mathrm{Im}(pq)}$ (from experiment $\tau_L \approx 5 \times 10^{-8}$ s, $\tau_S \approx 10^{-10}$ s.) The subscripts L and S refer to the long and short lives.

From lattice estimations of the bound state wave functions and other QCD modifications, p^2 and q^2 can be calculated by perturbation theory in the weak interaction. Fig.E.1 illustrates the fact that because some of the KM factors V_{is}, etc. are complex numbers, p and q are not equal. As a consequence neither $|K_L\rangle$ nor $|K_S\rangle$ is an eigenstate of CP. See Section (18.4).

E.2 B^o and $\bar{B}^o$

The neutral B meson pair B^o and $\bar{B}^o$ mix by the same mechanism as the neutral K mesons. The parameters m, Γ, p^2 and q^2 take, of course, different values.

For the B pair Im(pq) is much smaller than Γ so that the two mean lives are almost equal. There are two particles of different mass:

$$|B_L\rangle = N[p|B^o\rangle + q|\bar{B}^o\rangle],$$
$$|B_H\rangle = N[p|B^o\rangle - q|\bar{B}^o\rangle].$$

The subscripts L and H refer to their masses: light and heavy.

For $B^o\bar{B}^o$ mixing it is a fortunate circumstance that the top quark $q_i = t$, $\bar{q}_j = \bar{t}$ gives the dominant contribution to p^2 and q^2, p^2 is proportional to $(V_{tb}V_{td}^*)^2$ and q^2 is proportional to $(V_{tb}^*V_{td})^2$ (see Fig. E. 1) Calculations result in the expressions

$$\begin{aligned} p &= \sqrt{m_B m_t}\frac{G_F}{4\pi} f_B F_{tt} V_{tb} V_{td}^*, \\ q &= \sqrt{m_B m_t}\frac{G_F}{4\pi} f_B F_{tt} V_{tb}^* V_{td}. \end{aligned} \tag{E.5}$$

(Donoghue *et al.*, 1992, p. 395.)

All other contributions are smaller by factors of $(m_c/m_t)^2$, m_B is the B meson mass, $f_B \approx 0.3$ GeV is its 'leptonic decay constant' and F_{tt} is a dimensionless number, real to a very good approximation.

With F_{tt} real, $\mathrm{Im}(pq) = 0$, and B_L and B_H have the same mean life. Within experimental error this is seen to be so. Also $|p| = |q|$ and $p = |p|e^{i\beta}$ $q = |p|e^{-i\beta}$.

(See the unitarity triangle, Fig. 18.2). Hence

$$
\begin{aligned}
|B_L\rangle &= \frac{1}{\sqrt{2}}\left[e^{i\beta}\left|B^o\right\rangle + e^{-i\beta}\left|\bar{B}^o\right\rangle\right] \\
|B_H\rangle &= \frac{1}{\sqrt{2}}\left[e^{i\beta}\left|B^o\right\rangle - e^{-i\beta}\left|\bar{B}^o\right\rangle\right].
\end{aligned} \tag{E.6}
$$

A B_L meson or a B_H meson, at rest, develop independently with time

$$
\begin{aligned}
|B_L(t)\rangle &= |B_L(o)\rangle\, e^{-i(m-\Delta m/2)t-t/2\tau}, \\
|B_H(t)\rangle &= |B_H(o)\rangle\, e^{-i(m+\Delta m/2)t-t/2\tau}.
\end{aligned}
$$

After some algebra it then follows that an initial B^o or $\bar{B}^o$ develops in time into a mixture denoted by

$$
\begin{aligned}
|B^o_{phy}(t)\rangle &= \left[\cos\left(\frac{\Delta mt}{2}\right)|B^o\rangle + ie^{-2i\beta}\sin\left(\frac{\Delta mt}{2}\right)|\bar{B}^o\rangle\right]e^{-imt-t/2\tau} \\
|\bar{B}^o_{phy}(t)\rangle &= \left[ie^{2i\beta}\sin\left(\frac{\Delta mt}{2}\right)|B^o\rangle + \cos\left(\frac{\Delta mt}{2}\right)|\bar{B}^o\rangle\right]e^{-imt-t/2\tau}.
\end{aligned} \tag{E.7}
$$

If the meson decays at time t, to a final state $|f\rangle$ the decay amplitude for an initial B^o will be

$$
\langle f|B^o_{phy}(t)\rangle = \left[\cos\left(\frac{\Delta mt}{2}\right)A_f + ie^{-2i\beta}\sin\left(\frac{\Delta mt}{2}\right)\bar{A}_f\right]e^{-imt-t/2\tau}
$$

and an initial B^o

$$
\langle f|\bar{B}^o_{phy}(t)\rangle = \left[ie^{2i\beta}\sin\left(\frac{\Delta mt}{2}\right)A_f + \cos\left(\frac{\Delta mt}{2}\right)\bar{A}_f\right]e^{-imt-t/2\tau}. \tag{E.8}
$$

$A_f = \langle f|B^o_{phy}\rangle$ and $\bar{A}_f = \langle f|\bar{B}^o_{phy}\rangle$ are the amplitudes for the decays $B^o \to f$ and $\bar{B}^o \to f$. If the charge parity (CP) of f is $+1$ then it does not couple to the $CP = -1$ state $(B^o - \bar{B}^o)$; hence $A_f = \bar{A}_f$. The decay rates are then

$$
\begin{aligned}
\text{Rate}\left(B^o_{phy}(t) \to f\right) &= |A_f|^2 e^{-t/\tau}[1 + \sin(2\beta)\sin(mt)] \\
\text{Rate}\left(\bar{B}^o_{phy}(t) \to f\right) &= |A_f|^2 e^{-t/\tau}[1 - \sin(2\beta)\sin(mt)].
\end{aligned} \tag{E.9}
$$

If f has $CP = -1$ the same expression results but with the $+$ and $-$ signs interchanged.

At Cleo, Babar and Belle, B^o and $\bar{B}^o$ mesons are produced in pairs. If one undergoes a leptonic decay with a negative charge lepton it must have been a $\bar{B}^o$, its partner, at that instant is a B^o and it is the time dependence of this second decay that is measured.

Similarly a positive charge lepton identifies a B^o decay that leaves its partner an initial $\bar{B}^o$. This procedure is called tagging. The mass difference Δm and $\sin 2\beta$ are measured by tracking the time dependence of tagged mesons.

The formulae for p^2 and q^2 for K^o, $\bar{K}^o$ follow the same pattern as for B decays but the top quark contributions are highly suppressed by very small KM factors. c and u quarks contribute significantly and the simplicity for B mesons is lost.

References

Ahmad, Q. R. *et al.* (2002) *Phys. Rev. Lett.* **89**, 011301.

Altarelli, G. and Parisi, G. (1977). *Nucl. Phys.* **B126**, 298.

Anderson, P. W. (1963). *Phys. Rev.* **130**, 439.

Apollonio, M. *et al.* (2003) *Eur. Phys. J.* **C27**, 331.

Armstrong, T. A., Hogg, W. R., Lewis, G. M. *et al.* (1972). *Phys. Rev.* **D5**, 1640; *Nucl. Phys.* **B41**, 445.

Bali, G. S. and Schilling, K. (1993). *Phys. Rev.* **D47**, 661.

Bartelt, J., Csorna, S. E., Egyed, Z. *et al.* (1993). *Phys. Rev. Lett.* **71**, 4111.

Benvenuti, A. C., Bollini, D., Bruni, G. *et al.* (1989). *Phys. Lett.* **B223**, 490.

Booth, S. P., Henty, D. S., Hulsebos, A., Irving, A. C., Michael, C. and Stephenson, P. W. (1992). *Phys. Lett.* **B294**, 385.

CHARM II Collaboration (1994). *Phys. Lett.* **B335**, 246.

Cheng, T. P. and Li, L. F. (1984). *Gauge Theory of Elementary Particle Physics*. Oxford: Clarendon Press.

Close, F. (1979). *An Introduction to Quarks and Partons*. New York: Academic Press.

Coward, D. H., DeStaebler, H., Early, R. A. *et al.*, (1968). *Phys. Rev. Lett.* **20**, 292.

Dashen, R. and Gross, D. J. (1981). *Phys. Rev.* **D23**, 2340.

Davies, C. *et al.* (2004) *Phys. Rev. Lett.* **92**, 022001.

Davis, R. (1964) *Phys. Rev. Lett.* **12**, 303.

Donoghue, J. F., Golowich, E. and Holstein, B. R. (1992). *Dynamics of the Standard Model*. Cambridge: Cambridge University Press.

Dydak, F. (1990). In *Proceedings of the 1989 International Symposium on Lepton and Photon Interactions at High Energies*, ed. M. Riordan, p. 249. Singapore: World Scientific.

Eichten, E., Gottfreid, K., Kinoshita. T., Lane, K. D. and Yan, T. M. (1980). *Phys. Rev.* **D21**, 203.

Fero, M. J. (1994). In *Proceedings of the XXVII International Conference on High Energy Physics*, eds B. J. Bussey and I. G. Knowles, p. 399. Bristol: Institute of Physics Publishing.

Fukuda, Y. *et al.* (1996) *Phys. Rev. Lett.* **77**, 1683.

Gavrin, V. N. *et al.* (2003) *Nuc. Phys. B* (Proc. Supple.) 118.

Gross, D. J. and Wilczek, F. (1973) *Phys. Rev.* **D8**, 3633.

Gross, F. (1993). *Relativistic Quantum Mechanics and Field Theory*. New York: Wiley.

Hampel, W. *et al.* (1999) *Phys. Lett.* **B447**, 127.

Hansen, J. R. (1991) In *Proceedings of the 25th International Conference on High Energy Physics*, eds K. K. Phua and Y. Yamaguchi, p. 343. Singapore: World Scientific.
Hasenfratz, A. and Hasenfratz, P. (1985) *Ann. Rev. Nucl. Part. Sci.* **35**, 559.
Higgs, P. W. (1964) *Phys. Rev. Lett.* **13**, 508.
Hofstadter, R., Bumillar, F. and Yearian, M. R. (1958) *Rev. Mod. Phys.* **30**, 482.
Horn, R. A. and Johnson, C. R. (1985). *Matrix Analysis*. Cambridge: Cambridge University Press.
Itzykson, C. and Zuber, J. B. (1980) *Quantum Field Theory*. New York: McGraw-Hill.
Jarlskog, C. (1985) *Phys. Rev. Lett.* **55**, 1039.
Kinoshita, T. and Lindquist, W. B. (1990) *Phys. Rev.* **42**, 636.
Kobayashi, M. and Maskawa, J. (1973) *Prog. Theor. Phys.* **49**, 652.
Koks, F. W. J. and Van Klinken, J. (1976) *Nucl. Phys.* **A272**, 61.
Leader, E. and Predazzi, E. (1982) *Gauge Theories and the New Physics*. Cambridge: Cambridge University Press.
Mikheyev, S. P. and Smirnov, A. Yu. (1986) *Nuovo Cimento* **C9**, 17.
Mori, T. (1991). In *Proceedings of 25th International Conference on High Energy Physics*, eds K. K. Phua and Y. Yamaguchi, p. 360. Singapore: World Scientific.
Okun, L. B. (1982). *Leptons and Quarks*. Amsterdam: North-Holland.
Olive, D. I. (1997). In *Electron*, ed. M. Springford, p. 39. Cambridge: Cambridge University Press.
Particle Data Group (1996) *Phys. Rev.* **D54**, 1. (2006) W. M. Yao *et al. J. Phys.* G33 1.
Perkins, D. H. (1987) *Introduction to High Energy Physics*, 3rd edn. Menlo Park, CA: Addison-Wesley.
Politzer, H. D. (1973) *Phys. Rev. Lett.* **30**, 1346.
Pontecorvo, B. (1968) *Sov. Phys. JETP.* **26**, 984.
Prescott, C. Y. (1996) In *17th International Symposium on Lepton–Photon Interactions*, eds Z. P. Zheng and H. S. Chen, p. 130. Singapore: World Scientific.
Renton, P. B. (1996) In *17th International Symposium on Lepton–Photon Interactions*, eds Z. P. Zheng and H. S. Chen, p. 35. Singapore: World Scientific.
Salam, A. (1968). In *Elementary Particle Theory* (Nobel Symp. No. 8), ed. N. Svartholm. Stockholm: Almquist & Wiksell.
Skwarnicki, T. (1996) In *17th International Symposium on Lepton–Photon Interactions*, eds Z. P. Zheng and H. S. Chen, p. 238. Singapore: World Scientific.
t'Hooft, G. (1976) *Phys. Rev. Lett.* **37**, 8.
Treiman, S. B., Jackiw, R., Zumino, B. and Witten, E. (1985). *Current Algebra and Anomalies*. Singapore: World Scientific.
Van Dyck, R. S., Schwinberg, P. B. and Dehmelt, H. G. (1987). *Phys. Rev. Lett.* **59**, 26.
Weinberg, S. (1967). *Phys. Rev. Lett.* **19**, 1264.
Wolfenstein, L. (1978) *Phys. Rev.* **D17**, 2369.
Yang, C. N. and Mills, R. L. (1954) *Phys. Rev.* **96**, 191.

Hints to selected problems

Chapter 2

2.1 $a'_\mu = g_{\mu\rho}a'^\rho = g_{\mu\rho}L^\rho{}_\lambda a^\lambda = g_{\mu\rho}L^\rho{}_\lambda g^{\lambda\nu}a_\nu$. Hence $a'_\mu = L_\mu{}^\nu a_\nu$ where $L_\mu{}^\nu = g_{\mu\rho}L^\rho{}_\lambda g^{\lambda\nu}$. In particular, $L_0{}^1 = g_{00}L^0{}_1 g^{11} = -L^0{}_1$.

2.2 $a'^\mu = L^\mu{}_\nu a^\nu$. Multiply on the left by $L_\mu{}^\rho \cdot L_\mu{}^\rho a'^\mu = L_\mu{}^\rho L^\mu{}_\nu a^\nu = a^\rho$, or $a^\mu = a'^\nu L_\nu{}^\mu$. Similarly, $a_\mu = a'_\nu L^\nu{}_\mu$.

2.3 $\mathrm{d}\phi = \dfrac{\partial\phi}{\partial x^\mu}\mathrm{d}x^\mu = \dfrac{\partial\phi}{\partial x'^\nu}\mathrm{d}x'^\nu = \dfrac{\partial\phi}{\partial x'^\nu}L^\nu{}_\mu \mathrm{d}x^\mu$. Since the $\mathrm{d}x^\mu$ are arbitrary,

$$\frac{\partial\phi}{\partial x^\mu} = \frac{\partial\phi}{\partial x'^\nu}L^\nu{}_\mu.$$

This is a covariant vector field transformation (Problem 2.2).

2.4

$$\begin{aligned}\det\left(L_\mu{}^\nu\right) &= \det(g_{\mu\rho})\det\left(L^\rho{}_\lambda\right)\det(g^{\lambda\nu})\\ &= (-1)^2\det\left(L^\rho{}_\lambda\right).\end{aligned}$$

From (2.14), $\det(L_\mu{}^\nu)\det(L^\mu{}_\rho)\det\left(\delta^\nu{}_\rho\right) = 1$. The result follows.

2.6 Note that if $\det\mathbf{L}_1 = 1$ and $\det\mathbf{L}_2 = 1$ then $\det\mathbf{L}_1\det\mathbf{L}_2 = 1$.

2.7 $\delta'^\mu_\nu = L^\mu{}_\rho L_\nu{}^\lambda\delta^\rho_\lambda = L^\mu{}_\rho L_\nu{}^\rho = \delta^\mu_\nu$ using Problem 2.2.

2.8 Using (2.3), $\omega' = \omega\cosh\theta - k\sinh\theta$
$= \omega(\cosh\theta - \sinh\theta)$ since $\omega = k$
$= e^{-\theta}\omega$.
Since $v/c = \tanh\theta$, the result follows.

2.9 Jacobian is $\det(\partial x'^\mu/\partial x^\nu) = \det(L^\mu{}_\nu) = 1$.

2.10 The operation of space inversion can be written as $x_\mu{}' = P^\nu_\mu x_\nu$. Then the tensor $\varepsilon_{\mu\nu\lambda\rho}$, transforms as

$$\begin{aligned}\varepsilon'_{\mu\nu\lambda\rho} &= P^\alpha_\mu P^\beta_\nu P^\gamma_\lambda P^\delta_\rho\varepsilon_{\alpha\beta\gamma\delta}\\ &= \varepsilon_{\mu\nu\lambda\rho}\det\mathbf{P} = -\varepsilon_{\mu\nu\lambda\rho}.\end{aligned}$$

Chapter 3

3.1 Let $x_i (i = 1, ..., 3N)$ be the Cartesian coordinates of the particles. Since $x_i = x_i(q)$, $\dot{x}_i = (\partial x_i/\partial q_j)\dot{q}_j$. Then $T = (m/2)\dot{x}_i\dot{x}_i = (m/2)(\partial x_i/\partial q_j)(\partial x_i/\partial q_k)\dot{q}_j\dot{q}_k$.

3.2
$$\frac{\mathrm{d}E}{\mathrm{d}t} = \int \left[\dot{\phi}\frac{\partial}{\partial t}\left(\frac{\partial \mathcal{L}}{\partial \dot{\phi}}\right) + \frac{\partial \mathcal{L}}{\partial \dot{\phi}}\ddot{\phi} - \frac{\partial \mathcal{L}}{\partial \dot{\phi}}\ddot{\phi} - \frac{\partial \mathcal{L}}{\partial \phi'}\dot{\phi}' \right] \mathrm{d}x.$$

Integrate by parts the term $-(\partial\mathcal{L}/\partial\phi')(\partial\dot{\phi}/\partial x)$ and use (3.12).

3.4 Use orthogonality and the dispersion relation (3.20). Note that H and P^i form a contravariant four-vector $(H, \mathbf{P})$.

3.5 Varying ψ^*,

$$\begin{aligned}\delta S &= \int \delta\mathcal{L}\,\mathrm{d}t\,\mathrm{d}^3\mathbf{x} \\ &= \int \bigg[-(1/2\mathrm{i})\left(\delta\psi^*\frac{\partial\psi}{\partial t} - \frac{\partial(\delta\psi^*)}{\partial t}\psi\right) \\ &\qquad - (1/2m)\nabla(\delta\psi^*)\cdot\nabla\psi - \delta\psi^* V\psi \bigg]\,\mathrm{d}t\,\mathrm{d}^3\mathbf{x}.\end{aligned}$$

Integrating by parts the terms involving $\partial(\delta\psi^*)/\partial t$ and $\nabla(\delta\psi^*)$ gives

$$\delta S = \int \left[-(1/\mathrm{i})\frac{\partial\psi}{\partial t} + (1/2m)\nabla^2\psi - V\psi \right]\delta\psi^*\mathrm{d}t\,\mathrm{d}^3\mathbf{x}.$$

Since this is true for any $\delta\psi^*$, the integrand must vanish. Hence

$$\mathrm{i}\frac{\partial\psi}{\partial t} = -(1/2m)\nabla^2\psi + V\psi.$$

Chapter 4

4.1 $\mathcal{L} = -(1/4)F_{\mu\nu}F^{\mu\nu} - J^\mu A_\mu$. From (4.16), $F^{01} = -E_x = -F_{01}$, $F^{12} = -B_2 = F_{12}$, etc.

4.2 $\mathbf{A} \to \mathbf{A}' = \mathbf{A} - \nabla\chi$. We require $\nabla\cdot\mathbf{A}' = \nabla\cdot(\mathbf{A} - \nabla\chi) = f - \nabla^2\chi = 0$. The solution is

$$\chi(\mathbf{r}, t) = -\frac{1}{4\pi}\int \frac{f(\mathbf{r}', t)}{|\mathbf{r} - \mathbf{r}'|}\mathrm{d}^3\mathbf{r}'.$$

4.3
$$\begin{aligned}\tilde{F}_{01} &= (\varepsilon_{0123}F^{23} + \varepsilon_{0132}F^{32})/2 \\ &= (F^{23} - F^{32})/2 = (-B_x - B_x)/2 = -B_x, \text{ etc.}\end{aligned}$$

4.4
$$\begin{aligned}\mathbf{A} &= \frac{1}{\sqrt{2\omega V}}\left[(\varepsilon_x + \mathrm{i}\varepsilon_y)\mathrm{e}^{\mathrm{i}(kz-\omega t)} + (\varepsilon_x - \mathrm{i}\varepsilon_y)\mathrm{e}^{-\mathrm{i}(kz-\omega t)}\right] \\ &= \frac{1}{\sqrt{2\omega V}}\left[2\cos(kz - \omega t), -2\sin(kz-\omega t), 0\right], \\ \mathbf{E} &= -\frac{\partial\mathbf{A}}{\partial t} = \sqrt{\frac{2\omega}{V}}\left[\sin(\omega t - kz), -\cos(\omega t - kz), 0\right].\end{aligned}$$

By inspection, on any plane of fixed z, $\mathbf{E}$ rotates in a positive sense about the z-axis.

4.5 If the fields vanish at infinity, a term $\partial_i(A_0 F^{0i}) = \partial_\mu(A_0 F^{0\mu})$ does not contribute to the energy. Thus the energy density is not unique, and we may take

$$\begin{aligned} T^0_0 &= -F^{0\mu}\partial_0 A_\mu + \partial_\mu(A_0 F^{0\mu}) + \frac{1}{4}F_{\mu\nu}F^{\mu\nu} \\ &= -F^{0\mu}(\partial_0 A_\mu - \partial_\mu A_0) + \frac{1}{4}F_{\mu\nu}F^{\mu\nu}, \end{aligned}$$

since in free space $\partial_\mu F^{0\mu} = 0$ by (4.8),

$$= -F^{0\mu}F_{0\mu} + \frac{1}{4}F_{\mu\nu}F^{\mu\nu}.$$

4.6 $L = \frac{1}{2}m\dot{\mathbf{x}}^2 - q\phi + q\dot{\mathbf{x}}\cdot\mathbf{A}$, $p^i = (\partial L/\partial \dot{x}^i) = m\dot{x}^i + qA^i$ are the generalised momenta. The equation of motion $(\mathrm{d}p^i/\mathrm{d}t) = (\partial L/\partial x^i)$ is

$$m\ddot{x}^i + q(\partial A^i/\partial t) + q(\partial A^i/\partial x^j)\dot{x}^j = -q(\partial\phi/\partial x^i) + q\dot{x}^j(\partial A^j/\partial x^i),$$

giving

$$m\ddot{x}^i = q[-(\partial\phi/\partial x^i) - q(\partial A^i/\partial t)] - qF^{ij}\dot{x}^j$$

(noting $\partial^i = -\partial/\partial x^i$, and definition (4.6)). Taking $i = 1$,

$$\begin{aligned} m\ddot{x} &= q(E_x - F^{12}\dot{y} - F^{13}\dot{z}) \\ &= q(E_x + \dot{y}B_z - \dot{z}B_y), \end{aligned}$$

and similarly for the other components.

$$\begin{aligned} H(\mathbf{p},\mathbf{x}) &= p^i\dot{x}^i - L \\ &= \mathbf{p}\cdot(\mathbf{p}-q\mathbf{A})/m - [(\mathbf{p}-q\mathbf{A})^2/2m - q\phi + q(\mathbf{p}-q\mathbf{A})\cdot\mathbf{A}/m] \\ &= (\mathbf{p}-q\mathbf{A})^2/2m + q\phi. \end{aligned}$$

4.7 $\int L\,\mathrm{d}t = \int(\gamma L)\,\mathrm{d}\tau$, where $\mathrm{d}\tau = \mathrm{d}t/\gamma$ is Lorentz invariant (see (2.5); τ is the 'proper time'). Hence the result.

Chapter 5

5.3 Under the transformations (5.19) and (5.20),

$$\begin{aligned} \psi_R'^\dagger\psi_L' &= \psi_R^\dagger \mathbf{N}^\dagger\mathbf{M}\psi_L = \psi_R^\dagger\psi_L, \\ \psi_L'^\dagger\psi_R' &= \psi_L^\dagger \mathbf{M}^\dagger\mathbf{N}\psi_R = \psi_L'^\dagger\psi_R, \\ \psi_R'^\dagger\sigma^\mu\psi_R' &= \psi_R^\dagger\mathbf{N}^\dagger\sigma^\mu\mathbf{N}\psi_R = L^\mu{}_\nu\psi_R\sigma^\nu\psi_R, \\ \psi_L'^\dagger\tilde{\sigma}^\mu\psi_L' &= \psi_L^\dagger\mathbf{M}^\dagger\tilde{\sigma}^\mu\mathbf{M}\psi_L = L^\mu{}_\nu\psi_L^\dagger\tilde{\sigma}^\nu\psi_L, \\ \psi_R'\sigma^\mu\tilde{\sigma}^\nu\psi_L' &= \psi_R^\dagger\mathbf{M}^\dagger\sigma^\mu\mathbf{M}\mathbf{N}^\dagger\sigma^\nu\mathbf{N}\psi_L\ (\text{since }\mathbf{M}\mathbf{N}^\dagger = \mathbf{I}) \\ &= L^\mu{}_\lambda L^\mu{}_\rho\psi_R^\dagger\sigma^\lambda\tilde{\sigma}^\rho\psi_L,\ \text{etc.} \end{aligned}$$

5.4 Using (5.28), (5.31) becomes

$$\psi^\dagger\beta(\mathrm{i}\beta\partial_0 + \mathrm{i}\beta\alpha_i\partial_i - m)\psi = \psi^\dagger(\mathrm{i}\partial_0 + \mathrm{i}\alpha_i\partial_i - \beta m)\psi \text{ since } \beta^2 = \mathbf{I}.$$

5.6

$$\mathrm{i}\bar{\psi}\gamma^5\psi = \mathrm{i}(\psi_\mathrm{L}^\dagger, \psi_\mathrm{R}^\dagger)\begin{pmatrix} \mathbf{0} & \sigma^0 \\ \sigma^0 & \mathbf{0} \end{pmatrix}\begin{pmatrix} -\sigma^0 & \mathbf{0} \\ \mathbf{0} & \sigma^0 \end{pmatrix}\begin{pmatrix} \psi_\mathrm{L} \\ \psi_\mathrm{R} \end{pmatrix}$$
$$= \mathrm{i}(\psi_\mathrm{L}^\dagger\psi_R - \psi_\mathrm{R}^\dagger\psi_\mathrm{L}).$$

This is invariant under proper Lorentz transformations, but changes sign under the parity operation (5.27).

5.7 The results follow from the definitions (5.30) and (5.4).

Chapter 6

6.1

$$\psi_+^\dagger\psi_+ = \frac{1}{2}(\langle + \,|\, \mathrm{e}^{-\theta/2}, \langle + \,|\, \mathrm{e}^{\theta/2})\begin{pmatrix} \mathrm{e}^{-\theta/2}|+\rangle \\ \mathrm{e}^{\theta/2}|+\rangle \end{pmatrix}$$
$$= \frac{1}{2}[\mathrm{e}^{-\theta}\,\langle + \,|\, + \rangle + \mathrm{e}^{\theta}\,\langle + \,|\, + \rangle]$$
$$= \cosh\theta = \gamma = E/m.$$

From (6.14), probability of right-handed mode

$$= \frac{e^\theta}{e^\theta + e^{-\theta}} = \frac{e^\theta}{2\cosh\theta} = \frac{1}{2}\left(1 + \frac{v}{c}\right), \text{ since } \tanh\theta = \frac{v}{c}.$$

6.3

$$u_+^\dagger(\mathbf{p})u_+(\mathbf{p}) = \tfrac{1}{2}(e^\theta + e^{-\theta}) = \cosh\theta = E/m, \text{ etc.}$$
$$u_+^\dagger(\mathbf{p})u_-(\mathbf{p}) = 0 \text{ since } \langle + \,|\, - \rangle = 0.$$

Note that

$$\sigma\cdot\hat{\mathbf{p}}|+\rangle = |+\rangle \quad \text{and} \quad \boldsymbol{\sigma}\cdot\hat{\mathbf{p}}|-\rangle = -|-\rangle$$

implies

$$\boldsymbol{\sigma}\cdot(-\hat{\mathbf{p}})\,|+\rangle = -\,|+\rangle \quad \text{and} \quad \boldsymbol{\sigma}\cdot(-\hat{\mathbf{p}})\,|-\rangle = |-\rangle.$$

6.5 $|+\rangle$ and $|-\rangle$ are evidently normalised, and by direct substitution and the use of trigonometric identities, $\boldsymbol{\sigma}\cdot\mathbf{p}\,|+\rangle = |+\rangle, \boldsymbol{\sigma}\cdot\mathbf{p}\,|-\rangle = -|-\rangle$.

Chapter 7

7.1 This follows using the orthogonality properties of plane waves and those derived in Problem 6.3.

7.2 For example,

$$\psi_+^c = -\mathrm{i}\gamma^2\psi_+^* = \left(\mathrm{i}/\sqrt{2}\right)e^{-\mathrm{i}(pz-Et)}\begin{pmatrix} 0 & -\sigma^2 \\ \sigma^2 & 0 \end{pmatrix}\begin{pmatrix} \mathrm{e}^{-\theta/2} & |+\rangle \\ \mathrm{e}^{\theta/2} & |+\rangle \end{pmatrix}$$

and $\sigma^2\,|+\rangle = \mathrm{i}\,|-\rangle$, giving

$$\psi_+^c = \left(1/\sqrt{2}\right)\mathrm{e}^{-\mathrm{i}(pz-Et)}\begin{pmatrix} \mathrm{e}^{\theta/2} & |-\rangle \\ -\mathrm{e}^{-\theta/2} & |-\rangle \end{pmatrix}.$$

7.3 Under the parity operation,

$$\psi_L \to \psi_R, \tilde{\sigma}^\mu \partial_\mu \to \sigma^\mu \partial_\mu,$$

from (5.26) and (5.27). Under charge conjugation,

$$\psi_R \to i\sigma^2 \psi_L^*.$$

Hence under the combined operations,

$$i\psi_L^\dagger \tilde{\sigma}^\mu \partial_\mu \psi_L \to i\psi_L^T \sigma^2 \sigma^\mu \sigma^2 \partial_\mu \psi_L^* = -i\partial_\mu \psi_L^\dagger (\sigma^2 \sigma^\mu \sigma^2)^T \psi_L$$

(recall the – sign that must be introduced when spinor fields are interchanged). But $(\sigma^2\sigma^\mu\sigma^2)^T = \tilde{\sigma}^\mu$.

Finally, integrating by parts in the action yields the Lagrangian density $i\psi_L^\dagger \tilde{\sigma}^\mu \partial_\mu \psi_L$.

7.4 $\psi_R \to \psi_R' = \mathbf{N}\psi_R$ by (5.20).

$$i\sigma^2 \psi_R^* \to i\sigma^2 \mathbf{N}^* \psi_R^*.$$

But $\sigma^2 \mathbf{N}^* = \mathbf{M}\sigma^2$. This is true for $\mathbf{M}$ and $\mathbf{N}$ given by (5.24), and holds in general.

7.5 Varying $\mathbf{\Phi}^*$ in the action gives

$$\begin{aligned}\delta S &= \int \{-[(i\partial_\mu + qA_\mu)\delta\mathbf{\Phi}^*][(i\partial^\mu - qA^\mu)\mathbf{\Phi}] - m^2 \delta\mathbf{\Phi}^*\mathbf{\Phi}\}\, dt\, d^3\mathbf{x} \\ &= \int \delta\mathbf{\Phi}^* \{(i\partial_\mu - qA_\mu)(i\partial^\mu - qA^\mu)\mathbf{\Phi} - m^2\mathbf{\Phi}\}\, dt\, d^3\mathbf{x},\end{aligned}$$

after integrating by parts. Since this holds for any $\delta\Phi^*$, the Klein–Gordon equation follows.

7.6 If $\mathbf{\Phi} \to e^{i\alpha}\mathbf{\Phi}$ with $\alpha = \alpha(x)$ small,

$$\begin{aligned}(i\partial_\mu + qA_\mu)(e^{i\alpha}\mathbf{\Phi}) &= e^{i\alpha}(i\partial_\mu + qA_\mu)\mathbf{\Phi} - (\partial_\mu\alpha)e^{i\alpha}\mathbf{\Phi} \\ \delta S &= \int \{-(\partial_\mu\alpha)\mathbf{\Phi}^*[(i\partial^\mu - qA^\mu)\mathbf{\Phi}] + [(i\partial^\mu + qA^\mu)\mathbf{\Phi}^*](\delta_\mu\alpha)\mathbf{\Phi}\}\, dt\, d^3\mathbf{x} \\ &= \int \alpha(x)\partial_\mu\{\mathbf{\Phi}^*[(i\partial^\mu - qA^\mu)\mathbf{\Phi}] - [(i\partial^\mu + qA^\mu)\mathbf{\Phi}^*]\mathbf{\Phi}\}\, dt\, d^3\mathbf{x},\end{aligned}$$

after integrating by parts. Hence the current

$$j^\mu = i[\mathbf{\Phi}^*(\partial^\mu\mathbf{\Phi}) - (\partial^\mu\mathbf{\Phi}^*)\mathbf{\Phi}] - 2qA^\mu\mathbf{\Phi}^*\mathbf{\Phi}$$

is conserved, as is also qj^μ. (Note that $qj^\mu = -\partial\mathcal{L}/\partial A_\mu$ is the electromagnetic current.)

7.7 Verify by direct calculation, e.g. for positive helicity and taking $\mu = 3$,

$$\begin{aligned} qj^3 &= -e\psi^+\gamma^0\gamma^3\psi \\ &= -(e/2)\left(\mathrm{e}^{-\theta/2}\langle +|, \mathrm{e}^{\theta/2}\langle +|\right)\begin{pmatrix} -\sigma^3 & 0 \\ 0 & \sigma^3 \end{pmatrix}\begin{pmatrix} \mathrm{e}^{-\theta/2} & |+\rangle \\ \mathrm{e}^{\theta/2} & |+\rangle \end{pmatrix} \\ &= -e \sinh\theta, \text{ since } \sigma^3 |+\rangle = |+\rangle. \end{aligned}$$

7.8 This follows since the electric field lines are reversed in direction, $\mathbf{E} \to \mathbf{E}' = -\mathbf{E}$.

7.9 Assuming $\rho(\mathrm{t}) \to \rho'(\mathrm{t}') = \rho(-\mathrm{t})$, Maxwell's equations retain the same form if $\mathbf{E} \to \mathbf{E}' = \mathbf{E}, \mathbf{B} \to \mathbf{B}' = -\mathbf{B}, \mathbf{J} \to \mathbf{J}' = -\mathbf{J}$, or equivalently

$$\phi \to \phi' = \phi, \ \mathbf{A} \to \mathbf{A}' = -\mathbf{A}.$$

Taking the complex conjugate of (7.6) and multiplying on the left by $\gamma^1\gamma^3$ gives

$$\gamma^1\gamma^3[\gamma^{\mu *}(-\mathrm{i}\partial_\mu - qA_\mu) - m]\psi^* = 0.$$

Now

$$\begin{aligned} \gamma^1\gamma^3\left(\gamma^0\right)^* &= \gamma^1\gamma^3\gamma^0 = \gamma^0\gamma^1\gamma^3, \\ \gamma^1\gamma^3\left(\gamma^i\right)^* &= -\gamma^i\gamma^1\gamma^3 \quad \text{for } i = 1, 2, 3, \end{aligned}$$

and the result follows.

Chapter 8

8.3 If an $\mathrm{e}^+\mathrm{e}^-$ pair is created there is a frame of reference (the centre of mass frame) in which the total momentum of the pair is zero. The photon would also have zero momentum in this frame and hence zero energy: energy conservation would be violated.

Chapter 9

9.1 Conservation of energy gives $m_\pi = E_\mathrm{e} + E_\mathrm{v}$. Conservation of momentum gives $p_\mathrm{e} = p_\nu$. Also

$$E_\mathrm{v} = p_\mathrm{v}, \ E_\mathrm{e}^2 = p_\mathrm{e}^2 + m_\mathrm{e}^2, \ \upsilon_\mathrm{e} = p_\mathrm{e}/E_\mathrm{e}.$$

Hence

$$(m_\pi - p_\mathrm{e})^2 = E_\mathrm{e}^2 = p_\mathrm{e}^2 + m_\mathrm{e}^2, \quad p_\mathrm{e} = \frac{m_\pi^2 - m_\mathrm{e}^2}{2m_\pi}.$$

Then

$$\begin{aligned} E_\mathrm{e} &= m_\pi - p_\mathrm{e} = \frac{m_\pi^2 + m_\mathrm{e}^2}{2m_\pi}, \\ \frac{1}{2}\left(1 - \frac{\upsilon}{c}\right) &= \frac{1}{2}\left(1 - \frac{m_\pi^2 - m_\mathrm{e}^2}{m_\pi^2 + m_\mathrm{e}^2}\right) = \frac{m_\mathrm{e}^2}{m_\pi^2 + m_\mathrm{e}^2}. \end{aligned}$$

9.2 Final energy $E = E_e + E_\nu = E_e + p_e$

$$\frac{dE}{dp_e} = \frac{dE_e}{dp_e} + 1 = \frac{P_e}{E_e} + 1 = \frac{E_e + P_e}{E_e} = \frac{M_\pi}{E_e}.$$

9.3 Using Problem (9.1),

$$\left(1 - \frac{v_e}{c}\right) p_e{}^2 E_e = \frac{m_e{}^2}{4m_\pi{}^3} = \left(m_\pi{}^2 - m_e{}^2\right)^2,$$

with a similar expression for the μ leptons.

9.4 Since the pion is at rest, only the term $\partial\Phi/\partial t$ contributes. From (3.35), there is a factor in $\mathcal{L}_{\text{int}}$ arising from this:

$$\frac{1}{\sqrt{V}} \frac{(-im_\pi)}{\sqrt{2m_\pi}} a_0.$$

From Problem 6.5, the $\bar{\nu}$ factor is

$$\frac{1}{\sqrt{V}} d^\dagger_{\mathbf{p}'} \mathrm{e}^{\mathrm{i}(-\mathbf{p}'\cdot\mathbf{r})} \left|-\right\rangle_{\mathbf{p}'}.$$

From (6.24), the $e_L^\dagger$ factor is

$$\frac{1}{\sqrt{V}} \sqrt{\frac{m_e}{E_p}} b^\dagger_p \mathrm{e}^{\mathrm{i}(-\mathbf{p}\cdot\mathbf{r})} \frac{1}{\sqrt{2}} e^{-\theta/2} \left\langle+\right|_{\mathbf{p}}.$$

(Only this helicity term contributes.)

Integrating over volume gives $\mathbf{p}' = -\mathbf{p}$ and a volume factor V, so that, for a given $\mathbf{p}$,

$$\left\langle e_{\mathbf{p}}, \bar{\nu}_{-\mathbf{p}} \left|V(0)\right| \pi^-\right\rangle = \frac{(-\mathrm{i})}{\sqrt{V}} \sqrt{\frac{m_\pi}{2}} \sqrt{\frac{m_e}{E_e}} \frac{\alpha_\pi}{\sqrt{2}} \mathrm{e}^{-\theta/2}.$$

(Note that $\left|-\right\rangle_{-\mathbf{p}} = \left|+\right\rangle_{\mathbf{p}}$.)

Hence the transition rate s is obtained. The factor 4π in the density of states comes from summing over all directions of $\mathbf{p}$. Also $(E_e/m_e) = \cosh\theta$ and $\mathrm{e}^{-\theta}/\cosh\theta = (1 - \tanh\theta) = (1 - v/c)$.

9.7

$$G_F \approx \left(\frac{192\pi^3}{\tau m_\mu{}^5}\right)^{1/2} = 1.164 \times 10^{-5} (\mathrm{GeV})^{-2}.$$

9.8 The square of the centre of mass energy

$$s = (E_e + E_\nu)^2 - (\mathbf{p}_e + \mathbf{p}_\nu)^2$$

is Lorentz invariant. In the electron's rest frame

$$s = (m_e + E_\nu)^2 - p_\nu^2 = m_e^2 + 2m_e E_\nu.$$

9.9 The expression (9.8) contains the term

$$-2\sqrt{2} G_F g_{\mu\nu} e_L^\dagger \tilde{\sigma}^\mu \nu_{eL} \nu_{eL}^\dagger \tilde{\sigma}^\nu e_L.$$

The expression (9.15) contains the term

$$\left(G_{\mathrm{F}}/\sqrt{2}\right)g_{\mu\nu}\nu^{\dagger}_{\mu\mathrm{L}}\tilde{\sigma}^{\mu}\nu_{\mu\mathrm{L}}\bar{\psi}_{e}\gamma^{\nu}(c_{v}-c_{A}\gamma^{5})\psi_{\mathrm{e}}.$$

9.10

$$\frac{\tau\,(\mathrm{K}\to\mu\bar{\nu}_{\mu})}{\tau\,(\mathrm{K}\to\mathrm{e}\bar{\nu}_{\mathrm{e}})}=\frac{m_{\mathrm{e}}^{2}(m_{\mathrm{K}}^{2}-m_{\mathrm{e}}^{2})^{2}}{m_{\mu}^{2}(m_{\mathrm{K}}^{2}-m_{\mu}^{2})^{2}}=2.57\times10^{-5}$$

$$\frac{1}{\tau\,(\mathrm{K}\to\mu\bar{\nu}_{\mu})}=\frac{\alpha_{\mathrm{K}}{}^{2}}{4\pi}\left(1-\frac{v_{\mu}}{c}\right)p_{\mu}{}^{2}E_{\mu}\qquad\text{(cf. (9.3))},$$

where

$$\left(1-\frac{v_{\mu}}{c}\right)p_{\mu}^{2}E_{\mu}=\frac{m_{\mu}{}^{2}}{4m_{\mathrm{K}}{}^{2}}(m_{K}{}^{2}-m_{\mu}{}^{2})^{2}$$

(cf. Problem 9.3).

This gives $\alpha_{\mathrm{K}}=5.82\times10^{-10}\ \mathrm{MeV}^{-1}$, and $\alpha_{\pi}=2.09\times10^{-9}$ (text), giving $\alpha_{\mathrm{K}}/\alpha_{\pi}=0.28$.

9.11 Consider the decay $\tau^{-}\to\pi^{-}+\nu_{\tau}$. The term in $\mathcal{L}_{\mathrm{int}}$ that generates the decay is

$$\nu^{\dagger}_{\tau\mathrm{L}}\tilde{\sigma}^{\mu}\tau_{\mathrm{L}}\partial_{\mu}\Phi^{\dagger}.$$

Consider the τ to be at rest with its spin aligned along the z-axis, and the neutrino momentum to be $\mathbf{p}$. The pion momentum is then $(-\mathbf{p})$, and the interaction energy contains a term

$$\frac{\alpha_{\pi}}{\sqrt{V}}\frac{\mathrm{i}}{\sqrt{2E_{\pi}}}a^{\dagger}_{\pi}(-\mathbf{p})\,b^{\dagger}_{\nu}(\mathbf{p})\,b_{\tau}(0)\,\langle-|_{\mathrm{p}}\left(\sigma^{0}E_{\pi}-\boldsymbol{\sigma}\cdot\mathbf{p}\right)\frac{1}{\sqrt{2}}\begin{pmatrix}1\\0\end{pmatrix}.$$

Now $\langle-|_{\mathbf{p}}(\sigma^{0}E_{\pi}-\boldsymbol{\sigma}\cdot\mathbf{p})=\langle-|_{\mathbf{p}}(E_{\pi}+p_{\nu})=\langle-|_{\mathbf{p}}\,m_{\tau}$, and from Problem 6.5, $\langle-|_{\mathbf{p}}=(-\sin(\theta/2)\mathrm{e}^{\mathrm{i}\phi},\ \cos(\theta/2))$ where θ and ϕ are the polar angles of $\mathbf{p}$. Hence

$$\langle\pi_{-\mathbf{p}},\nu_{\mathbf{p}}\,|V|\,\tau\rangle=-\frac{\alpha_{\pi}}{\sqrt{V}}\frac{\mathrm{i}}{\sqrt{2E_{\pi}}}m_{\tau}\frac{1}{\sqrt{2}}\sin(\theta/2)\,\mathrm{e}^{\mathrm{i}\phi}.$$

The decay rate is

$$\frac{1}{\tau}=2\pi\int|\langle f|\,V\,|i\rangle|^{2}\,p\,(m_{\tau})\,\mathrm{d}\Omega$$

where

$$p\,(m_{\tau})=\frac{V}{(2\pi)^{3}}\frac{\left(m_{\tau}{}^{2}-m_{\pi}{}^{2}\right)^{2}}{4m_{\tau}{}^{2}}\frac{E_{\pi}}{m_{\tau}},$$

and the angular integration gives a factor 2π.

Chapter 10

10.1 The term $-(m^{2}/2\phi_{0}{}^{2})\sqrt{2}\phi_{0}\chi\psi^{2}$ links the χ and ψ fields, and $m=m_{\chi}/\sqrt{2}$. Since the ψ particles are massless, the final energy $E=2p$, and the density of states factor for the decay is

$$\rho(E) = \frac{V}{(2\pi)^3} 4\pi p^2 \frac{dp}{dE} \quad \text{where} \quad \frac{dp}{dE} = \frac{1}{2},$$

and the factor 4π comes from the angular integration.

In the matrix element $\langle \mathbf{p}, -\mathbf{p}|V|\chi \text{ at rest}\rangle$, the χ field gives a factor $1/\sqrt{2m_\chi}$ from the expansion (3.21), and each of the ψ fields gives a factor $1/\sqrt{2p}$. Hence

$$2\pi|\langle \mathbf{p}|V|i\rangle|^2\rho(E) = 2\pi \frac{m_\chi{}^4}{8\phi_0{}^2} \frac{1}{2m_\chi} \frac{1}{4p^2} \frac{4\pi p^2}{(2\pi)^3} \frac{1}{2}$$
$$= \frac{m_\chi}{128\pi}\left(\frac{m_\chi}{\phi_0}\right)^2.$$

10.2 The decay of an isolated vector boson requires a term in $\mathcal{L}_{\text{int}}$ linear in A_μ. There is a term $(\sqrt{2}\phi_0 q^2)A_\mu A^\mu h$ that allows the decay of the scalar boson if energy conservation can be satisfied, i.e. $m_h = \sqrt{2}m > 2\left(\sqrt{2}q\phi_0\right)$.

Chapter 11

11.1 The term $\mathbf{UWU}^\dagger$ satisfies $(\mathbf{UWU}^\dagger)^\dagger = \mathbf{UWU}^\dagger$ and $\text{Tr}(\mathbf{UWU}^\dagger) = \text{Tr}(\mathbf{U}^\dagger\mathbf{UW}) = \text{Tr}(\mathbf{W}) = 0$.

Noting that $(\hat{\boldsymbol{\alpha}} \cdot \boldsymbol{\tau})^2 = \mathbf{I}$ and $(\partial_\mu\alpha^j)\alpha^j = 0$ since $\alpha^j\alpha^j = 1$, the term $(2i/g_2)(\partial_\mu\mathbf{U})\mathbf{U}^\dagger$ may be written as a linear combination of the matrices τ^j with real coefficients. Each τ^j is Hermitian and has zero trace.

11.3 The last term may be written as $(g_2{}^2\phi_0{}^2/4)(W_\mu{}^1W^{1\mu} + W_\mu{}^2W^{2\mu})$, and in the absence of electromagnetic fields the term that precedes it can be handled similarly. There are therefore two independent fields each with mass $g_2\phi_0/\sqrt{2}$ (cf. Section 4.9).

11.4 The interaction Lagrangian density (11.32) contains a term $g_2{}^2/\sqrt{2})hW_\mu^- W^{+\mu}$ coupling the h field and the charged W fields.

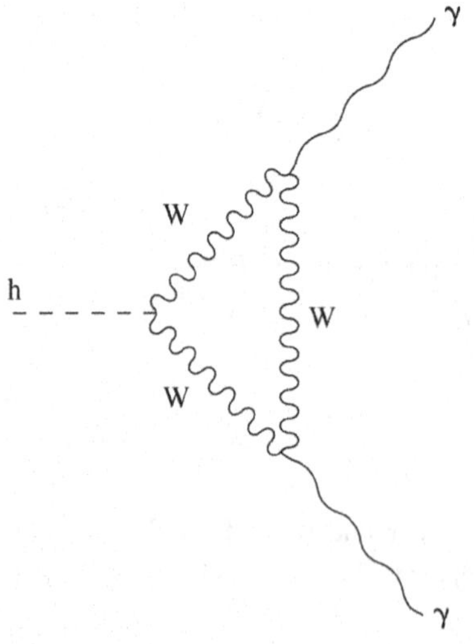

11.5 Consider

$$\mathbf{U} = \cos\alpha\mathbf{I} + \mathrm{i}\sin\alpha\boldsymbol{\tau}\cdot\hat{\boldsymbol{\alpha}}\ (\text{see B.9}).$$

Then

$$\mathbf{U}^* = \cos\alpha\mathbf{I} - \mathrm{i}\sin\alpha(\tau^1\hat{\alpha}^1 - \tau^2\hat{\alpha}^2 + \tau^3\hat{\alpha}^3)$$

and

$$\tau^2\mathbf{U}^* = [\cos\alpha\mathbf{I} + \mathrm{i}\sin\alpha(\tau^1\hat{\alpha}^1 + \tau^2\hat{\alpha}^2 + \tau^3\hat{\alpha}^3)]\tau^2$$

using

$$\tau^2\tau^1 = -\tau^1\tau^2,\ \tau^2\tau^3 = -\tau^3\tau^2.$$

Hence

$$\mathrm{i}\tau^2\mathbf{U}^* = \mathbf{U}(\mathrm{i}\tau^2) \quad \text{and} \quad \mathrm{i}\tau^2 = \begin{pmatrix} 0 & 1 \\ -1 & 0 \end{pmatrix}.$$

The result follows.

11.6 Using (B.9).

$$\begin{aligned}\mathbf{U} &= \cos\alpha\mathbf{I} + \sin\alpha(\sin\phi\tau^1 + \cos\phi\tau^2)\\ &= \begin{pmatrix} \cos\alpha & \mathrm{i}\sin\alpha(\sin\phi - \mathrm{i}\cos\phi) \\ \mathrm{i}\sin\alpha(\sin\phi + \mathrm{i}\cos\phi) & \cos\alpha \end{pmatrix}.\end{aligned}$$

Chapter 12

12.2 Take the two fields to be

$$\mathbf{L} = \begin{pmatrix} L_1 \\ L_2 \end{pmatrix}.$$

To maintain local gauge invariance, the dynamical term in the Lagrangian density must be $\mathbf{L}^\dagger\tilde{\sigma}^\mu\mathrm{i}(\partial_\mu + \mathrm{i}(g_2/2)\mathbf{W}_\mu)\mathbf{L}$.
There are terms which mix L_1 and L_2, for example,

$$\begin{aligned}&-(g_2/2)L_1{}^\dagger\tilde{\sigma}^\mu(W_\mu{}^1 - \mathrm{i}W_\mu{}^2)L_2\\ &= -(g_2/2)L_1{}^\dagger\tilde{\sigma}^\mu L_2 W_\mu{}^\dagger.\end{aligned}$$

The operator $W_\mu{}^\dagger$ destroys electric charge e, so that to conserve charge $L_1{}^\dagger\tilde{\sigma}^\mu L_2$, must create charge e.

12.3 The Higgs particle at rest has zero momentum and zero angular momentum. Hence the e^+ and e^- have opposite momentum. If they had opposite helicities, they would have to carry orbital angular momentum with a component $+1$ or -1 along their direction of motion, to conserve angular momentum. This is not possible since $\mathbf{p}\cdot(\mathbf{r}\times\mathbf{p}) = 0$.

The final density of momentum states is

$$\rho(E) = \frac{V}{(2\pi)^3}4\pi p_e{}^2\frac{\mathrm{d}p_e}{\mathrm{d}E}.$$

The final energy $E = 2E_e$,where $E_e^2 = m_e^2 p_e^2$. Hence

$$\frac{dp_e}{dE} = \frac{1}{2}\frac{dp_e}{dE_e} = \frac{E_e}{2p_e}, \quad \text{and} \quad p(E) = \frac{V}{(2\pi)^2} p_e E_e.$$

The interaction term in (12.9) is $-(c_e\sqrt{2})h\bar{\psi}\psi$. From (6.24) and (3.21), this gives

$$\langle f|V|i\rangle = \frac{1}{\sqrt{V}}\frac{1}{\sqrt{2m_H}}\frac{m_e}{E_e}[\bar{\mu}_+(\mathbf{p})v_+(-\mathbf{p})]$$

or

$$[\bar{\mu}_-(\mathbf{p})v_-(-\mathbf{p})].$$

Now $\bar{\mu}_\pm(\mathbf{p})v_\pm(-\mathbf{p}) = \sinh\theta$, and $E_e/m_e = \cosh\theta$. Hence the decay rate to positive helicities is

$$2\pi|\langle f|V|i\rangle|^2\rho(E) = 2\pi\frac{c_e{}^2}{2}\frac{1}{2m_H}\tanh^2\theta\frac{1}{(2\pi)^2}p_e E_e.$$

Also $\tan\theta = v_e/c = p_e/E_e$ and $E_e = m_H/2$. The decay rate to negative helicities is the same, and the result follows.

12.4 Since $c_\tau > c_\mu > c_e$ (see (12.13)) the decay to $\tau^+\tau^-$ dominates in the leptonic partial width. Also, since the Higgs mass is much greater than the τ mass, $v_\tau \approx c$. Hence

$$\frac{\Gamma}{m_H} \approx \frac{c_\tau^2}{16\pi} = \frac{1}{16\pi}\left(\frac{m_\tau}{\phi_0}\right)^2.$$

Chapter 13

13.1 In the rest frame of the W, and neglecting the lepton mass, $\mathbf{p}_l = -\mathbf{p}_v$, $E_l = p_l = M_w/2$, and $p_l{}^2 = M_w{}^2/4 = p_x{}^2 + p_y{}^2 + p_z{}^2$. Taking the x-axis to be the beam direction, the mean square transverse momentum is

$$\overline{p_x{}^2} + \overline{p_y{}^2} = (2/3)p_l{}^2 = M_w{}^2/6.$$

13.2 From (12.23), the Z_μ is produced by right-handed electron fields with a coupling $e\tan\theta_w = 2e\sin^2\theta_w/\sin(2\theta_w)$ and by left-handed fields with a coupling $-e\cos(2\theta_w)/\sin(2\theta_w)$. In head-on collisions at high energies the right-handed component of the electron (positron) has positive (negative) helicity. Hence the total spin is $+1$ along the electron beam direction. The spin of the left-handed components is opposite. For unpolarised beams the left-handed and right-handed components are equally populated, and the result follows.

13.3 Consider the decay $W^- \to e^- + \bar{\nu}_e$ in the W^- rest frame. With no loss of generality we may take the W^- to have $J = 1$, $J_z = 0$ (see Section 4.9). The interaction Lagrangian density responsible for the decay is (from (12.15) and (12.16))

$$\mathcal{L} = -(g_2/\sqrt{2})j^3 W_3^-.$$

If the electron has momentum **p**, the neutrino has momentum $-\mathbf{p}$. Neglecting the electron mass (see Problem 6.5) the matrix element for the decay is

$$\langle f|\,V\,|i\rangle = \frac{g_2}{\sqrt{2}}\frac{1}{\sqrt{2M_\mathrm{w}V}}\langle -|\,\sigma^3\,|+\rangle\,.$$

(Recall $\boldsymbol{\sigma}\cdot\mathbf{p}\,|-\rangle = -\,|-\rangle\,, \boldsymbol{\sigma}\cdot(-\mathbf{p})\,|+\rangle = -\,|+\rangle$.) Also, from Problem 6.6, $\langle -|\,\sigma^3\,|+\rangle = -\sin\theta \mathrm{e}^{\mathrm{i}\phi}$. The decay rate is

$$\Gamma = 2\pi\int |\,\langle f|\,V\,|i\rangle\,|^2 \mathrm{d}\Omega\frac{V}{(2\pi)}{}^3 p_e{}^2\frac{\mathrm{d}p_\mathrm{e}}{\mathrm{d}E}$$

where $\mathrm{d}p_\mathrm{e}/\mathrm{d}E = 1/2$, $p_\mathrm{e} = M_\mathrm{w}/2$, giving

$$\Gamma = \frac{g_2{}^2}{48\pi}M_\mathrm{w} = \frac{G_\mathrm{F}M_\mathrm{w}{}^3}{6\pi\sqrt{2}},\ \text{by (12.22)}.$$

The decay rate for $Z\to\nu\bar{\nu}$ requires a similar calculation, with M_w replaced by M_z and the coupling constant $g_2/\sqrt{2}$ replaced by $e/\sin 2\theta_\mathrm{w} = g_2/2\cos\theta_\mathrm{w} = g^2M_\mathrm{z}/2M_\mathrm{w}$. (We have used (12.23), (11.38) and (11.37a).) Then

$$\Gamma(Z\to\nu\bar{\nu}) = \frac{G_\mathrm{F}M_\mathrm{z}^3}{12\pi\sqrt{2}}.$$

There are two terms in (12.23) contributing to $\Gamma(Z\to \mathrm{e}^+\mathrm{e}^-)$, yielding

$$\Gamma(Z\to e^+e^-) = \Gamma(Z\to\nu\bar{\nu})[(2\sin^2\theta_\mathrm{w})^2 + (\cos 2\theta_\mathrm{w})^2].$$

13.4 83.86 MeV.

Chapter 14

14.3 Under an $SU(2)$ transformation, and from Appendix A.2

$$(\Phi^\mathrm{T}\varepsilon L)\to(\Phi^\mathrm{T}U^\mathrm{T}\varepsilon UL)$$

$$\begin{aligned} U^\mathrm{T}\varepsilon U &= \begin{bmatrix} U_{AA} & U_{BA}\\ U_{AB} & U_{BB}\end{bmatrix}\begin{bmatrix} 0 & 1\\ -1 & 0\end{bmatrix}\begin{bmatrix} U_{AA} & U_{AB}\\ U_{BA} & U_{BB}\end{bmatrix} = \begin{bmatrix} 0 & \mathrm{Det}(U)\\ -\mathrm{Det}(U) & 0\end{bmatrix}\\ &= (\mathrm{Det}(U))\varepsilon\\ &= \varepsilon,\ \text{since Det(U)} = 1.\ \text{Hence}\,(\mathbf{\Phi}^\mathrm{T}U^\mathrm{T}\varepsilon UL) = (\mathbf{\Phi}^\mathrm{T}\varepsilon L)\end{aligned}$$

14.4 From (11.23),

$$\mathbf{\Phi} = \begin{pmatrix} 0\\ \phi_0 + h/\sqrt{2}\end{pmatrix}.$$

Inserting this in (14.6) gives the coupling terms

$$-(1/\sqrt{2})\sum[G^d_{ij}d^\dagger_{\mathrm{L}i}d_{\mathrm{R}j}h + \text{Hermitian conjugate}.$$

Similar terms arise from (14.9) and (14.10). Using the true quark masses these become

$$-(1/\sqrt{2}\phi_0)\sum[m^d_i(d^\dagger_{\mathrm{L}i}d_{\mathrm{R}i} + d^\dagger_{\mathrm{R}i}d_{\mathrm{L}i}) + m^u_i(u^\dagger_{\mathrm{L}i}u_{\mathrm{R}i} + u^\dagger_{\mathrm{R}i}u_{\mathrm{L}i})]h.$$

The coupling to the top quark is

$$c_t = \frac{m_t}{\sqrt{2}\phi_0} \approx \frac{180\,\text{GeV}}{\sqrt{2} \times 180\,\text{GeV}} \approx 0.7.$$

14.5 For $\text{K}^+ \to \mu^+ + \nu_\mu$, the terms

$$s_{\text{L}}{}^{\dagger}\tilde{\sigma}^{\mu}u_{\text{L}}V_{us}^{*} \text{ from } j^{\mu}, \ \nu_{\mu\text{L}}{}^{\dagger}\tilde{\sigma}^{\mu}\mu_{\text{L}} \text{ from } j^{\mu\dagger}$$

contribute in the second order of perturbation theory. (See (a).)

(a)

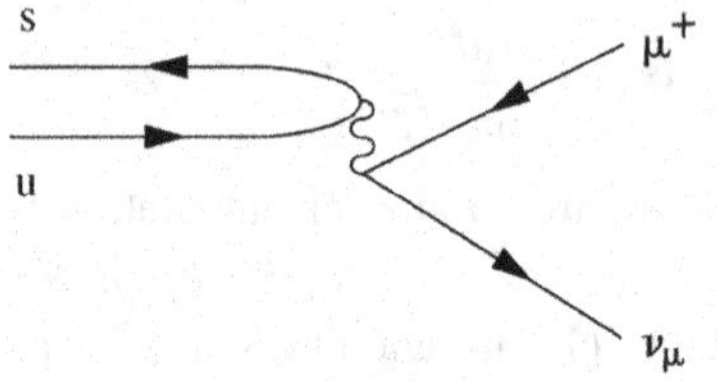

(b)

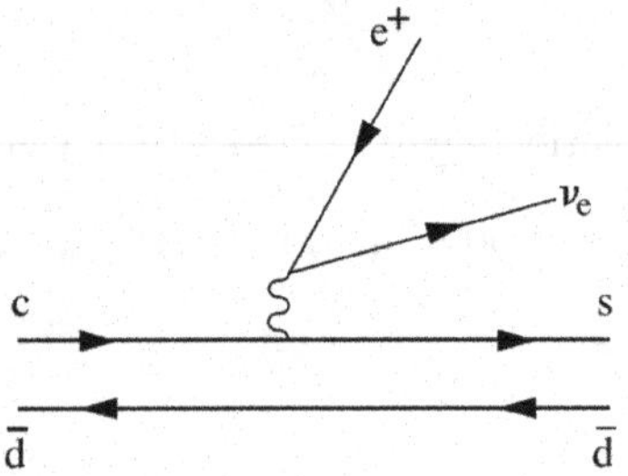

(c)

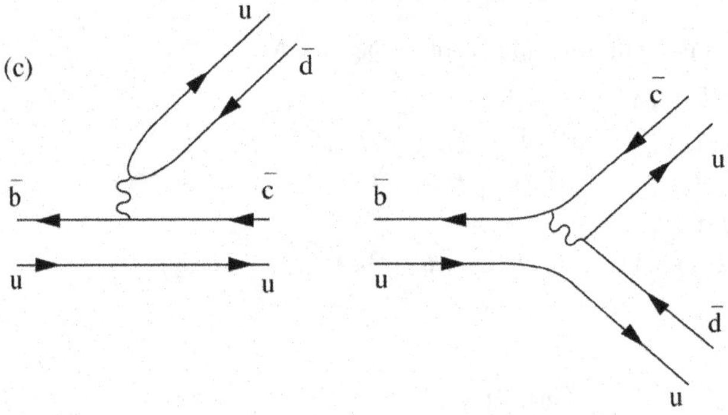

For $\text{D}^+ \to K^0 + e^+ + \nu_{\text{e}}$,

$$s_{\text{L}}^{\dagger}\tilde{\sigma}^{\mu}c_{\text{L}}V_{cs}^{*} \text{ from } j^{\mu}, \ \nu_{\text{eL}}^{\dagger}\tilde{\sigma}^{\mu}e_{\text{L}} \text{ from } j^{\mu\dagger}. \text{ (See (b).)}$$

For $\text{B}^{\dagger} \to \bar{\text{D}}^0 + \pi^{\dagger}$,

$$b_{\text{L}}^{\dagger}\tilde{\sigma}^{\mu}c_{\text{L}}V_{cb}^{*} \text{ from } j^{\mu}, \ u_{\text{L}}^{\dagger}\tilde{\sigma}^{\mu}d_{\text{L}}V_{ud} \text{ from } j^{\mu\dagger}. \text{ (See (c).)}$$

14.6

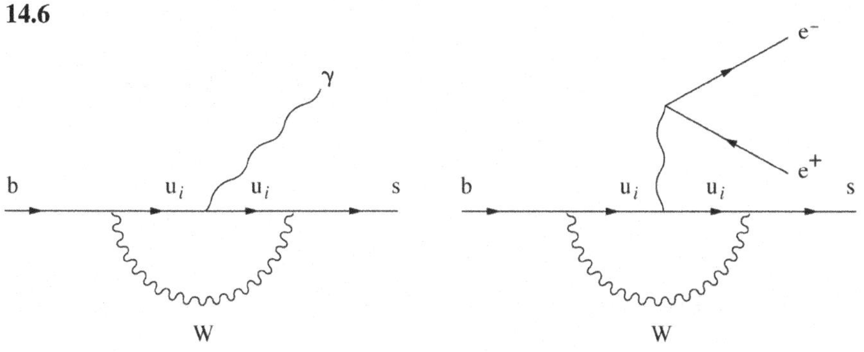

The quark labelled u_i can be u, c or t.

Chapter 15

15.1 The decay rate for $Z \to d\bar{d}$ of (15.3) can be compared with the decay rate for $Z \to e^+e^-$ of (13.3), calculated in the answer to Problem 13.3. Comparing the interaction Lagrangian densities (12.23) and (14.14), the term in the left-handed coupling $\cos 2\theta_W = 1 - 2\sin^2\theta_W$ is replaced by $(1 - (2/3)\sin^2\theta_W)$, and in the right-handed coupling $2\sin^2\theta_W$ is replaced by $(2/3)\sin^2\theta_W$. Including a colour factor of 3 and replacing $\sin^2\theta_W$ by $(1/3)\sin^2\theta_W$ in the rate (13.3) gives the rate (15.3).

Similarly for $Z \to u\bar{u}$. Comparing (12.23) with (14.14), $\sin^2\theta_W$ is replaced by $(2/3)\sin^2\theta_W$.

The decay rate $W^+ \to u_i\bar{d}_j$ of (15.6) can be compared with the rate $W^+ \to e^+\nu_e$ of (13.2) calculated in the answer to Problem 13.3. Comparing the interactions (12.18) and (14.20), $g_2/\sqrt{2}$ is replaced by $eV_{ij}/\sqrt{2}\sin\theta_W = g_2V_{ij}/\sqrt{2}$. Including the colour factor of 3, the rate (15.6) follows from the rate (13.2).

Chapter 16

16.1

$$\begin{aligned} \mathrm{G}_{\mu\nu} &= \partial_\mu \mathbf{G}_\nu - \partial_\nu \mathbf{G}_\mu + \mathrm{i}g(\mathbf{G}_\mu \mathbf{G}_\nu - \mathbf{G}_\nu \mathbf{G}_\mu) \\ &= (\partial_\mu \mathbf{G}^a_\nu - \partial_\nu \mathbf{G}^a_\mu)(\lambda_a/2) \\ &\quad + \mathrm{i}(g/4)\Big(G^b_\mu G^c_\nu \lambda_b\lambda_c - G^c_\nu G^b_\mu \lambda_c\lambda_b\Big), \end{aligned}$$

and

$$(\lambda_b\lambda_c - \lambda_c\lambda_b) = 2\mathrm{i} f_{bca}\lambda_a \quad \text{(see (B.27))}.$$

Hence

$$G_{\mu\nu} = [(\partial_\mu G^a_\nu - \partial_\nu G^a_\mu) - g f_{abc} G^b_\mu G^c_\nu](\lambda_a/2).$$

16.2 These are the terms in (16.9) cubic and quadratic in the G fields.

16.3 Variation of G^a_ν gives

$$\delta S = \int \left[-(1/2)G^{a\mu\nu}\delta G^a_{\mu\nu} - g\sum_f \bar{\mathbf{q}}_f\gamma^\nu \delta G^a_\nu(\lambda_a/2)\mathbf{q}_f \right] d^4x,$$

and

$$-(1/2)G^{a\mu\nu}\delta G^a_{\mu\nu} = -G^{a\mu\nu}\partial_\mu(\delta G^a_\nu) + gG^{c\mu\nu}G^b_\mu\delta G^a_\nu f_{cba}.$$

(There are two equal contributions to the right-hand side.) Integrating by parts gives

$$\delta S = \int \left[\partial_\mu G^{a\mu\nu} - gG^{c\mu\nu}G^b_\mu f_{abc} - g\sum_f \bar{\mathbf{q}}_f\gamma^\nu(\lambda_a/2)\mathbf{q}_f \right] \delta G^a_\nu \, d^4x$$

$(f_{cba} = -f_{abc})$.

Since the δG^a_ν are arbitrary (16.14) is obtained.

16.4

$$Q^2/4m^2 = e^{12x^2/e^2} = e^{3\pi/\alpha} = 10^{560}.$$
$$2m \sim 1\,\text{MeV},\ Q^2 \sim 10^{560}\,(\text{MeV})^2.$$

16.5 Take $\mathbf{Q}\cdot\mathbf{r} = Qr\cos\theta$ and $d^3\mathbf{Q} = Q^2 dQ\, d(\cos\theta)d\phi$ where (Q, θ, ϕ) are the polar coordinates of $\mathbf{Q}$, with $\mathbf{r}$ taken to be (0, 0, r).

Chapter 18

18.1

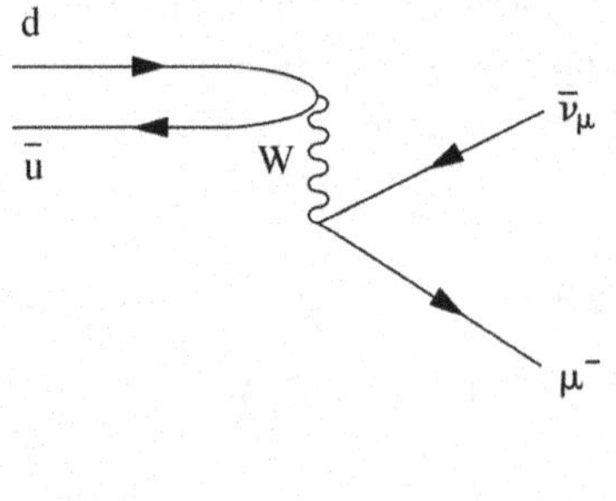

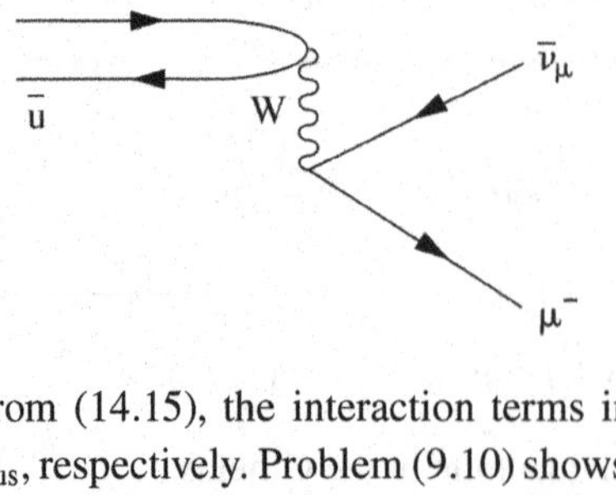

From (14.15), the interaction terms in $\bar{u}dW^+$ and $\bar{u}sW^+$ contain factors V_{ud} and V_{us}, respectively. Problem (9.10) shows $\alpha_K/\alpha_\pi \approx 0.28$. Setting this equal to V_{us}/V_{ud} gives $\sin\theta_{12} \approx 0.27$.

18.2 The internal wave function of two pions at $\mathbf{r}_1$ and $\mathbf{r}_2$ in an S state is a function of only $|\mathbf{r}_1-\mathbf{r}_2|$ and $|\mathbf{r}_1-\mathbf{r}_2|$ is invariant under both C and P. Hence

$$CP\left|\pi^0\pi^0\right\rangle = \left|\pi^0\pi^0\right\rangle \quad \text{and} \quad CP\left|\pi^+\pi^-\right\rangle = \left|\pi^+\pi^-\right\rangle.$$

18.3 The internal wave function of three pions at $\mathbf{r}_1$, $\mathbf{r}_2$, $\mathbf{r}_3$, depends only on two relative coordinates, say $\mathbf{r}_{12} = \mathbf{r}_2 - \mathbf{r}_1$ and $\mathbf{r}_{23} = \mathbf{r}_3 - \mathbf{r}_2$. To be invariant under rotations ($J=0$) the internal wave function can be a function of only three scalars: $\mathbf{r}_{12}\cdot\mathbf{r}_{12}$, $\mathbf{r}_{12}\cdot\mathbf{r}_{23}$, and $\mathbf{r}_{23}\cdot\mathbf{r}_{23}$. These are invariant under C and P. Since the intrinsic parity of the π^0 is negative,

$$CP\left|\pi^0\pi^0\pi^0\right\rangle = -\left|\pi^0\pi^0\pi^0\right\rangle.$$

18.4 The area of the triangle formed by the origin and the points $\mathbf{r}_1 = (x_1, y_1, 0)$ and $\mathbf{r}_2 = (x_2, y_2, 0)$ is

$$\begin{aligned}(1/2)|\mathbf{r}_1\times\mathbf{r}_2| &= (1/2)|x_1y_2 - x_2y_2)| \\ &= (1/2)|\mathrm{Im}(z_1^* z_2)|,\end{aligned}$$

where $z_1 = x_1 + \mathrm{i}y_1$, $z_2 = x_2 + \mathrm{i}y_2$. Hence the area of the unitary triangle is

$$(1/2)|\mathrm{Im}(V_{ud}^* V_{ub} V_{cd} V_{cb}^*)| = J/2.$$

18.5 All the complex numbers z_i are transformed to $z_i^! = \mathrm{e}^{\mathrm{i}(\theta_d-\theta_b)} z_i$ and the triangle is rotated through an angle $(\theta_d - \theta_b)$.

Chapter 19

19.2 (a) $(U_{\beta j}^* U_{\alpha j} U_{\beta i} U_{\alpha i}^*) = (U_{\beta i}^* U_{\alpha i} U_{\beta j} U_{\alpha j}^*)^*$ hence
$\mathrm{Im}(U_{\beta j}^* U_{\alpha j} U_{\beta i} U_{\alpha i}^*) = -\mathrm{Im}(U_{\beta i}^* U_{\alpha i} U_{\beta j} U_{\alpha j}^*)$.

(b) Since $\mathbf{U}$ is unitary,
$\sum_i F_{\beta\alpha ij} = \mathrm{Im}(\partial_{\alpha\beta} U_{\beta j} U_{\alpha j}^*) = \mathrm{Im}(|U_{\alpha j}|^2) = 0.$
As two examples $F_{\beta\alpha 12} + F_{\beta\alpha 32} = 0$ and $F_{\beta\alpha 13} + F_{\beta\alpha 23} = 0$.
Hence $F_{\beta\alpha 12} + F_{\beta\alpha 23} = F_{\beta\alpha 31}$.

(c)

$$\sum_{i>j} F_{\mu e ij}\sin\left(\frac{\Delta m_{ij}^2 L}{2E}\right) = -J\left[\sin\left(\frac{\Delta m_{21}^2 L}{2E}\right) + \sin\left(\frac{\Delta m_{32}^2 L}{2E}\right) - \sin\left(\frac{(\Delta m_{21}^2 + \Delta m_{32}^2)L}{2E}\right)\right]$$

and the result follows.

Chapter 21

21.1 Let $\left(\mathrm{i}\sigma^2\nu^*\right)^\dagger \sigma^\mu \partial_\mu \left(\mathrm{i}\sigma^2\nu^*\right) = E$
Inserting explicit spinor indices

$$E = \nu_i \sigma_{ij}^2 \sigma_{jk}^\mu \sigma_{kl}^2 \partial_\mu \nu_l^*, \quad \text{(repeated indices summed).}$$

But from the algebra of Pauli matrices $\sigma^2_{ij}\sigma^{\mu}_{jk}\sigma^2_{kl} = \tilde{\sigma}^{\mu}_{li}$. Taking account of the anticommuting spinor fields $E = -\partial_\mu \nu^*_l \tilde{\sigma}^{\mu}_{li} \nu_i$. and discarding a total derivative that makes no contribution to the action

$$E = \nu^*_l \tilde{\sigma}^{\mu}_{li} \partial_\mu \nu_i = \nu^\dagger \tilde{\sigma}^\mu \partial_\mu \nu.$$

21.2 Inserting explicit spinor indices

$$\nu^{\mathrm{T}}_\alpha \sigma^2 \nu_\beta = \nu_{\alpha i}\sigma^2_{ij}\nu_{\beta j} = -\nu_{\alpha i}\sigma^2_{ji}\nu_{\beta j} = \nu_{\beta j}\sigma^2_{ji}\nu_{\alpha i} = \nu^{\mathrm{T}}_\beta \sigma^2 \nu_\alpha.$$

21.3 From (21.15)

$$U^M_{\beta j} U^{M*}_{\alpha j} = U^D_{\beta j} e^{i\Delta j} U^{D*}_{\alpha j} e^{-i\Delta j} = U^D_{\beta j} U^{D*}_{\alpha j}.$$

Appendix A

A.1 The equation holds for $\alpha\beta \ldots \nu = 1, 2, \ldots, n$. Interchanging, say, α and β is equivalent to interchanging column i with column j, and gives the same sign change.

A.3 $\mathbf{M} = (\mathbf{M} + \mathbf{M}^\dagger)/2 + i(\mathbf{M} - \mathbf{M}^\dagger)/2i$. $(\mathbf{M} + \mathbf{M}^\dagger)/2$ is Hermitian, as is $(\mathbf{M} - \mathbf{M}^\dagger)/2i$. $\mathbf{A}$ and $\mathbf{B}$, and hence $\mathbf{M}$, can be diagonalised by the same transformation if and only if

$$\mathbf{AB} - \mathbf{BA} = \mathbf{0}, \text{ i.e. } (\mathbf{M} + \mathbf{M}^\dagger)(\mathbf{M} - \mathbf{M}^\dagger) - (\mathbf{M} - \mathbf{M}^\dagger)(\mathbf{M} + \mathbf{M}^\dagger) = \mathbf{0}$$

or

$$\mathbf{M}^\dagger\mathbf{M} - \mathbf{M}\mathbf{M}^\dagger = \mathbf{0}.$$

(This condition is satisfied if $\mathbf{M}$ is unitary.)

A.4 Since $(\mathbf{MM}^\dagger)^\dagger = \mathbf{MM}^\dagger$, we can find $\mathbf{U}_1$ such that $\mathbf{U}_1(\mathbf{MM}^\dagger)\mathbf{U}_1^\dagger = \mathbf{M}_D{}^2$. $\mathbf{M}_D{}^2$ has diagonal elements ≥ 0, since $\mathbf{M}_D{}^2 = \mathbf{U}_1\mathbf{M}(\mathbf{U}_1\mathbf{M})^\dagger$. Thus we can choose $\mathbf{M}_D$ with real diagonal elements ≥ 0. If none are zero, $\mathbf{M}_D$ can be inverted. We may then define

$$\mathbf{H} = \mathbf{U}_1{}^\dagger\mathbf{M}_D\mathbf{U}_1 = \mathbf{H}^\dagger, \quad \text{and} \quad \mathbf{V} = \mathbf{H}^{-1}\mathbf{M}.$$

Hence

$$\begin{aligned}
\mathbf{VV}^\dagger &= \mathbf{H}^{-1}\mathbf{MM}^\dagger\mathbf{H}^{-1} \text{ since } (\mathbf{H}^{-1})^\dagger = \mathbf{H}^{-1} \\
&= \mathbf{H}^{-1}\mathbf{U}_1{}^\dagger\mathbf{M}_D{}^2\mathbf{U}_1\mathbf{H}^{-1} \\
&= \mathbf{U}_1{}^\dagger\mathbf{M}_D{}^{-1}\mathbf{U}_1\mathbf{U}_1{}^\dagger\mathbf{M}_D{}^2\mathbf{U}_1\mathbf{U}_1{}^\dagger\mathbf{M}_D{}^{-1}\mathbf{U}_1 \\
&= \mathbf{I}, \text{ since } \mathrm{U}_1\mathrm{U}_1{}^\dagger = \mathbf{I}.
\end{aligned}$$

Thus $\mathbf{V}$ is unitary, as is $\mathbf{U}_1\mathbf{V} = \mathbf{U}_2$.
Finally, $\mathbf{M} = \mathbf{HV} = \mathbf{U}_1{}^\dagger\mathbf{M}_D\mathbf{U}_1\mathbf{V} = \mathbf{U}_1^\dagger\mathbf{M}_D\mathbf{U}_2$.

Appendix B

B.1 A unitary transformation, $\mathbf{H} \to \mathbf{H}' = \mathbf{VHV}^\dagger = \mathbf{H}_\mathrm{D}$, say, also diagonalises each term of $\mathbf{U}$ and hence

$$\mathbf{U} \to \mathbf{U}' = \mathbf{VUV}^\dagger = \mathbf{U}_\mathrm{D} = \exp(i\mathbf{H}_\mathrm{D}).$$

$$\det \mathbf{U} = \det \mathbf{U}_\mathrm{D} = \prod_n \exp \mathrm{i}(\mathbf{H}_\mathrm{D})_{nn}$$
$$= \exp\left[\mathrm{i}\sum_n (\mathbf{H}_\mathrm{D})_{nn}\right] = \exp[\mathrm{iTr}\, \mathbf{H}_\mathrm{D}].$$

But $\mathrm{Tr}\mathbf{H}_D = \mathrm{Tr}\mathbf{H}$. Hence if $\mathrm{Tr}\,\mathbf{H} = 0$, $\det \mathbf{U} = 1$.

B.2 The $SU(2)$ matrices corresponding to $R_{01}(\theta)$ and $R_{02}(\theta)$ are respectively

$$\begin{pmatrix} \cos(\theta/2) & \mathrm{i}\sin(\theta/2) \\ \mathrm{i}\sin(\theta/2) & \cos(\theta/2) \end{pmatrix} \quad \text{and} \quad \begin{pmatrix} \cos(\theta/2) & \sin(\theta/2) \\ -\sin(\theta/2) & \cos(\theta/2) \end{pmatrix}$$

and the correspondence can be checked directly.

B.3 From equation (B.5), using (B.12) and Problem B.2, $R(\psi, \theta, \phi)$ corresponds to the product

$$\begin{pmatrix} \mathrm{e}^{\mathrm{i}\psi/2} & 0 \\ 0 & \mathrm{e}^{-\mathrm{i}\psi/2} \end{pmatrix} \begin{pmatrix} \cos(\theta/2) & \sin(\theta/2) \\ -\sin(\theta/2) & \cos(\theta/2) \end{pmatrix} \begin{pmatrix} \mathrm{e}^{\mathrm{i}\phi/2} & 0 \\ 0 & \mathrm{e}^{-\mathrm{i}\phi/2} \end{pmatrix}.$$

B.4 Under a Lorentz transformation, $\mathbf{l} \to \mathbf{l}' = \mathbf{Ml}$, $\mathbf{r} \to \mathbf{r}' = \mathbf{Nr}$.
Hence

$$\begin{aligned} \mathbf{l}^\dagger \tilde{\sigma}^\mu \sigma^\nu \mathbf{r} &\to \mathbf{l}^\dagger \mathbf{M}^\dagger \tilde{\sigma}^\mu \sigma^\nu \mathbf{N}\mathbf{r} \\ &= \mathbf{l}^\dagger \mathbf{M}^\dagger \tilde{\sigma}^\mu \mathbf{M}\mathbf{N}^\dagger \sigma^\nu \mathbf{N}\mathbf{r} \quad \text{since } \mathrm{MN}^\dagger = \mathbf{I} \\ &= \mathbf{l}^\dagger L^\mu{}_\lambda \tilde{\sigma}^\lambda \mathbf{L}^\nu{}_\rho \sigma^\rho \mathbf{r} \quad \text{from (B.17) and (B.18)} \\ &= L^\mu{}_\lambda L^\nu{}_\rho (\mathbf{l}^\dagger \tilde{\sigma}^\lambda \sigma^\rho \mathbf{r}). \end{aligned}$$

It is easy to verify that

$$\tilde{\sigma}^\mu \sigma^\nu + \tilde{\sigma}^\nu \sigma^\mu = \begin{cases} 0 & \text{if } \mu \neq \nu, \\ 2 & \text{if } \mu = \nu = 0, \\ -2 & \text{if } \mu\nu = i;\ i = 1, 2, 3. \end{cases}$$

B.5 Equation (B.10) gives

$$\begin{aligned} \mathbf{X}(x) &= x^i \sigma^i \\ \mathbf{X}'(x') &= x'^i \sigma^i = R^i_j x^j \sigma^i. \end{aligned}$$

Also $\mathbf{X}' = \mathbf{UXU}^\dagger = \mathbf{U}x^j\sigma^j\mathbf{U}^\dagger$. The x^j are arbitrary. Hence $\mathbf{U}\sigma^j\mathbf{U}^\dagger = \mathrm{R}^i_j\sigma^i$. Multiplying on the left by σ^k and taking the trace,

$$\mathrm{Tr}(\sigma^k \mathbf{U} \sigma^j \mathbf{U}^\dagger) = R^i{}_j \mathrm{Tr}(\sigma^k \sigma^i).$$

Now

$$\mathrm{Tr}(\sigma^k \sigma^i) = \begin{cases} 2 & \text{if } k = i, \\ 0 & \text{if } k \neq i. \end{cases}$$

Hence the result.

B.6 From (B.17), $\mathbf{M}^\dagger \tilde{\sigma}^\mu \mathbf{M} = L^\mu{}_\lambda \tilde{\sigma}^\lambda$. Multiplying on the left by $\tilde{\sigma}^\nu$ and taking the trace, the result follows, since

$$\mathrm{Tr}(\tilde{\sigma}^\nu \tilde{\sigma}^\lambda) = \begin{cases} 2 & \text{if } \lambda = \nu, \\ 0 & \text{if } \lambda \neq \nu. \end{cases}$$

Appendix C

C.2 The ground state is given by $a|0\rangle = 0$, or $(X + \mathrm{i}P)|0\rangle = 0$. In the Schrödinger representation. $P = -\mathrm{i}\,\mathrm{d}/\mathrm{d}X$, so that $(X + \mathrm{d}/\mathrm{d}X)\psi_0 = 0$, giving $\psi_0 = A\mathrm{e}^{-X^2}/2$, where the constant A is determined by normalisation.

C.3

$$\begin{aligned} N_i b_i{}^\dagger|0\rangle &= b_i{}^\dagger b_i b_i{}^\dagger|0\rangle \\ &= b_i{}^\dagger(1 - b_i{}^\dagger b_i)|0\rangle = b_i{}^\dagger|0\rangle. \end{aligned}$$

Appendix D

D.1

$$\begin{aligned} Q^2 &= (\mathbf{p} - \mathbf{p}')^2 - (E - E')^2 \\ &= (p^2 - E^2) + (p'^2 - E'^2) - 2\mathbf{p}\cdot\mathbf{p}' + 2EE'. \end{aligned}$$

But $E^2 = p^2 + m^2$, $E'^2 = p'^2 + m^2$, so that, neglecting electron masses,

$$Q^2 = -2pp'\cos\theta + 2EE' = 2EE'(1 - \cos\theta) = 4EE'\sin^2(\theta/2).$$

The energy and momentum of the recoil proton are given by $E_p = M + E - E'$, $\mathbf{P} = \mathbf{p} - \mathbf{p}'$; also $E_\mathrm{p}{}^2 = M^2 + P^2$. Hence

$$\begin{aligned} Q^2 &= p^2 - (E - E')^2 \\ &= (M + E - E')^2 - M^2 - (E - E')^2 \\ &= 2M(E - E') \end{aligned}$$

so that (D.3) follows.

D.3

$$\begin{aligned} Q^2 &= 2EE'(1 - \cos\theta) \\ \nu &= E - E' \\ \mathrm{d}Q^2\mathrm{d}\nu &= \frac{\partial(Q^2, \nu)}{\partial(\cos\theta, E')}\mathrm{d}(\cos\theta)\mathrm{d}E' \end{aligned}$$

where the Jacobian of the transformation is

$$\begin{vmatrix} -2EE' & 2E(1 - \cos\theta) \\ 0 & -1 \end{vmatrix} = 2EE'.$$

Hence the result.

Index

Printed in the United States
by Baker & Taylor Publisher Services